THE MAMMALS OF LUZON ISLAND

THE MAMMALS OF LUZON ISLAND

Biogeography and Natural History of a Philippine Fauna

Lawrence R. Heaney,
Danilo S. Balete, and Eric A. Rickart
*Illustrated by Velizar Simeonovski
and Andria Niedzielski*

JOHNS HOPKINS UNIVERSITY PRESS BALTIMORE

Printed in China on acid-free paper
9 8 7 6 5 4 3 2 1

Johns Hopkins University Press
2715 North Charles Street
Baltimore, Maryland 21218-4363
www.press.jhu.edu

Library of Congress Cataloging-in-Publication Data

Heaney, Lawrence R.
The mammals of Luzon Island : biogeography and natural history of a Philippine fauna / Lawrence R. Heaney, Danilo S. Balete, Eric A. Rickart ; illustrated by Velizar Simeonovski and Andria Niedzielski.
pages cm
Includes bibliographical references and index.
ISBN 978-1-4214-1837-7 (hardcover : alk. paper) — ISBN 978-1-4214-1838-4 (electronic) — ISBN 1-4214-1837-1 (hardcover : alk. paper) — ISBN 1-4214-1838-X (electronic) 1. Mammals—Philippines—Luzon. I. Balete, Danilo S., 1960– II. Rickart, Eric A. III. Title.
QL729.P5H434 2016
599.09599'1—dc23 2015008660

A catalog record for this book is available from the British Library.

Special discounts are available for bulk purchases of this book. For more information, please contact Special Sales at 410-516-6936 or specialsales@press.jhu.edu.

Johns Hopkins University Press uses environmentally friendly book materials, including recycled text paper that is composed of at least 30 percent post-consumer waste, whenever possible.

Contents

Preface

Tropical islands have held great attraction for biologists for centuries. The reason seems to lie both in the often exceptionally great concentration of unique biodiversity on such islands and in the implicit questions that arise from that unique diversity: what species are present on the island, how and when did they arrive, why are there so many highly distinctive species, and how has the unique geological history of the islands influenced the biota that is present?

Biologists are an inherently curious lot; the bigger and more complex the questions, the happier we seem to be. That curiosity is what drew the authors of this volume to studies of the highly diverse mammals of the Philippines. The best available evidence at hand indicates that Luzon has the greatest concentration of unique species of mammals of any place in the world. We found the attraction of trying to understand why that is the case to be wonderfully compelling.

Biologists are also explorers, attracted to the unknown and to a sense of discovery. In our case, organizing teams of field researchers and porters from local communities to trek into rugged mountains in isolated parts of the country has an undeniable romantic appeal, and we confess that it was a large part of the attraction for us. And from the first time any of the three of us captured species that no biologist had seen before, the surge of adrenalin was powerful.

We also are strongly motivated to share what we have learned—to share the excitement of exploration and discovery, and to share the new knowledge gained. Formal teaching in a classroom or speaking to a crowd in an auditorium are fine ways to impart much of the excitement and knowledge, but writing can be the best way to share the information as widely as possible, and most biologists love to write. Further, we feel a strong compulsion to help protect the wonderful animals we have studied and the places where we have studied them. We firmly believe that protecting these animals and places will work to the good of people everywhere, both those who are so fortunate as to have them in their backyard, and those who dream, as we once did, of green forests on steep mountains where the unknown awaits.

This book is intended to answer to all of these parts of being a biologist. It is designed to be useful as a field guide, a classroom textbook, and a source of information for anyone who needs or simply wishes to know more about mammalian diversity on Luzon. It is not a comprehensive compilation of all of what is known regarding every aspect of each species, or the sum total of the data that we have gathered. Rather, it is a summary of what we know at this time about the evolutionary origin and ecological maintenance of mammalian diversity on Luzon, from a biogeographical perspective. For some additional details, we refer the reader to the online version of the Synopsis of Philippine Mammals, available at www.fieldmuseum.org/synopsis-philippine-mammals/. We offer our regrets that this volume does not include information about the wonderful marine mammal fauna of the Philippines; aside from the Synopsis, we recommend Alava et al. (1993), Dolar et al. (2006), and Tan (1995) as sources.

Acknowledgments

This book is the result of 15 years of focused effort to document the extraordinary mammal fauna of Luzon, but it is also the product of far longer periods for all of us. Heaney benefited greatly from early encouragement, and opportunities to study mammals and their patterns of distribution and diversity, provided by Tom McIntyre, Richard Thorington Jr., Hank Setzer, Charles Handley, and Jim Mead at the US National Museum of Natural History; their generosity with time and knowledge set him securely on the road to field-oriented studies of mammalian diversity, including an appreciation for the mammals of Southeast Asia. Rickart's initial interest in mammals stems from his early association with Charles Remington and his son Eric, who revealed the joys of field exploration, the value of close observation, and the powerful attraction of natural history and science as lifelong pursuits. Heaney and Rickart both studied under Robert Hoffmann, whose gentle demeanor, love for learning, and deep interest in biogeography and mammals has provided a lifetime of inspiration. Balete was inspired by many teachers and colleagues, especially Lipke Holthuis and Boris Sket, and benefited immeasurably from the generosity of Anton Cornelius Jacobus Burgers.

Our early studies of biogeography and mammals in the Philippines, beginning in the 1980s, were encouraged, supported, and inspired by a great many people: foremost among them were Angel C. Alcala, Pablo Alphonso, Pedro Alviola, Mike Carleton, Pedro Gonzales, Karl Hutterer, Karl Koopman, Guy Musser, and Richard W. Thorington Jr. Silliman University provided our base of operations in the Philippines during much of this period and made us so welcome that it came to feel like home.

The 12 years of field work on which this book is primarily based would not have been possible without the strong support of the Philippine Department of Environment and Natural Resources (DENR), especially the Protected Areas and Wildlife Bureau (recently renamed the Biodiversity Management Bureau). We offer our most sincere thanks to all of the current and former staff, especially to Angel C. Alcala, T. Mundita Lim, Cora Catibog-Sinha, Luz Gonzales, Samuel Peñafiel, Wilfredo Pollisco, Josie DeLeon, Carlo Custodio, Marlynn Mendoza, and Anson Tagtag. A great many other staff members of the DENR in Regions 1, 2, 3, 4, 5, and the Cordillera Autonomous Region gave us their help and support; they are far too numerous to name, but to all of them we offer our gratitude. We thank the staff of the National Commission for Indigenous People (NCIP) for their support in obtaining the necessary permission for our studies.

We have also benefited from a long, close, and highly productive association with the National Museum of the Philippines, especially Jeremy Barns, Anna Labrador, Angel Bautista, Julie Barcelona, Arvin Diesmos, Pete Gonzales, and Sweepea Veluz; their efforts to bring the study of natural history in the Philippines to a new and greater level will have benefits for generations to come.

Our first studies on Luzon (Mt. Isarog, in 1988), which took place while Heaney benefited from a position as a Smithsonian Institution Research Fellow in the Division of Mammals at the US National Museum of Natural History, played a major role in inspiring this book; support from the US National Science Foundation made the field work possible. Our collaborators during that field season—Ruth Utzurrum, Myrissa Lepiten, Mylene Laranjo, and Leoning Tagat from Silliman University, as well as Steve Goodman, Doug Samson, and Dave Schmidt—deserve much credit for what followed. At the University of the Philippines (Diliman, Los Baños, and Baguio campuses) we have enjoyed close collaboration with Grace Ambal, Phillip Alviola, Leonard Co, Aloy Duya, Liza Duya, Nancy Ibuna, Mandy Mijares, Perry Ong, Sol Pedregosa, Phil Piper, Aris Reginaldo, and many others. The staff of the Haribon Foundation, especially Blas Tabaranza Jr., Genevieve Gee, and Myrissa Lepiten-Tabao, all made substantial contributions, including collaborating on our studies of the mammals of Balbalasang-Balbalan National Park. The current and former staff of Conservation International–Philippines in Quezon City, especially Perry Ong, Tom Brooks, and Romy Trono, provided us not only with enthusiastic collaborators but also with a base of operations for over a decade; to all of them we are grateful.

Our field research has depended on skilled and energetic support from a remarkable team over the years, especially from Nonito Antoque, Boying Fernandez, and Bernard Malaga in the early years, with Joel Sarmiento, Cardo Buenviaje, and Ronel Plutado providing expert assistance in many recent years. We would have accomplished little without them. At each of our study areas, we received generous hospitality and friendship; to all of our hosts, friends, cooks, porters, and guides, we owe a vast debt. We are thankful to local governmental units at provincial, municipal, and barangay levels, and to the Protected Area Management Boards for each national park, for showering us with good will and hospitality.

The phylogenetic portions of our studies have been carried out in collaboration with Sharon Jansa, Scott Steppan, Manuel Ruedi, and the members of their lab groups; to all we are grateful. Museum studies of morphology have been made possible by many people, especially including Karl Koopman, Guy Musser, and Rob Voss at the American Museum of Natural History; Mike Carleton, Linda Gordon, Jeremy Jacobs, Helen Kafka, Charley Potter, Dave Schmidt, Jean Smith, and Richard Thorington Jr. at the US National Museum of Natural History; and Phil Myers and Barry O'Connor at the University of Michigan Museum of Zoology. Our colleagues Rafe Brown, Jake Esselstyn, Paul Heideman, Nina Ingle, Guy Musser, Trina Roberts, Jodi Sedlock, Bob Timm, and Joe Walsh have provided encouragement and enthusiasm, as well as expert advice and information.

At the Field Museum, Anna Goldman, Andria Niedzielski, John Phelps, Mary Beth Prondzinski, Michi Schulenberg, Bill Stanley, and Betty Strack have all provided extensive assistance with the cataloging and preparation of our specimens and other matters. John Bates, Rudiger Bieler, Lori Breslauer, Barry Chernoff, Lance Grande, Mary Johnson, Bob Martin, and Steve Strohmeier did much to help us surmount obstacles; to all we are grateful. Trish DeCoster and Andria Niedzielski served as project assistants for several years each and deserve much of the credit for bringing this book to fruition. The beautiful illustrations that grace this volume have been prepared by Velizar Simeonovski and Andria Niedzielski. For reading and making helpful suggestions on draft chapters, we thank Mike Douglas, Jake Esselstyn, Phil Piper, Jodi Sedlock, Alex Stuart, and especially Robert Hall, whose comments on two successive drafts of chapter 4 greatly improved the content. An anonymous reviewer of the entire manuscript also improved the final product. Vince Burke, our editor at Johns Hopkins University Press, encouraged us to develop the issues covered in the book more extensively than initially planned, and it is much improved as a result. We thank Kathleen Capels, Glen Burris, and Courtney Bond for sharp-eyed attention to detail, improving our garbled syntax, and creating an accurate and visually appealing final product; they do great work.

Terry Horton, Ben Heaney, and Shelby and Sam Rickart have tolerated our absences and absorption in our studies for decades, have shared our enthusiasm at our adventures and discoveries, and have tolerated our dejection when things have not gone as we hoped. We are deeply grateful for their love and support.

A special thanks for support of the project is owed to Robin and Richard Colburn and members of the Brown family, especially Barbara and Roger; without their steadfast encouragement, much of the recent work in the field and museum could not have been accomplished. This book was made possible through the generous support of:

The Barbara Brown Fund for Mammal Research
The Negaunee Foundation
Ellen Thorne Smith Fund (Field Museum)
The Marshall Field Fund (Field Museum)
The Women's Board (including the Field Dreams program / Field Museum)
The Grainger Foundation Inc.

Department of Environment and Natural Resources–Biodiversity Management Bureau

Diliman Science Research Foundation

University of the Philippines

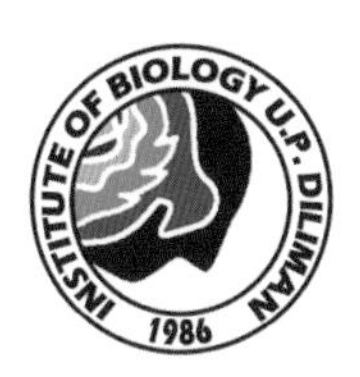

Institute of Biology, University of the Philippines Diliman

Biodiversity Research Laboratory, Institute of Biology, University of the Philippines Diliman

National Museum of the Philippines

Philippine Tropical Forest Conservation Foundation

Part I

THE BIOGEOGRAPHY OF DIVERSITY

1

The Mammals of Luzon, 1895–2012

In the Beginning: 1895

When an adventurous young British ornithologist named John Whitehead arrived on Luzon in 1893, he had high hopes of discovering many previously unknown species of birds in the Philippines to send back to the British Museum. He succeeded spectacularly during his travels over the next three years, obtaining the first specimen of the huge Philippine eagle (which was named *Pithecophaga jefferyi* in honor of his father, Jeffery) and many others. Although numerous collectors of birds had preceded him, and the avian fauna of this region was known to be rich in distinctive endemic species, it was still poorly known in many parts of the geographically complex and rugged archipelago. Whitehead had found a fertile ground for scientific exploration (Dickinson et al., 1991).

In contrast, Whitehead did not expect to find many mammals of interest. Just a bit over a decade earlier, one of the founders of biogeography (and codiscoverer of natural selection), Alfred Russel Wallace, had published his great work, *Island Life* (1880), in which he said that 288 species of birds were then known from the Philippines, but only 21 species of mammals. In what seems an almost dismissive tone, he added that "no doubt several other [mammals] remain to be discovered" (Heaney, 2013a). Only one of the mammals (*Phloeomys*, the giant cloud rat) represented a genus limited to the Philippines, a poor showing compared with the rich mammal fauna of Borneo and much of the rest of the Malay Archipelago, where Whitehead had traveled and collected natural-history specimens from 1885 to 1888. Wallace ascribed this lack of distinctiveness to land-bridges from Borneo throughout the Philippines and surmised that the poverty of mammalian species was due to "a great amount of submersion in recent times," based on "large tracts of elevated coral-reefs"; the mammals that might once have been present, he believed, had been driven to extinction by a great reduction in island area in the recent past (Wallace, 1880:362).

Whitehead almost certainly knew of Wallace's writings and can hardly have anticipated much of note among any mammals he collected in the course of his ornithological explorations. So, near the end of his trip in December 1894, when he headed east from the Ilocos coast up to the crest of the Central Cordillera on horseback, with a string of mules, his mind most likely had remained fixed on birds. In his notes published afterward (Whitehead, 1899), he described "a weary tramp that lasted six days, through an absolutely treeless country, and so depressed by the useless-looking landscape that I nearly turned back. However, . . . one morning while riding ahead of my baggage, on rounding a steep cliff, I came in sight of oak- and pine-forests. . . . After a good deal of hard work, we reached

FIG. 1.1. Steep slopes on the western side of the Central Cordillera of Luzon are burned frequently, giving rise to hot, dry mountainsides covered by tough grasses and scattered pine trees (*Pinus kesiya*).

the summit [of Mt. Data, at 7000–8000 feet] . . . on the morning of the 14th of January [1895] at 9 A.M.; and I was delighted to find the ground in shady places still covered with frost. On Data I remained camping out for thirty days, during which time the collections made were both large and remarkable for interesting birds and mammals" (figs. 1.1 and 1.2).

We do not know with certainty what went through Whitehead's mind as his time on Mt. Data passed, but it is not hard to guess at the astonishment he must have felt when the local men he hired as collectors first brought him a giant rodent (approaching 1 m in length) covered with long, lustrous black fur from its nose to the tip of its long tail (fig. 1.3). The fur was so thick that it easily shed water from the frequent fog and rain Whitehead experienced, and the rodent's short, broad hind feet allowed it to grasp limbs as it climbed through the treetops along the moss-covered branches. The men brought him smaller rodents as well, some with bold stripes on their backs and short, sturdy fore-

FIG. 1.2. Mossy forest in the Central Cordillera supports a profusion of life, with ferns, orchids, and moss growing on the ground and on trees. Vines link the canopies of trees, and a thick layer of partially decomposed plant matter forms a water-absorbent cushion on the ground. This mossy forest in Balbalasang, Kalinga Province, is similar to what John Whitehead saw on Mt. Data.

FIG. 1.3. The native mice that John Whitehead found on Mt. Data astounded both him and the most knowledgeable mammalogist of his day, Oldfield Thomas. A: *Crateromys schadenbergi*. B: *Chrotomys whiteheadi*. C: *Rhynchomys soricoides*. D: *Apomys datae*. E: *Batomys granti*. F: *Carpomys phaeurus*.

limbs and forepaws that implied an animal that could dig well—but for what? Yet other small rodents had large, robust hind limbs and slender hind feet, with a snout drawn out to a distant tip where a tiny mouth had delicate, pincer-like incisors. Still others had broad heads, sturdy muzzles, and long furry tails. Nothing resembled anything he had seen before, on Borneo or elsewhere.

In his initial lack of enthusiasm for Philippine mammals, Whitehead was joined by the British Museum's eminent mammalogist, Oldfield Thomas, who worked during the height of the British Empire, describing many hundreds of new species from around the world. Thomas, whose publications are notable for their absence of expressed emotion, appears to have exceeded his capacity for reserve when he received Whitehead's collection. "Little . . . could have been expected from the expedition further than the discovery of a few fresh species of genera known to inhabit the [Philippines]. . . . But in the great northern island of [Luzon] Mr. Whitehead has made a most wonderful and unexpected discovery, that of a new and peculiar Mammal-fauna inhabiting the Luzon highlands . . . no less than six new genera and eight new species were discovered in the island, a proportion of novelty that has perhaps never been equaled in the history of Mammal-collecting" (Thomas, 1898:377–378). Thomas thought the giant fur-covered rodent (which he named as *Crateromys schadenbergi*) and the one with the thin, elongated snout (*Rhynchomys soricoides*) were related to animals from Celebes (= Sulawesi, in Indonesia), the one with stripes (*Chrotomys whiteheadi*) to animals from Australia and New Guinea, and the previously known giant cloud rats (*Phloeomys*), which Whitehead also found on Mt. Data, "is so isolated that I can make no suggestion as to what is its nearest ally" (p. 378). He did not even mention the ones with long furry tails and sturdy muzzles (*Carpomys*). One can easily imagine Thomas scratching his head, exclaiming over the "most wonderful and unexpected discovery."

Return to Mt. Data, 2006

When our field team returned to Mt. Data in 2006, a fully paved, major highway from Manila reached Baguio, a metropolitan area of over a million people near the south end of Central Cordillera. After checking in with the regional office of the Department of Environment and Natural Resources, we boarded a small, open-air bus that went directly to the mountain, traveling along the heavily used, partially paved highway that zigzags through the Central Cordillera, passing over Mt. Data's shoulder on the way to Bontoc, the rapidly growing capital city of Mountain Province.

We settled into a comfortable bunkhouse that was the former headquarters for a logging company but had been converted to the local headquarters for Mt. Data National Park, which officially covered 5512 ha (one hectare is about 2.5 acres; Ong et al., 2002). We quickly learned that the park was covered primarily by commercial vegetable farms; the rich volcanic soil is perfect for raising cabbage, cauliflower, radishes, green beans, and other vegetables that will grow only in cool climates, and the entire park lies near to the highway (fig. 1.4). A few patches of shrubby second growth along steep ravines, scattered pine trees, and a fenced patch of partially logged forest of about 80 ha was all that remained of the forest that Whitehead had seen (Heaney et al., 2006b). The two species of giant cloud rats (*Phloeomys* and *Crateromys*) were long gone from the mountain, overhunted for meat and for skins to be made into hats for the tourist trade, and the dwarf cloud rats (*Carpomys*, with long furry tails and sturdy muzzles) and long-snouted *Rhynchomys* had disappeared as well. We were not surprised to see that non-native pest rats (Oriental house rats, *Rattus tanezumi*; and spiny ricefield rats, *R. exulans*) were abundant at the edges of vegetable fields and in grassy areas that were burned frequently (in addition to the house rats in our bunkhouse), but they were scarce in the secondary forest and along the ravines.

But the mountain retained some remarkable surprises, well over 100 years after Whitehead's visit. Two species of native forest mice (*Apomys*) were common in the secondary forest and ravines, and a small relative of the cloud rats (*Batomys granti*) was still present in the secondary forest. Striped earth mice (*Chrotomys*) were abundant in the patch of secondary forest and even in shrubby ravines at the edges of fields, where they dug for earthworms, their primary food. Even in the cool, wet climate at ca. 2300 m elevation (about 8000 feet),

FIG. 1.4. The top of Mt. Data is a broad plateau that was covered by mossy forest in 1895 but was mostly converted to commercial vegetable farms when we studied mammals there in 2006. The boundary of the remaining 80 ha fragment of mossy forest is fenced, with cabbage fields up to its edge.

FIG. 1.5. Surprises remained on Mt. Data even in 2006. A mouse (*Soricomys montanus*) we saw scampering about the shrubby ravines during the daytime proved to be a previously unknown species that is a member of a new genus we described in 2012.

we captured five species of bats in our nets, the first time four of them had been documented so high up in the mountains, and the first time the fifth species (*Nyctalus plancyi*) was documented anywhere in the Philippines (Heaney et al., 2012). And scurrying about during the daytime in the patch of forest and shrubby areas in the ravines were small, dark-furred, shrew-like mice that fed on small earthworms and other soft-bodied invertebrates—members of a previously unknown genus of rodents that we named *Soricomys* (fig. 1.5; Balete et al., 2012).

Lessons from Mt. Data

From these two stories, set apart by more than a century, emerge a set of general features that define much of what characterizes the mammal fauna of Luzon, and of the Philippines in general. They also represent the primary topics we will deal with in this book, organized as chapters.

The mammal fauna of Luzon Island, once thought to be quite poor, is astoundingly diverse, with over 110 native species: there are giant cloud rats (and giant fruit bats), tiny tree-mice (and one of the world's smallest bats), striped earthworm mice (that feed on earthworms), bats with ears as long as their bodies, and a host of others. Most bats are fairly closely related to those in other parts of Southeast Asia (Heaney, 1991), but most native small mammals, such as the cloud rats and earthworm mice that Whitehead captured and Thomas described for the scientific world, are members of entire branches on the tree of life that live nowhere else.

The processes that have produced that diversity can be summarized as climatic, topographic, and geological, each operating in concert with the others. Luzon lies in the tropics, with extensive areas once covered by lowland tropical forest. But much of the island is mountainous and, as Whitehead recognized when he described finding frost on the ground at the top of Mt. Data in January, climatic variation along the elevational gradient is great. Habitats change in concert with this climatic variation, and also with soil types that range from volcanic to nutrient-poor ultramafic soils and to highly rugged karst over limestone. In chapter 2, we describe the plant communities of each kind of habitat, with emphasis on the aspects that influence the mammals that live there.

It was in the Central Cordillera, the highest and largest mountain chain on Luzon, where John Whitehead had his greatest success in finding mammals previously unknown to biologists. But Luzon is made up of many isolated small mountain ranges, with intervening lowlands. We have learned that each of these isolated ranges has a unique set of mammal species, and the taller the mountains, the more mammal species are present. It is this combination of isolated mountains and the increase in diversity with higher elevations that accounts for Luzon having what may be the greatest concentration of endemic mammal diversity of any place in the world. Chapter 3 documents and explains the ecological processes that influence these patterns, based on our recent research.

Alfred Russel Wallace hinted at the potential complexity of Luzon's geological history when he speculated about uplift of old coral reefs, land-bridges to Asia, and periods of higher seas, but information was highly limited and fragmentary at that time, and (unusually for him), he misjudged the real history. With the possible exception of the Palawan group, the Philippine islands were never connected by land to the Asian mainland; they have existed entirely as true oceanic islands. Sea level, however, did fluctuate greatly during the ice ages of the past 2.6 million years, and that impacted the history of connections between some islands within the Philippines (Heaney, 1985; Piper et al., 2011). In chapter 4, we summarize this geological history, explaining how the Philippines were formed largely as a result of continental drift, with subduction zones that result in the formation of the volcanoes that dominate Luzon's geological history and current topography. We also discuss what little is known about the fossil mammals of Luzon, including current efforts to peer directly into the depths of the diverse Luzon adaptive radiations.

Bats were very difficult to capture in Whitehead's time, and he sent only a few back to the British Museum. But we know now that the bats of Luzon are highly diverse, with at least 57 species presently known. Most of them are fairly closely related to species outside of the Philippines, with only a few small endemic clades (i.e., sets of closely related species that live only within the Philippines). This is a very different pattern of diversity than we see in non-flying mammals: most of the 50 or more non-flying native mammals are members of just two "branches on the tree of life" of rodents that live only in the Philippines (Jansa et al., 2006). Clearly, most of the species of non-volant mammals on Luzon have arisen by speciation within the island. Recent DNA-based studies illustrate the remarkable extent of morphological and ecological diversity that has resulted (e.g., Justiniano et al., 2015). In chapter 5, we provide more details and discuss why the bats and non-flying mammals show such different patterns of distribution.

Conventional wisdom currently holds that native animals on oceanic islands such as Luzon are highly vulnerable to invasive alien species, such as *Rattus rattus* and its relatives. Our studies on Mt. Data and elsewhere on Luzon have demonstrated that the "invasive" non-native small mammals often are unable to invade forest habitat, even when the forest has been degraded by human activity; native species appear to be competitively superior to non-native rodents except in the most disturbed areas, such as in buildings and agricultural fields. We discuss this surprising result in chapter 6 and offer an explanation as to why Luzon is different and why there may be many other islands in Indo-Australia where the Luzon pattern also exists.

Habitat destruction, overhunting, extinction of native species, and invasion by non-native pest rats are familiar stories—but they are by no means the only significant aspects of these stories from Mt. Data, nor should they be taken as an inevitable outcome. Luzon Island has an area of about 103,000 km^2 (a bit larger than Indiana at 94,300 km^2), and a human population of about 42 million people (Indiana has roughly 6.6 million). With great disparity in income and opportunity among the island's people, and with substantial ethnic diversity, Luzon has been subject to a complex set of factors that have produced an enormous loss of natural habitat through logging, burning, mining, and agricultural expansion, but it has also retained extensive forest in some rugged and remote areas, especially those controlled by traditional indigenous people. Hunting pressure is intense, not only on the few large mammals—deer and wild pigs—but also on some large rodents and many bats, leading to varying levels of endangerment for many species. But we have also found that isolation and traditional management practices have afforded effective protection in some places. In chapter 7, we describe the extent of current problems and threats and offer recommendations on what actions may prevent extinction, as well as lead to improvements in effective conservation. We also discuss the geographic patterns of mammalian diversity and endemism and the implications they have for management.

The results of our research described in this volume have immediate relevance in understanding the evolutionary origin and ecological maintenance of mammalian diversity on Luzon, but they also have profound implications for current theory in island biogeography. In chapter 8, we summarize the pertinent points from our studies and show how they call for similar studies elsewhere to test new models of island biogeography.

Mammals of Luzon Project, 2000–2012

The questions that drive scientific research arise in many ways, often as the result of simple curiosity about unexpected observations. The Mammals of Luzon Project on which this book is largely based had its primary origin in conversations around cook-fires in 2000. We had gone to the Central Cordillera to obtain additional specimens of the fauna that Whitehead discovered, for use in genetic and morphological studies of the evolutionary relationships of the diverse native rodents (e.g., Heaney et al., 2011a; Jansa et al., 2006). We also hoped to add to what little was known about their ecology and conservation status. Rather than go initially to Mt. Data, which we had been told had been largely clear cut for conversion to vegetable farms, we went 80 km north to the small village of Balbalasang, high in the northern Central Cordillera, a remote area designated as a national park where there were high peaks and extensive areas of forest. Our hopes were confirmed: we documented 13 native species of small mammals, along a transect from about 900 m elevation (in lowland forest) to 2150 m (in mossy forest), in a place of serene beauty protected by the traditional management practices of the Banao people (fig. 1.6; Heaney et al., 2000, 2003). As expected from data elsewhere in the Philippines (Heaney, 2001), diversity increased from 3 species in the lowland forest at 900 m elevation to 10 in the mossy forest at 2000 m and above, and the distinctive genera discovered by Whitehead increased in diversity and abundance with increasing elevation (Rickart et al., 2011a). We had previously conducted a similar elevational transect on Mt. Isarog, in the southern peninsula of Luzon, and found a similar pattern of increasing species richness and abundance with increasing elevation (Heaney et al., 1999; Rickart et al., 1991). Of the eight species on Mt. Isarog, four were local endemic species (now tallied as five, due to new taxonomic studies; Balete et al., 2015) in the same genera as those in the Central Cordillera. The three that were shared by the two areas were the only three that occurred in the lowlands as well as mountains. Two other rather large mountain ranges—the Zambales and Northern Sierra Madre—had just one locally endemic species each.

This left us scratching our heads: it made sense that the pattern of diversity was similar on the two mountains, and that species that occur in lowlands would be widespread, but why was it *just those two places* where more than a single locally endemic species was known? Luzon is highly mountainous, with many isolated mountain ranges and individual mountains (fig. 1.7); what made the Cordillera and Mt. Isarog different? The answer quickly became obvious: with a few almost trivial exceptions, *no one had looked anywhere else* (Heaney, 2004b; Heaney and Mallari, 2000; Heaney et al., 2000). A quick count of likely prospects showed at least six mountain ranges or individual mountains that rose to over 1500 m and were isolated by lowland areas, which meant that each was effectively a "sky island," consisting of an area of the montane or mossy forest favored

FIG. 1.6. A morning view from the village of Balbalasang, Kalinga Province, where we had our first introduction to the full breadth of the Luzon mammal fauna. Successive waves of ridges rise progressively higher from the village, which lies at about 850 m elevation, to the highest peaks at over 2300 m.

by the native small mammals, surrounded by a "sea" of lowland forest where the native species did not occur (now often converted to agriculture or urban areas).

So, when we completed our survey in Balbalasang, we began a series of standardized surveys elsewhere on Luzon. One of our first surveys in 2004, on Mt. Tapulao in the Zambales Mountains, demonstrated a similar pattern of increasing diversity and endemism with increasing elevation, and we discovered five new species in the process (Balete et al., 2007, 2009; Heaney et al., 2011a). With that positive reinforcement, we conducted surveys in five additional highland areas on Luzon,

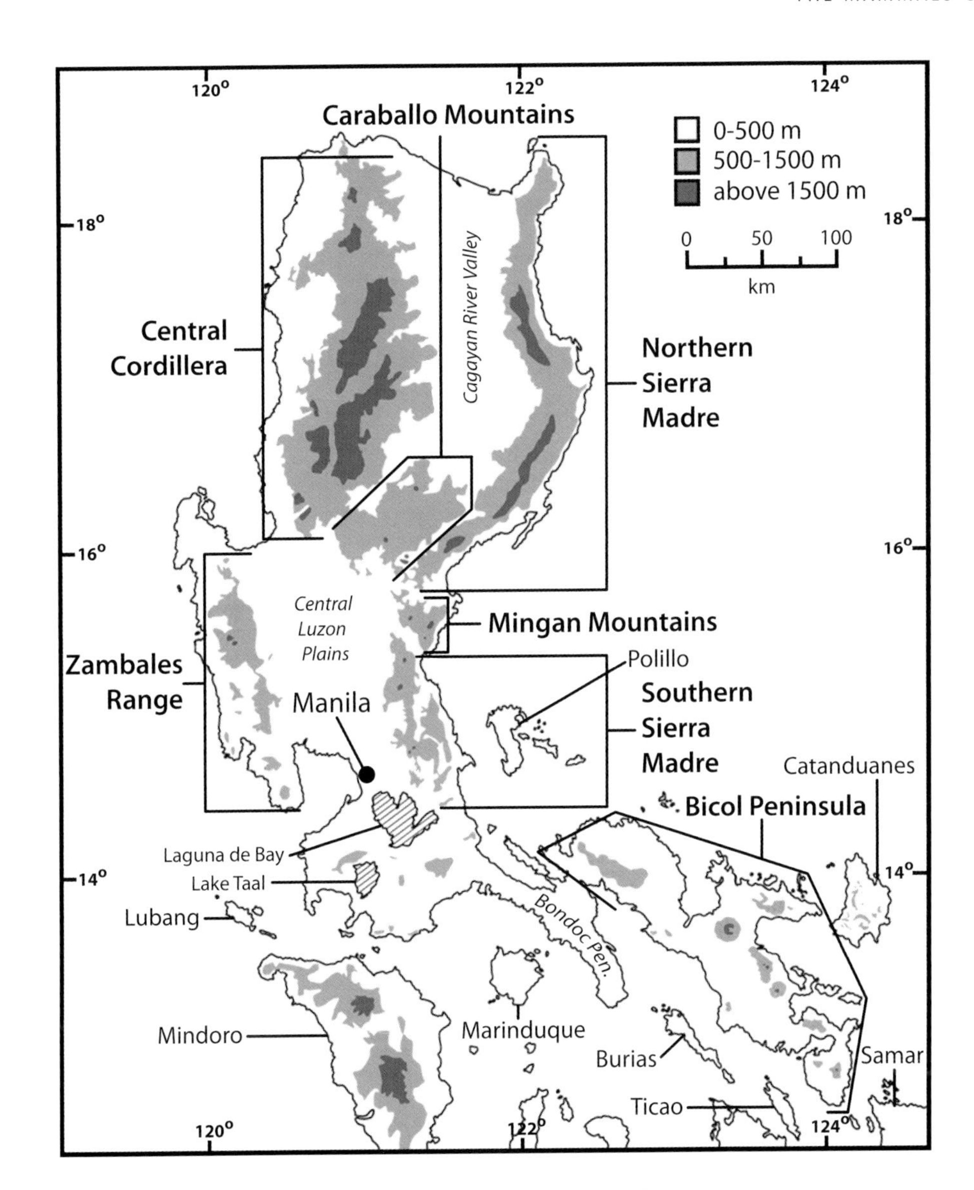

FIG. 1.7. Primary topographic features of Luzon mentioned in this volume, plus nearby smaller islands.

later adding two more as we recognized that the intervening lowlands did not need to be more than about 10 km wide to mark the boundaries between distinctive highland areas (fig. 1.8; Heaney, 2004b).

Over the course of the Luzon project from 2000 to 2012, we estimate that we spent about 37 team-months in the field, with six to eight people usually making up the team. In each study area, we began our survey at the lowest elevation where forest was present, usually at 600–800 m, and proceeded up the mountain to as close to the peak as we could reach. We sampled at intervals of about 200 m elevation along the transect; depending on the height of the mountain, that usually meant four to eight sampling localities, each requiring about a week of sampling. The best way to determine the number of species present was to track the number of trap-nights and net-nights (our measures of sampling intensity; one trap or one mist net set for one night is counted as a trap-night or a net-night). We learned by experience that it takes about 600–1000 trap-nights to

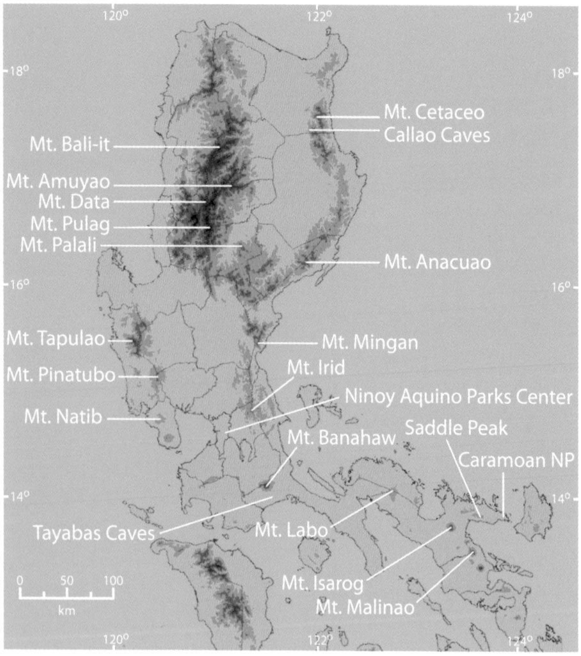

FIG. 1.8. Locations of our primary study areas on Luzon during our 2000–2012 field surveys. We attempted to survey the mammals in every mountain or mountain chain that is surrounded by lowlands and reaches to at least 1400 m elevation, as well as in several lowland areas and one urban area.

reach the point at which we were unlikely to catch additional species of small mammals at a given locality (Rickart et al., 1991, 2011a), so we adopted that range as a minimum target for each locality. We baited traps with two very different types of bait, intended to appeal to animals that feed on seeds or on invertebrates: either lightly fried slices of coconut coated with peanut butter, or live earthworms. We set traps both on the ground surface and in trees; the latter were set in places to which we could safely climb, and so rarely exceeded 5 m above the ground. Our species-accumulation curves typically reached a plateau at each locality, indicating that the sampling had captured all (or nearly all) species that were present (e.g., Balete et al., 2009, 2011, 2013a, 2013b; Heaney et al., 2013a, 2013b; Rickart et al., 1991, 2011a, 2013). At each sampling locality, we ranked the mammals and some features of the biotic community (e.g., tree species, moss, ants, earthworms, etc.), based on their relative abundance (i.e., the number per unit area), as "abundant," "common," "uncommon," "scarce," or "absent"; we use those terms throughout this volume, especially in making comparisons among localities.

Giant fruit bats (also called flying foxes) fly quite high and are difficult to catch, but the highly diverse small to medium-sized fruit bats are fairly easy to capture (they lack echolocation abilities; i.e., no "sonar"), and we know quite a lot about most of them. We set 25 net-nights as a limit for catching representatives of most of these fruit bat species in an area (Heideman and Heaney, 1989). Insectivorous bats, on the other hand, have sensitive echolocation. We caught many and have learned much about them, but we soon decided that we would not make the extensive and specialized effort to sample them thoroughly along elevational gradients, except in special cases (e.g., Sedlock et al., 2011). As a result, while we present all of the information we have about the distribution, ecology, and general natural history of insectivorous bats on Luzon in this book, we have not focused on their diversity patterns, and much basic, essential information remains to be learned about them.

Scientific studies always require evidence to support any conclusions. In the case of our taxonomic and, more broadly, evolutionary and biogeographic studies, voucher specimens are required: these are preserved specimens deposited in permanent research collections where they are available to support future research. From them we collect data that can be examined by other scientists to verify the accuracy of our data and our subsequent conclusions (e.g., descriptions and measurements of cranial anatomy, as well as reproductive patterns and stomach contents). Our voucher specimens have been deposited at the Field Museum of Natural History (FMNH) for initial study; half are being permanently deposited at the National Museum of the Philippines as we complete our taxonomic studies. We obtained permits for this research from local governments, tribal groups, and the Department of Environment and Natural Resources, and we followed all requirements established in Philippine law for research on wildlife.

Our studies of the specimens and data we collected from 2000 to 2012 are continuing; this book is a progress report and certainly not an end point. But we can readily summarize a few of the most striking outcomes that we have published elsewhere and develop more extensively in the later parts of this volume.

We captured five species of bats on Mt. Data, representing three of the seven families of bats on Luzon. These are among the very highest elevational records for all of these families; most bat species live in the lowlands or low foothills, and species richness declines steadily with an increase in elevation (Heaney and Rickart, 1990; Heaney et al., 1989; Utzurrum, 1998). Most species are fairly widespread within the Philippines, with very few locally endemic species—most occur either throughout much of the archipelago, or in one of the faunal regions discussed in chapter 4; Greater Luzon is one such faunal region (Heaney, 1985, 1986). Only two species of bats (*Desmalopex leucopterus* and *Otopteropus cartilagonodus*) are currently recognized as being restricted to Luzon. The reason seems straightforward: because most bat species are widespread at low elevation, there is little opportunity for reduced gene flow, and therefore not much chance for genetic isolation and speciation within a given island, even one as large as Luzon. Thus the story of the bats' diversification (chapter 5) is one that is told on a larger geographic scale than a single island, even one as large as Luzon (Heaney and Rickart, 1990; Heaney and Roberts, 2009).

The contrast between bats and the non-volant mammals is striking in many ways. There were 28 native non-volant species known in 2000; as of 2014, at the time of this writing, we have formally described and named 18 new species, and we are in the process of describing 10 more, an abrupt increase of 28 new species (fig. 1.9). This brings the total for Luzon to 56 species—an increase of 100% in the known fauna (Heaney et al., submitted), an extraordinary record of discovery, perhaps not equal to Whitehead's, but quite remarkable for the current century. Some of these new species represent two new genera, *Soricomys* and *Musseromys* (Balete et al., 2012; Heaney et al., 2014b). All of the new species are members of two clades (i.e., branches on the tree of life) that live only in the Philippines, and primarily on Luzon: the cloud rats and the earthworm mice (chapter 5). As we suspected when we began our intensive surveys on Luzon, we documented that there were not just four mountain ranges on Luzon where there were unique (i.e., endemic) species, there were nine—every isolated mountain or mountain range, as defined above, is a unique center of mammalian diversity. The evolutionary proliferation of these small mammals on Luzon, given its modest size relative to continents, is quite stunning.

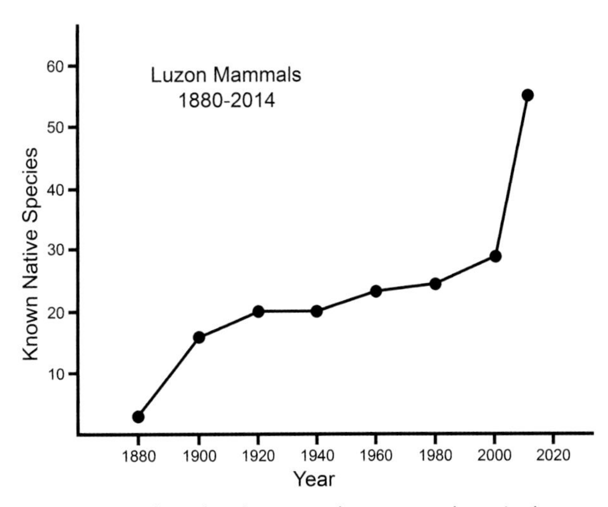

FIG. 1.9. Number of native non-volant mammal species known to occur on Luzon, in 20-year increments, from 1880 to 2000 and from 2000 to 2014.

The rest of part I represents our effort to tell the story of mammalian diversity on Luzon from the inferences we can draw from ecological aspects of current distributions, the geological history of the archipelago, and DNA-based studies of the fauna. This story is far from complete, and a vast amount of research remains to be done. But it is also true that we have learned a great deal, and the main features of that story are visible to an extent much greater than that of just 20 years ago. Part II of this book allows readers to become familiar with the evolution and ecology of individual species, including how to identify them in the field, along with an assessment of their conservation status. We hope this volume will motivate some readers to conduct new and more insightful research on all of these issues. Our intent is to be explicit about what we do and do not know, both in the interest of accuracy and in the hope that others will recognize the vast opportunities that remain for enlarging and enhancing the world's understanding and appreciation for the wonderful mammal fauna of Luzon, and thereby increasing the likelihood of its survival into the future.

2

Climate and Habitats

Luzon lies squarely in the tropics, with its northern coast reaching to about 18.6° N, well below the Tropic of Cancer, which marks the northern limit of the tropics at 22.5° N. The Pacific Ocean lies to the east, and the South China Sea lies to the west, between Luzon and the Asian continent, so that Luzon is surrounded by tropical seas. But Luzon extends over a full six degrees of latitude, from 12.6° N to 18.6° N, equivalent to the distance from the north coast of Spain to the south coast of England, or from Washington, DC, to Bangor, Maine, and the elevation varies from sea level to 2930 m on Mt. Pulag. It should therefore come as no surprise that there is considerable variation in its climate.

Temperature and Rainfall

Typhoons form one of the most prominent and dramatic elements of Luzon's climate. The number of tropical storms, which includes typhoons, that pass through Philippine territory and influence its weather ranges from about 15 to 30 each year. Many of these never directly strike land, veering off to the north and only brushing the east coast of Luzon (fig. 2.1). The number of typhoons that do pass over the islands increases from south to north: they are rare over southern Mindanao, moderately common in the central Visayan islands, and most common in central and northern Luzon. The area influenced by a given typhoon varies greatly, depending not only the path of the storm but also its diameter, which can range from as little as about 100 km up to about 500 km. Each of these typhoons brings strong winds near its center (over 300 km/hr in the strongest), causing direct, extensive damage to natural vegetation (fig. 2.2) as well as to buildings, agricultural areas, and roads (chapter 6). The amount of rain can be prodigious; a typhoon that struck Baguio City in 1911 dropped 1.2 m of rain in a 24-hour period, and nearly twice that amount over the full four days that its presence was felt. Some estimates place typhoons as generating 25%–30% of the mean annual rainfall on Luzon (Manalo, 1956).

Although Luzon lies in the wet tropics, seasonality is pronounced. This seasonality is not based on substantial variation in temperature, however; for example, in Manila, mean monthly high temperature fluctuates only from about 24°C to 28°C (fig. 2.3). While this may feel like a major change to people accustomed to little variation, it is quite minor to people from Chicago, where the coldest month (January) averages 0°C and the warmest (July) is 29°C. Instead, seasonality in Manila is imposed mostly by variation in rainfall, which reaches a mean monthly low of 10 cm in February and a high of 42 cm in June (fig. 2.3). Together, the dry months of January through April in Manila typically produce only a bit more rainfall than the single month of July.

The timing and extent of seasonality vary considerably within Luzon, however, depending on the loca-

FIG. 2.1. Typhoon tracks in the vicinity of the Philippines from 2009 to 2013. Based on maps in the JAXA/EORC Tropical Cyclone Database (2014).

FIG. 2.2. Lowland forest on Mt. Malinao (15 May 2007), damaged by two typhoons in the previous two years. Leaves were stripped away and many trees were blown over, opening the canopy, allowing fast-growing plants to spring up and increasing within-habitat diversity for the animals that live there.

tion. From the southern tip of the island near Legaspi up to Lucena and Infanta, there are no months that are especially dry (i.e., all have above about 15 cm of rainfall), though there are always some periods of the year that are wetter than others (fig. 2.3). Over central and northern Luzon, seasonality is always conspicuous, with some portions of the Zambales and Ilocos coasts in west-central and northwestern Luzon (from Iba to Laoag) having several months each year with virtually no rain. This extended period of dry weather results in many trees in the lowlands dropping their leaves during the dry season, much as trees in the north temperate zone (such as in Chicago) do in the winter. And although temperature varies little overall, the months of highest temperature usually occur during the late dry season, so the combination of clear, rainless skies and elevated temperatures produce very dry conditions, with rivers shrinking and agricultural fields becoming unproductive.

Clear evidence of two other, crucially important patterns in the climate of Luzon is shown in figure 2.3. Although temperatures in cities in the mountains, such as Baguio and Bontoc, are no more variable than those in the lowlands, it is always cooler in the mountains. Additionally, while the annual pattern of rainfall may be very similar in nearby lowland and mountain cities (compare Vigan and Baguio), the mountains receive vastly more rainfall.

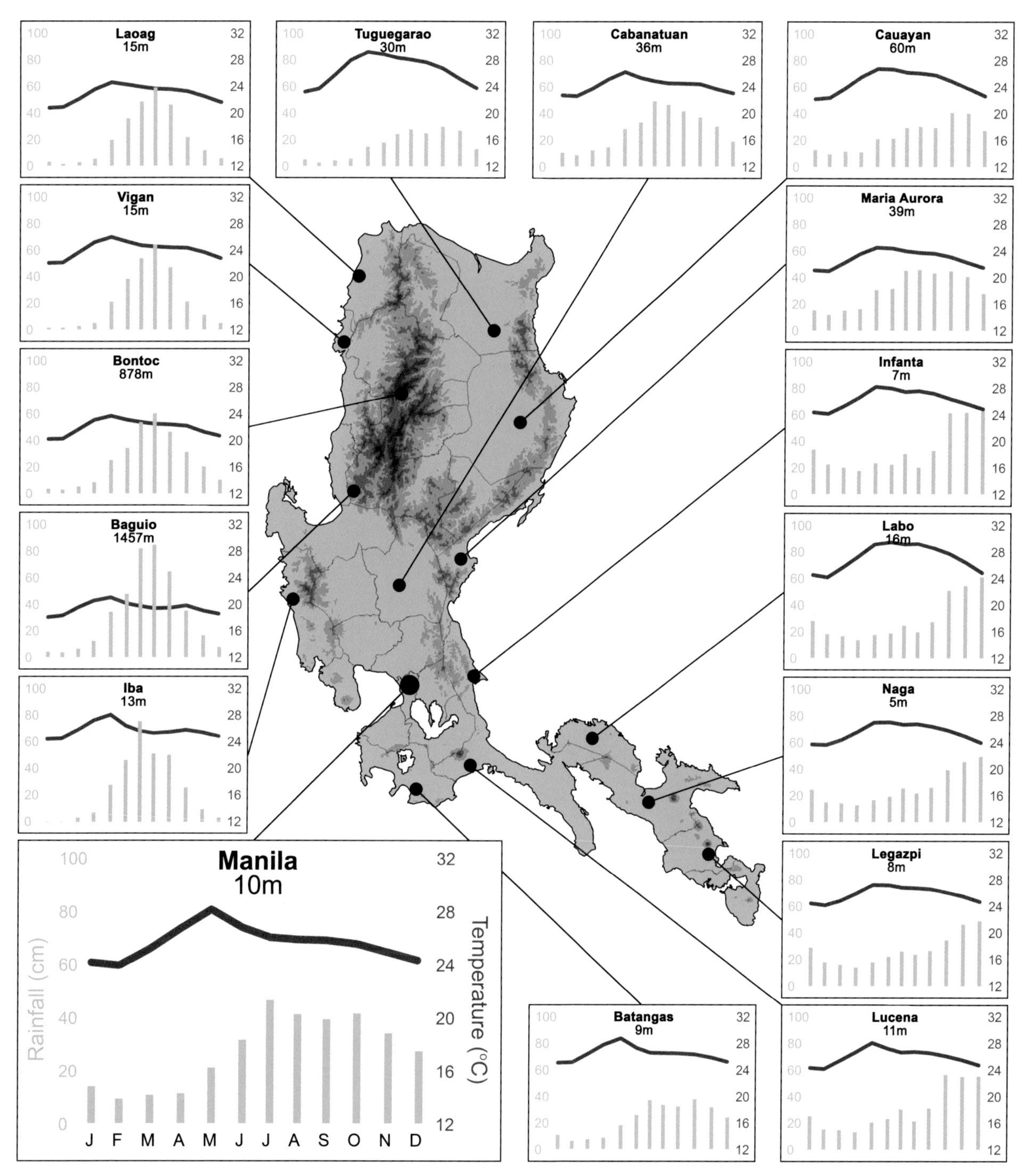

FIG. 2.3. Mean monthly temperature and rainfall at 16 locations on Luzon. Redrawn from figures in the World Bank Group's Climate Change Knowledge Portal (2014).

The first of these differences—the decline in temperature with increasing elevation—is evident in mountainous areas all over the globe. As air rises along a mountainside, it cools at a regular rate; in the Philippines, with its typically humid air, the rate is about 6°C of cooling per 1000 m rise in elevation. Thus, on a day when it is 30°C near sea level in Manila, it is likely to be around 21°C in Baguio, which lies at about 1500 m. This change in temperature was evident at our camp sites on Luzon. Although we measure daily high and low temperatures only for five days to two weeks while we sample at a given locality, the data we have collected show precisely this pattern: on average, temperature declines at 6°C per 1000 m from the lowest camps where we have data (about 900 m elevation) up to the highest at about 2600 m (fig. 2.4). There is variation in this graph because of the shortness of our sampling periods, but the trend is remarkably consistent. Since Baguio, at 1500 m, has a February mean low temperature of 10°C, we should expect that Mt. Data, at 2300 m, should have February lows averaging about 5°C. In light of that, John Whitehead's report of seeing frost on the ground is not surprising, and we should expect all of the higher mountain peaks in the Central Cordillera to have their temperatures drop to near freezing on a regular basis.

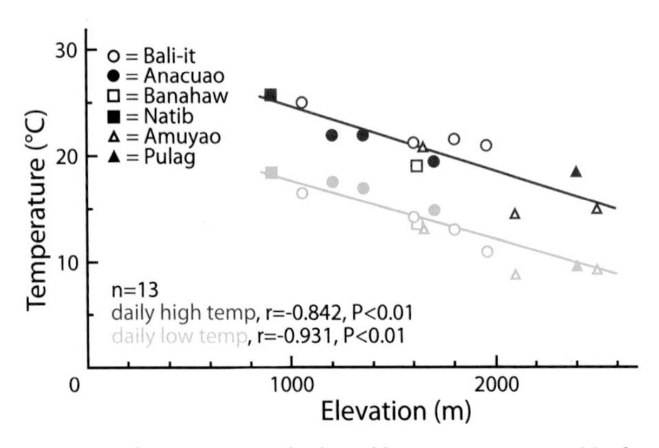

FIG. 2.4. Change in mean high and low temperature at 13 of our study sites on Luzon; each point is the mean of 5–15 days of measurements. Based on data in faunal reports cited here and in unpublished field notes (Division of Mammals Archives, FMNH).

The second difference between Vigan and Baguio—the difference in rainfall—is also strongly associated with elevation. As the air cools with increasing elevation, it also decreases in its ability to hold moisture in a gaseous state: in other words, the humidity in the air condenses, and droplets of water form. Initially, this takes the form of clouds or fog, but if there are enough droplets, they may aggregate into raindrops. Thus rainfall tends to increase with increasing elevation, with some variation due to local circumstances (mostly from the amount of humidity in the air, but wind speed and the steepness of the terrain also contribute). As is evident in figure 2.5, there is considerable variation in mean annual rainfall in the lowlands. At sea level along the Pacific (eastern) coast of Luzon, where typhoons make their first approach to land, rainfall ranges from about 2.5 to 3.5 m per year. Along the conspicuously drier Ilocos (western) coast, annual rainfall ranges from about 2.1 to 2.5 m; the Cagayan Valley lowlands are similarly dry (1.5–2.5 m), and perhaps slightly drier in places. The lower amounts of rainfall in the Cagayan Valley and on the Ilocos coast are due in part to being in a rain shadow; typhoons drop much of their moisture as they rise over the Sierra Madre and Central Cordillera, leaving the air masses less humid overall.

The most striking part of figure 2.5 is associated with the Central Cordillera. In that area, information is available about rainfall at many elevations, and it is apparent that there is a significant correlation between elevation and rainfall: the higher the location on the mountain, the more rain falls. At 500 or 1000 m elevation, the amount is not much different than on nearby lowlands, but at 2000 m, the average is over 3 m per year, and in some places at 2500 m (such as Atok), rainfall averages about 5 m per year (Manalo, 1956). Even this understates the amount that can fall at high elevation in a wet year; for example, on Mt. Isarog (at Balete's field camp at 1650 m elevation) in the year 1993–94 (which had an unusually high numbers of typhoons), at least 8.8 m of rain fell (Heaney et al., 1999). While 5 m per year may be the highest long-term average on most high mountains on Luzon, it is likely that every

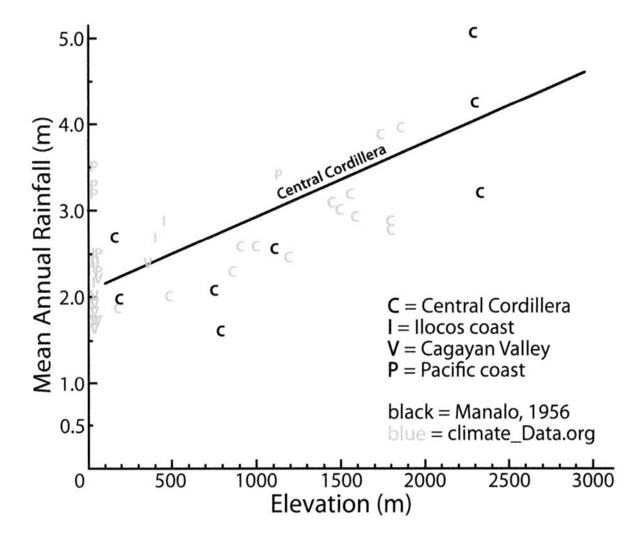

FIG. 2.5. Correlation of mean annual rainfall with elevation at 21 locations in the Central Cordillera (as shown in the regression line), in the Cagayan Valley, and on the Pacific coast and the Ilocos coast of northern Luzon. Data from http://en.climate-data.org/ and Manalo (1956).

year there are some mountains (different ones each year, depending on typhoon tracks) that receive 8 m of rain, and perhaps occasionally more. We suspect that the highest amounts fall on the high peaks of the Sierra Madre but have no data to test that prediction. By any standard, the high peaks on Luzon are *wet*.

These data on rainfall explain why streams and rivers are so important to agriculture on Luzon. Much of the rainfall in any given province falls on the high mountains, even though they have a proportionately small area. These high-elevation watersheds thus serve as crucial sources of water for agriculture, household use, and business. Because these mountains are steep, they are somewhat prone to natural landslides, and when the vegetation and soil that serve as a natural sponge to hold water are removed by agriculture, logging, or mining, it becomes all but inevitable that there will be massive landslides and floods.

Another measure of moisture availability is mean annual cloud cover (fig. 2.6). This figure shows in dramatic and obvious fashion the extent of areas that receive large amounts of rainfall, and even more clearly the areas that are often immersed in fog. Fog is a prominent part of the ecology of the mountains of Luzon; we have spent many soggy days in camp when no rain fell but fog was continuous and heavy. The largest area of heavy cloud cover is in northeastern Luzon; from the Caraballo Mountains and the adjacent Northern Sierra Madre up to the northeastern tip of Luzon at Cape Engano, cloud cover is frequent, especially over the mountains themselves. The narrow strip of lowland between the Mingan Mountains and the southern edge of the Northern Sierra Madre (just west of Baler) is clearly visible, suggesting that there may be little opportunity for species restricted to montane or mossy forest to cross between the two mountain chains, and a similar area of low cloud cover extends between the Mingan Mountains and the Southern Sierra Madre. We will return to a discussion of these two climatic barriers several times in this volume.

Other areas of high cloud cover are easily identified in figure 2.6: the Central Cordillera is not continuously connected by areas of frequent cloud cover, and the Zambales Mountains are easily visible but with variable levels of connection. Several isolated areas of heavy cloud cover are evident in the Bicol Peninsula, and a broad gap separates the Southern Sierra Madre from the nearest cloud-covered portions of the Bicol Peninsula. For mammals and other organisms that occur in montane and mossy forest, with their distinctive vegetation and soil conditions, the presence of frequent cloud cover, with the rainfall and fog that accompany it, may play a major role in influencing moisture and temperature conditions that shape patterns of distribution, abundance, and endemism.

Habitats

The combination of declining temperature and increasing rainfall along elevational gradients creates a clear set of environmental conditions that determines much of the variation in habitat on Luzon. Except in areas with special soil conditions (see below), there are three primary types of fairly distinct habitat (fig. 2.7). These descriptions are based largely on Fernando et al.

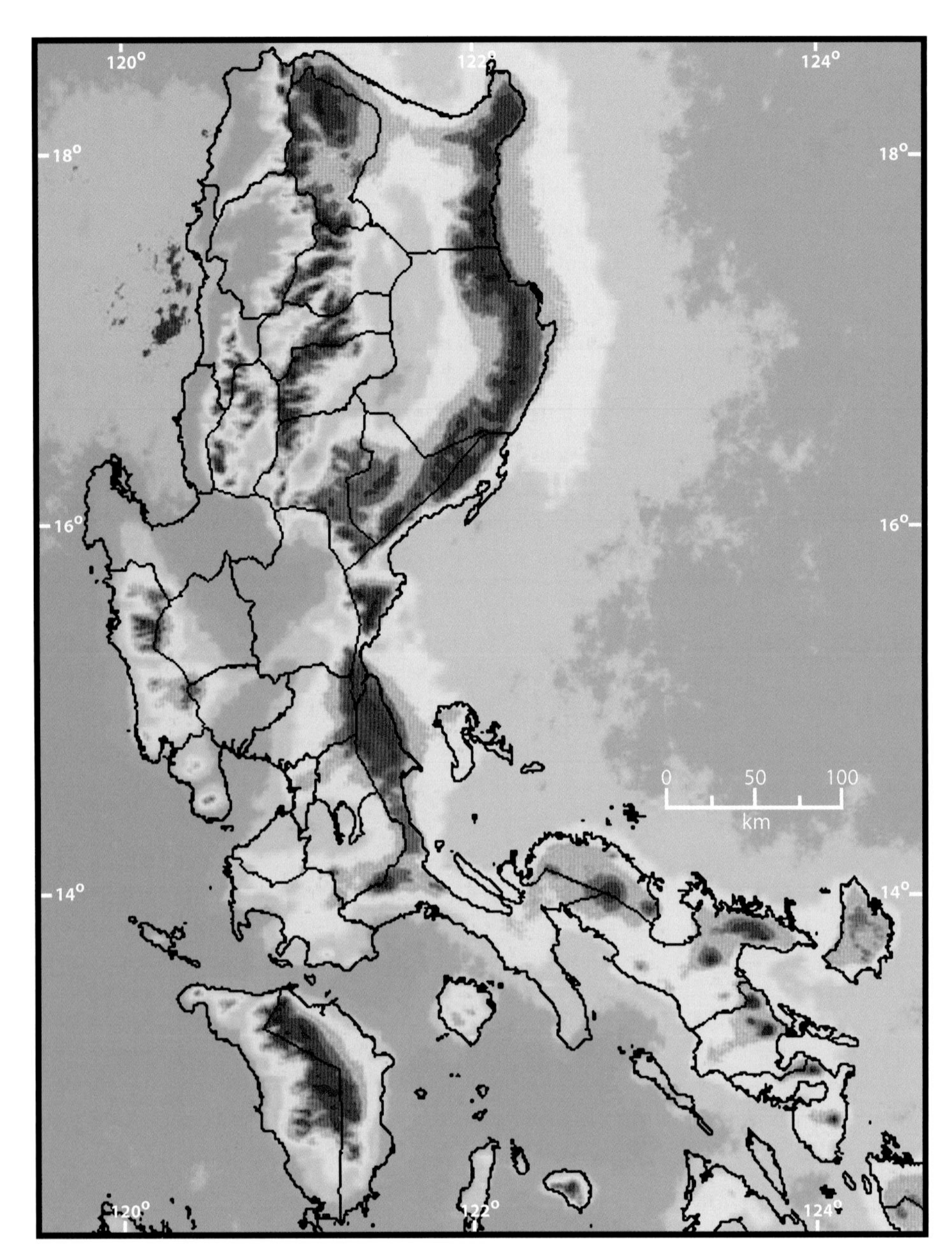

FIG. 2.6. Mean annual cloud cover over Luzon and nearby islands, generated from 67 months (January 2008–July 2013). The deeper shades of red indicate the heaviest annual cloud cover; orange, yellow, and green indicate progressively lower amounts of cloud cover. Based on images from Terra and Aqua satellites operated by the US National Oceanographic and Atmospheric Administration (2014).

(2008) and our publications on elevational transects (e.g., Heaney, 2011, 2013b).

From sea level to about 900 m elevation in most places, the primary natural vegetation on Luzon consists of lowland tropical rainforest. The most abundant trees are the dipterocarps (members of the Family Dipterocarpaceae, sometimes referred to as Philippine mahogany), including *Anisoptera*, *Hopea*, *Parashorea*, and *Shorea*. These grow to great height—at least 40 m on moist lowland soils and over 25 m in the foothills—and often are 1 meter or more in diameter. They are the source of beautiful and highly valuable wood, so they have been the subject of intensive logging. Figs (*Ficus*, Moraceae), *Elaeocarpus* (Elaeocarpaceae), *Dracontomelon* (Anacardiaceae), *Canarium* (Burseraceae), and many others also are common trees. Vines, including

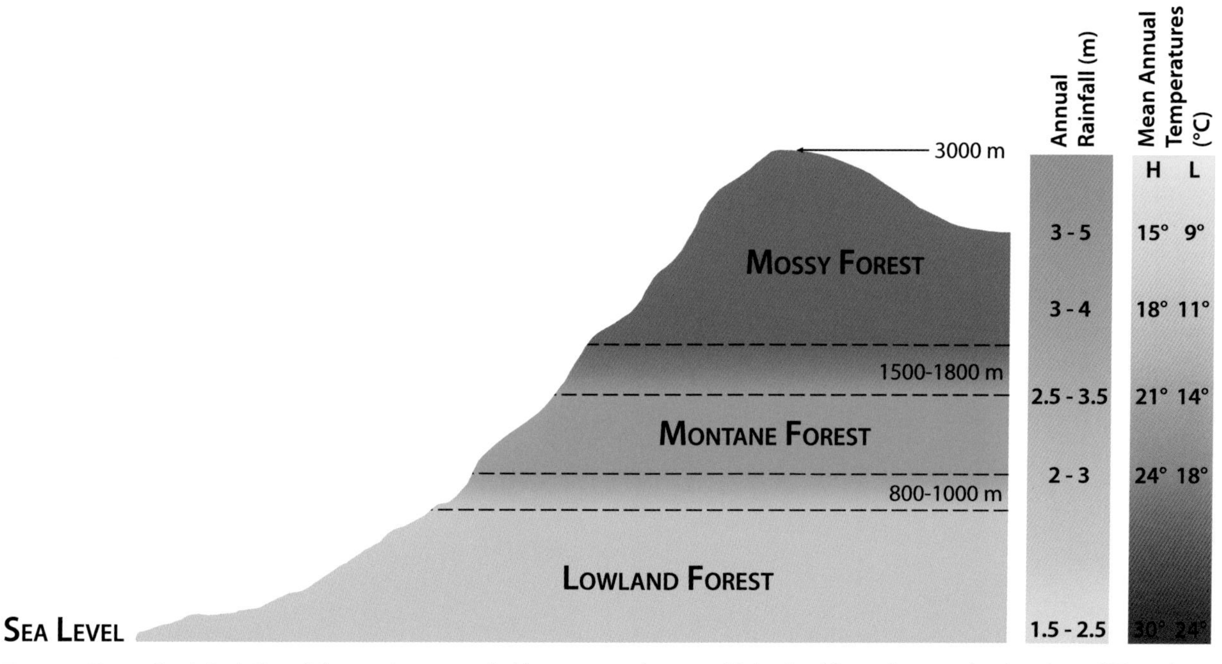

FIG. 2.7. Generalized depiction of the most common habitat types on Luzon, with lowland forest from sea level to about 900 m elevation, montane forest from 900 m to 1600 m, and mossy forest from 1600 m to the highest peaks. The elevation at which a transition takes place between these types varies, based on local rainfall patterns, exposure to wind and fog, and steepness of the terrain. Annual rainfall in some areas along the Pacific coast is probably higher than shown. Based on Heaney and Regalado (1998), updated with additional data from figs. 2.3 and 2.4.

rattan (*Calamus*), pepper (*Piper*), climbing bamboo (*Dinochloa*), and *Tetrastigma* snake up into the canopy. Epiphytes are fairly common, usually ferns, orchids, and aroids (e.g., *Alocasia*), but moss is rather scarce. In mature forest, there is a continuous canopy that catches nearly all of the sunlight, so the forest floor is rather dark and open, since there is too little light to allow much plant growth. As a result, ground cover in mature lowland forest (fig. 2.8) usually consists of scattered ferns, including tree ferns (*Cyathea*), gingers (*Zingiber* spp.), and palms (including *Caryota* and *Pinanga*). Because most lowland forest on Luzon has been logged, we have worked almost exclusively in regenerating forest that has smaller trees and a more open canopy, allowing thicker ground cover to grow. In such places, fast-growing plants (such as wild bananas, *Musa*) are often plentiful. Because of the abundance of termites and bacteria, little wood or leaf-litter builds up, leaving the surface of the ground relatively bare. Ants are abundant, and they quickly consume any edible fruits, seeds, or invertebrates they can find (including the baits in mammal traps). Small mammals are typically low in species richness and abundance, but both fruit-eating and insect-eating bats are diverse and common.

With increasing elevation, the temperature declines and rainfall increases, with gradually increasing impact on the plants and animals (fig. 2.7). By about 800 m elevation, with mean daily high temperatures reaching no more than 25°C and annual rainfall consistently at about 2.5 m or more, a transition begins into forest vegetation that is quite different from the lowland forest, and by about 1000 m elevation the appearance and composition of the forest has changed greatly (fig. 2.9). In this montane forest (sometimes called "lower montane forest"), dipterocarps are usually uncommon or absent, with oaks (*Lithocarpus*), myrtles (e.g., *Syzygium* and *Tristaniopsis*), laurels (e.g., *Actinodaphne*, *Cinnamomum*, and *Litsea*), and members of the tea family (e.g.,

FIG. 2.8. Lowland forest, 550 m elevation on Saddle Peak, Camarines Sur Province, 27 February 2008.

Eurya and *Camellia*) usually being the most common trees; some gymnosperms (e.g., *Agathis* and *Dacrycarpus*) and/or figs (*Ficus*) are sometimes present, usually in low numbers. Canopy height rarely exceeds 25 m, and on steep hillsides or ridgetops it may be lower. Tree trunks and branches are often coated with a layer of moss; epiphytic ferns, begonias, and orchids are common; and vines such as rattan (*Calamus*), pandans (*Freycinetia*), pepper (*Piper*), and climbing bamboo (*Dinochloa*) are common to abundant. Because the canopy tends to be more open than in lowland forest, the forest floor has more vegetation, made up of ferns and herbaceous plants. Tree ferns (*Cyathea*) and shrubs (e.g., *Melastoma, Medinilla*, and *Psychotria*) are often common. Termites are virtually absent, so the forest floor typically has at least a thin layer of leaf-litter and humus, with scattered fallen tree limbs and trunks; ants are scarce. In the humus and logs, earthworms, millipedes, and fungi are increasingly abundant with higher elevations. Small mammals also become more diverse and abundant with increasing elevation, but bats show the reverse pattern (chapter 3).

FIG. 2.9. Montane forest. A: 1625 m elevation on Mt. Banahaw, 5 April 2005. B: 1000 m elevation on Mt. Natib, 23 March 2005.

FIG. 2.10. Mossy forest, ca. 2100 m elevation on Mt. Bali-it, February 2003.

Usually between 1500 m and 1800 m elevation—as temperatures reach mean annual highs of 21°C or less, rainfall reaches 3 m or more, and fog becomes common—another transition takes place: from montane to mossy forest (sometimes called "upper montane forest"; fig. 2.10). Oaks (*Lithocarpus*) and gymnosperms (*Agathis, Dacrycarpus, Dacrydium, Falcatifolium,* and *Podocarpus*) are often the most abundant taxa, though laurels (*Actinidaphne, Cinnamomum,* and *Litsea*), elaeocarps (*Elaeocarpus*), and myrtles (*Leptospermum, Syzygium,* and *Tristaniopsis*) are usually common as well. Canopy height is rarely more than 15 m and not infrequently is as low as 5 m. The trees are usually festooned with thick layers of moss that hang from branches, and even leaves may have moss growing on them. Some other epiphytes also grow profusely, including orchids, *Medinilla, Melastoma, Vaccinium,* and ferns. Moss and leaf-litter cover the ground above a layer of humus that may vary in depth from 10 cm to over a meter, providing habitat for earthworms and other soil-living invertebrates in abundance. Fragments of acorn shells (*Lithocarpus* seed coats) and the tiny piles of dirt left by earthworms often are quite common. Ground plants are often grow profusely, including ferns, ground orchids, and liverworts; shrubs (*Medinilla* and sometimes *Rhododendron*) are common; tree ferns and vines (bamboos, pandans, and pitcher-plants) are often abundant. Bats are uncommon and represented by only a few species, but small mammals reach their greatest diversity and abundance (chapter 10).

Scattered places on Luzon have extensive areas with limestone as the primary rock underlying the surface; these usually occur in the lowlands. This limestone may be dense, hard rock or little more than poorly consolidated coral reef. Limestone slowly dissolves when exposed to rainfall, which is always slightly acidic, and because the hardness of the limestone varies, the flowing water gradually creates cavities that grow over time. The result is a jagged, rocky surface, referred to as karst (fig. 2.11), with many holes that allow water to drain away rapidly, so the surface is quite dry most of the time, even in places with substantial annual rainfall. The trees and other plants that grow over limestone must therefore be highly tolerant of dry conditions,

FIG. 2.11. Forest over limestone. A: 400 m elevation, Tayabas, Quezon Province, 8 November 2008. B: 140 m elevation, Caramoan National Park, Camarines Sur Province, 29 May 2008.

and plant communities here are typically quite different from those on nearby soils that do not lie over limestone. The most common tree species in these locales is molave (*Vitex parviflora*). and leguminous trees (e.g., *Afzelia*, *Albizia*, and *Intsia*) often are also common, with an understory of figs (*Ficus*), *Melastoma*, and palms (*Calamus*, *Caryota*, and *Pinanga*). The soil is usually quite thin, trees are widely spaced, and we have found ant populations to typically be very high.

Small mammals in karstic areas tend to be scarce, with one dramatic exception. Limestone may not retain water well because of the many cavities that lie below the surface, but those cavities—especially the large ones—provide the locations for the densest populations of mammals on Luzon. Prior to human disturbance, most limestone caves supported huge colonies of bats; some still do, but many have been severely impacted by overhunting. Both species richness and abundance of bats in limestone areas are the highest we have documented (chapter 7).

At scattered places around Luzon, there are geological rock units known as ophiolites (chapter 4) that produce ultramafic soil. Forest growing on ultramafic soil (i.e., soil rich in heavy metals—often including magnesium, iron, and nickel—but low in silica, potassium, calcium, and phosphorus) is often quite distinctive, with plants that are rare or absent elsewhere. The woody vegetation in this habitat typically has thick, rather leathery leaves, and often the trees are stunted, to the point that herbaceous ground plants and grasses may be taller than the trees (fig. 2.12). Some trees that grow in ultramafic soils are nickel accumulators, including *Phyllanthus* (Phyllanthaceae) and *Rinorea* (Violaceae). Other trees that grow in this habitat include

FIG. 2.12. Forest over ultramafic soil, 1100 m elevation on Mt. Anacuao, Aurora Province, 31 March 2010.

Calophyllum (Calophyllaceae), *Gymnostoma* (Casuarinaceae), *Dillenia* (Dilleniaceae), and several myrtles (*Leptospermum*, *Syzygium*, and *Tristaniopsis*). Podocarps (*Agathis*, *Dacrydium*, and *Falcatifolium*) are also present, especially at higher elevations. The ground vegetation commonly includes many genera of ferns; carnivorous plants, such as pitcher-plants (*Nepenthes*) and sundews (*Drosera*); as well as ground orchids (*Paphiopedilum*) and monocots (*Dianella* and *Patersonia*). Few studies of mammals have been conducted in this habitat, but apparently both species richness and density tend to be low. Birds and insects also generally occur at low density, and bats may as well, but we have too few data to be sure of this.

Pine forest is widespread on Luzon and attracts much attention, because it is highly visible in the vicinity of Baguio, which is a major tourist destination. Pines originally grew only where the terrain was very steep and rocky, at moderate to high elevations, but frequent fires set by humans in the Cordillera and Zambales Mountains have allowed pines to spread widely. In places that are regularly burned (usually every few years), the understory consists of grasses, mostly cogon (*Imperata cylindrica*) and *talahib* (*Saccharum spontaneum*), with ferns (especially bracken) growing in relatively moist places (fig. 1.1). When no fires occur, the forest vegetation begins to regenerate, with plants such as *Melastoma* quickly invading, followed by the diverse broad-leafed plants that occur in montane or mossy forest. In many places in the Cordillera, we have conducted our studies in mature montane and mossy forest where a few quite large pines persist; these are the remnants from the time burning ceased, perhaps a century or more ago (Kowal, 1966). Pines may occur as low as 500 m elevation or as high as 2800 m, but they are most common from about 1000 m to 2000 m. The species in the Central Cordillera is *Pinus kesiya*, and in the Zambales it is *P. merkusii*. As John Whitehead noted long ago, the open, frequently burned pine forest has very few native mammals or birds, but non-native pest rats are usually common; when pine forest regenerates into montane or mossy forest, the native mammals move back in (chapter 6).

3

Discovering Diversity: Topography and Elevational Diversity Patterns

Most people on Luzon live along the coast or in relatively flat lowland areas, so they think of Luzon as being composed of lowlands. A large portion of the island, however, is covered by hills and mountains that ascend to nearly 3000 m elevation (about 10,000 feet; fig. 1.7), and much of the environmental variation on the island is associated with this topographic diversity (chapter 2). That environmental and topographic diversity is also closely associated with the patterns of diversity and distribution of the mammal fauna.

For many years it was widely believed—and frequently stated in textbooks—that maximum biodiversity occurs in lowland tropical rainforest. Surprisingly, in the past two decades it has become increasingly clear that although many groups of organisms have their greatest diversity (and often abundance) in the lowlands, many others are most diverse in the cooler, wetter habitats on mountainsides, and some of these groups are entirely absent from the lowlands (Heaney, 2001; Heaney et al., 1989; McCain, 2005; Rahbeck, 1995). For example, old-growth lowland forests on Luzon are dominated by dipterocarp trees, but this family is virtually absent above 1100 m elevation; above about 900 m, oaks, laurels, and myrtles are among the most common trees (chapter 2). Obviously, we should not assume that every group of organisms is common or diverse in lowland tropical forest, and it is precisely this geographic variation among groups of organisms that adds greatly to the total biodiversity in any tropical area, including Luzon.

Species Richness along Elevational Gradients

Among mammals, we see two very different patterns of elevational diversity. Bats are certainly among the groups that are most diverse in the lowlands. A plot of documented records of the eight common, easily captured, medium-sized to small fruit bats (Family Pteropodidae; fig. 3.1; chapter 11) shows this clearly. Only one species, *Otopteropus cartilagonodus*, is most abundant above 1000 m elevation and is virtually absent below 500 m. All of the other species are most abundant in the lowlands, below 1000 m. One, *Ptenochirus jagori*, ranges from sea level up to 2000 m and is a generalist with respect to habitat (chapter 11). Another species, *Macroglossus minimus*, feeds heavily on the flowers, nectar, and fruits of both wild and domestic bananas (*Musa* spp.). Since wild bananas spring up quickly in areas where landslides have occurred, and people grow domestic bananas at surprisingly high elevations, we have found these little flower bats at places up to about 1700 m elevation, but they are far more common in the lowlands.

Other families of bats show a similar pattern of greatest diversity and abundance in lowland habitats. For example, the seven species of roundleaf bats (Family Hipposideridae) on Luzon all occur in the lowlands, including the three that are endemic to the Philippines (*Coelops hirsutus*, *Hipposideros antricola*, and *H. pygmaeus*; fig. 3.2). One species (*H. diadema*) has occasionally been found as high as 2500 m elevation, but even they are rare above 1000 m. Among other families of bats (chapter 11), only 2 (*Falsistrellus petersi* and *Nyctalus plancyi*) out of more than 30 species in the Vespertilionidae appear to be restricted to high-elevation areas (Heaney et al., 2012); the other 28 all occur primarily or entirely in the lowlands.

The contrast between the distribution patterns of bats and of the native small mammals is dramatic. For example, of the 12 species of cloud rats on Luzon (chapter 10), only 3 occur in the lowlands; the rest occur only above about 1400 m elevation (fig. 3.3), and the most diverse communities of cloud rats occur between 2000 m and 2500 m. Although *Batomys* and *Phloeomys* spend some time on the ground, all of the species are primarily arboreal; the cool, wet, mossy forest canopy

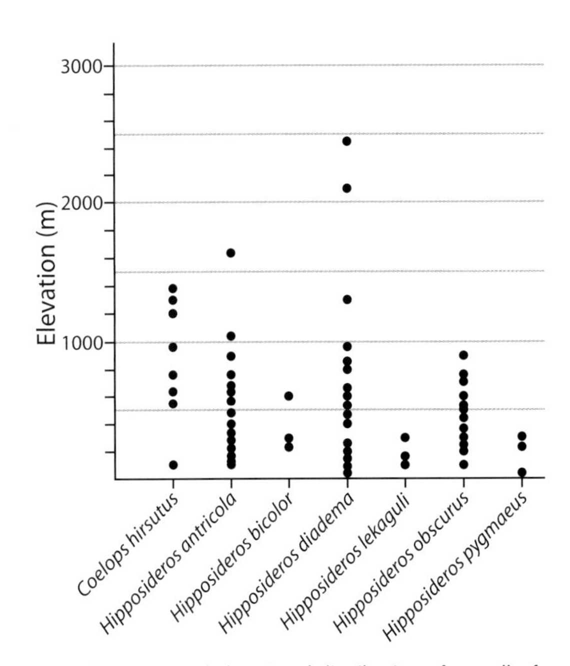

FIG. 3.2. Documented elevational distribution of roundleaf bats (Family Hipposideridae) on Luzon Island. Each dot corresponds to a given elevational record; thus one dot may represent many specimens from several localities. Based on museum specimens examined by the authors.

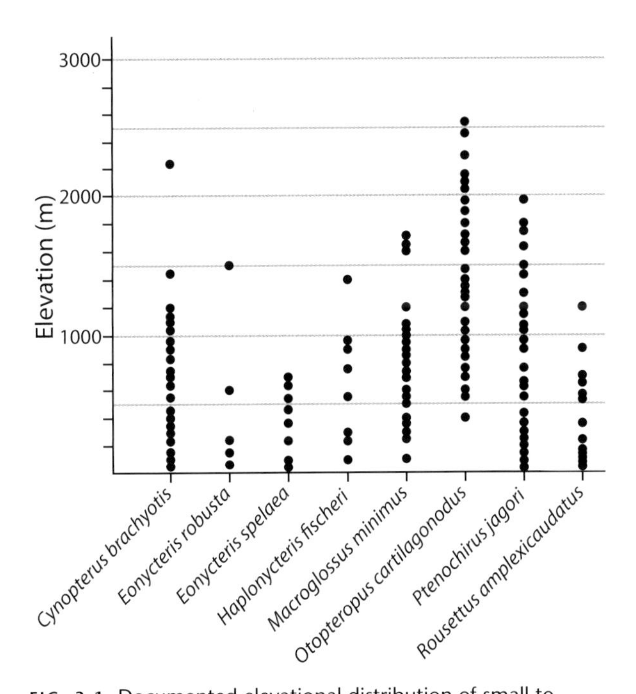

FIG. 3.1. Documented elevational distribution of small to medium-sized fruit bats (Family Pteropodidae) on Luzon Island. Each dot corresponds to a given elevational record; thus one dot may represent many specimens from several localities. Based on museum specimens examined by the authors.

clearly provides the primary habitat for these animals. All of the cloud rats are strictly herbivorous, with the giant cloud rats (*Phloeomys* and *Crateromys*) feeding on tender young leaves, bamboo shoots, and some fruits, while the smaller species (*Musseromys*) mostly eat seeds (chapter 10).

A similar pattern is evident in the other large group of small mammals endemic to Luzon, the earthworm mice (chapter 10). There are 30 species of earthworm mice currently recognized on Luzon, with more likely to be described when ongoing studies are completed. The most thoroughly studied are the large forest mice, subgenus *Megapomys* of the genus *Apomys* (Heaney et al., 2011a; Justiniano et al., 2015). We present more details about them in chapter 5, which focuses on speciation; in this chapter, we are concerned only with their elevational distribution. These large forest mice are among the most abundant native mice on Luzon. They are active on the surface of the ground, rarely climbing more than 1 m up the trunks of large trees,

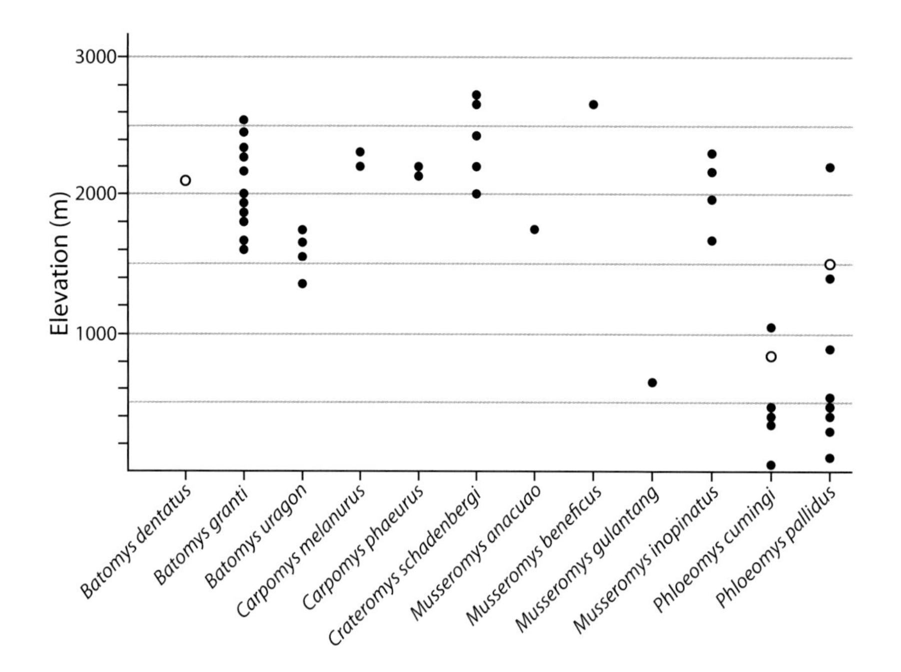

FIG. 3.3. Documented elevational distribution of cloud rats on Luzon Island. Each dot corresponds to a given elevational record; thus one dot may represent many specimens from several localities. Open circles represent approximate elevations. Based on museum specimens examined by the authors.

and—uniquely among the earthworm mice—they feed on seeds and fruits in addition to worms. Most of the species occur only above 500 m elevation; of 11 species on Luzon, only 2 (*A. sacobianus* and *A. zambalensis*) extend down to about 300 m (fig. 3.4). Surprisingly, where *A. sierrae*, the forest mice that are widespread in the Northern Sierra Madre, occur on Luzon itself, they have been documented down to about 600 m elevation, but on the small island of Palaui these mice live at elevations less than 200 m. Similarly, *A. lubangensis*, a species that is endemic to Lubang Island (a small, deepwater island southwest of Manila Bay), also occurs at low elevations. This suggests the presence of some ecological difference between small islands such as Lubang and Palaui and large islands such as Luzon that allows "normally" montane species to extend downslope, nearly to sea level. This is a tantalizing topic about which we have almost no information—certainly another subject for future research.

Another pattern is evident in figure 3.4: most of the mountain ranges, regardless of their area, have two or three species of the subgenus *Megapomys* on them. In every instance, there is some elevational overlap between the two species, and we have documented them as living syntopically (i.e., with overlapping home ranges), but there is always some substantial difference in their ranges, with one species occurring mostly at higher elevations and the other lower down.

Aside from the species of *Apomys* shown in figure 3.4, figure 3.5 shows the elevational records for all of the other earthworm mice on Luzon. Again, only a few species occur in forest at low elevation. *Apomys microdon* occur widely in secondary and mature lowland and montane forest from sea level up to about 2000 m elevation; this is a generalist species, eating invertebrates and seeds on the ground and in trees. *Chrotomys mindorensis* feed only on soft-bodied invertebrates, including earthworms, snails, and millipedes. This species lives only on the ground and often digs into the soil to catch its prey. Although it is a specialist in its feeding, it is a habitat generalist, living in agricultural areas, shrubby fallow fields, and mature forest (Alviola et al., 2011; Balete et al., 2009; Rickart et al., 2013). The animal noted in figure 3.5 as *Apomys* sp. is an as-yet-unnamed species from an area of lowland forest over limestone near Peñablanca, Cagayan Province (chapter 12); this hints at the potential importance of karst as a specialized habitat (chapter 7), another topic about

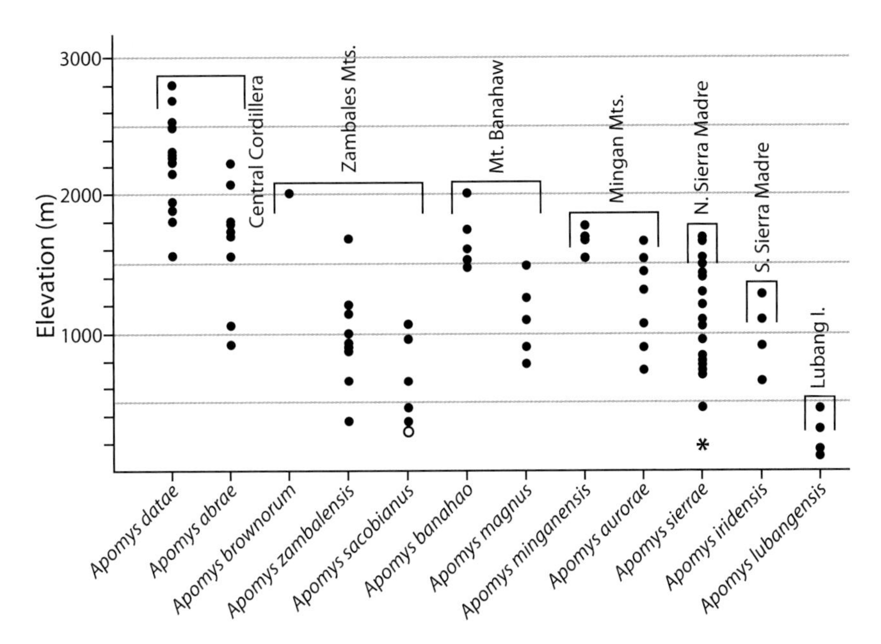

FIG. 3.4. Documented elevational distribution of large forest mice (subgenus *Megapomys* of the genus *Apomys*) on Luzon and two adjacent islands. One dot corresponds to a given elevational record; thus each dot may represent many specimens from several localities. Open circles represent approximate elevations. The asterisk represents the occurrence of *A. sierrae* on Palaui, a small island in shallow water adjacent to northeastern Luzon. *A. lubangensis* occur only on Lubang Island. Based on museum specimens examined by the authors.

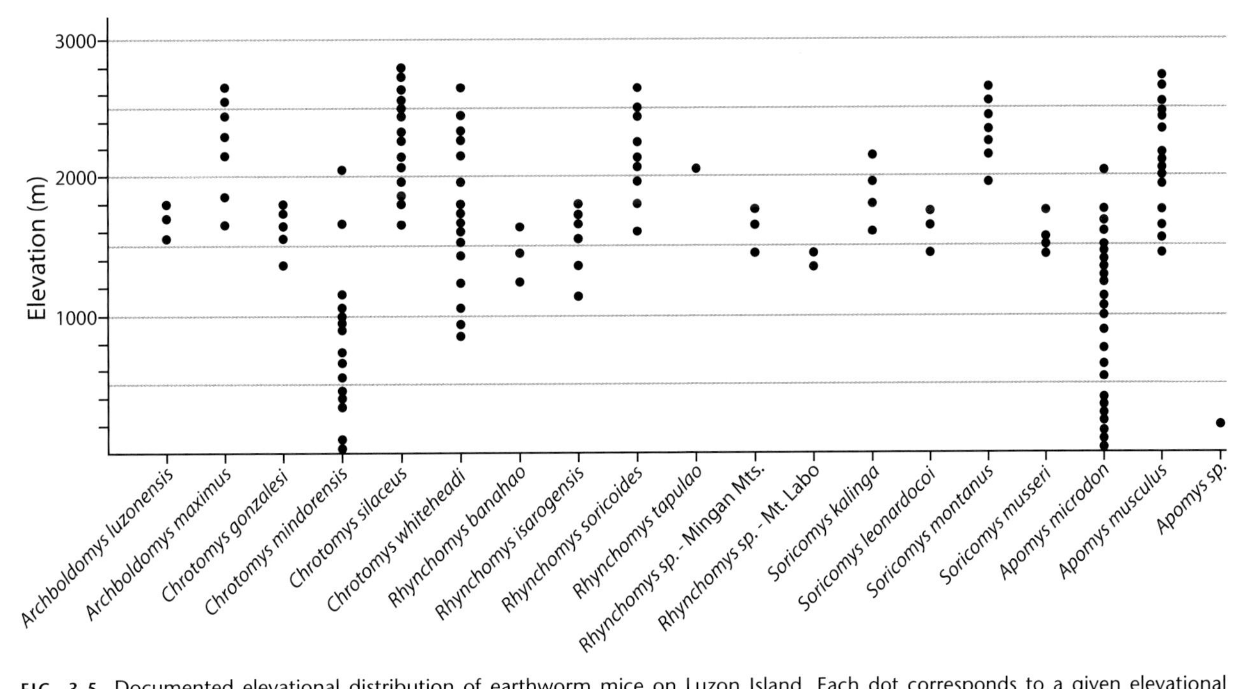

FIG. 3.5. Documented elevational distribution of earthworm mice on Luzon Island. Each dot corresponds to a given elevational record; thus one dot may represent many specimens from several localities. Based on museum specimens examined by the authors.

which we have limited information but that is deserving of future research.

All of the rest of the species of earthworm mice in figure 3.5 occur only above 1100 m elevation, and quite a few only above 1500 m. With the exception of the more omnivorous species of *Apomys*, they all feed primarily on soft-bodied soil invertebrates, especially earthworms. These invertebrates are scarce in lowland forest but are common to abundant in montane and mossy forest. Up to seven species of earthworm mice may occur within a hectare of mossy forest in the Central Cordillera (Rickart et al., 2011a): two *Chrotomys* dig for their prey within the layer of humus; one or two shrew-mice species (one each of *Archboldomys* and *Soricomys* on Mt. Amuyao) forage in the abundant leaf-litter; a species of *Rhynchomys* makes its trails on the forest floor, which it patrols for earthworms; two large *Apomys* species feed at the surface of the ground (and include some seeds and fruits in their diet); and one *Apomys* species hunts in the canopy for earthworms and seeds (the epiphytes contain small pockets of soil in which earthworms live).

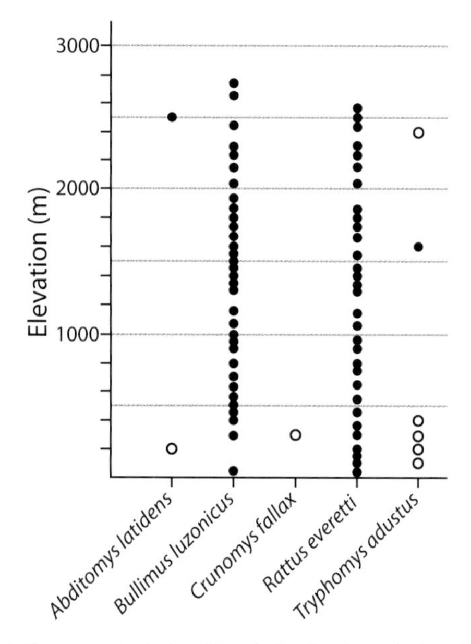

FIG. 3.6. Documented elevational distribution of New Endemic native mice on Luzon Island. Each dot corresponds to a given elevational record; thus one dot may represent many specimens from several localities. Open circles represent approximate elevations. Based on museum specimens examined by the authors.

The far-less-diverse New Endemic native rodents (chapter 10) show quite a different pattern (fig. 3.6): the two well-known species (*Bullimus luzonicus* and *Rattus everetti*) occur from sea level to over 2500 m elevation and most often are associated with disturbed habitat. The poorly known *Abditomys latidens* and *Tryphomys adustus* have similarly large ranges; only *Crunomys fallax*, still known only from a single specimen after more than 125 years since its discovery by John Whitehead, appear to have a restricted elevational range, at just a few hundred meters above sea level.

Seven species of non-native small mammals have established populations on Luzon (Singleton et al., 2008). Three of these—the house shrew (*Suncus murinus*), spiny ricefield rat (*Rattus exulans*), and Asian house rat (*R. tanezumi*)—all have broad elevational ranges (fig. 3.7). All of these species are usually associated with heavily disturbed habitat, including agricultural areas and buildings, but all of them also occur in regenerating natural vegetation and thus have the potential for extensive interactions with the native small mammals (chapter 6). House mice (*Mus musculus*) are common in buildings and occasionally in some lowland crops, but scarce elsewhere; most records are from 500 m elevation or below, though they occur in urban areas up to at least 1500 m elevation. On Luzon, *R. argentiventer* have only been recorded in ricefields near Los Baños (Barbehenn et al., 1973), and wharf rats (or Norway rats, *R. norvegicus*) have been found only in urban areas and a few lowland agricultural areas. *Rattus nitidus*, which are a common agricultural pest in southern China and the Himalayas (Smith and Xie, 2008), have been found on Luzon only in a few places in the Central Cordillera, and none during our studies in the past decade. Variable squirrels (*Callosciurus finlaysonii*) were introduced to Manila from Thailand in the 1960s; they now occur widely in urban and suburban areas and are spreading rapidly. Overall, these data indicate that most non-native species are highly restricted to human habitations and agricultural areas (chapter 6).

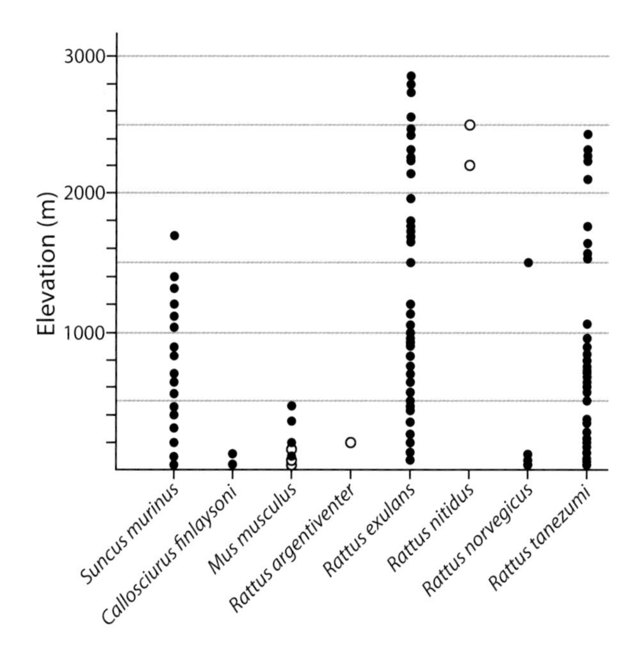

FIG. 3.7. Documented elevational distribution of non-native small mammals on Luzon Island. Each dot corresponds to a given elevational record; thus each dot may represent many specimens from several localities. Open circles represent approximate elevations. Based on museum specimens examined by the authors.

The final set of species encompasses the large mammals—monkeys, civets, wild pigs, and deer (some of which are arguably "medium mammals"; we use the term "large mammals" for these five species for convenience). All of these species are heavily hunted and have been eliminated from many places where they once occurred (chapters 7 and 9). The records in figure 3.8 are from specimens in museum research collections, some of which date from the early 1900s, and from our published and unpublished field notes. Thus these should be viewed as historical elevational ranges, rather than current ranges.

All of these large mammals originally occurred in lowland forest, and two of them—macaques (*Macaca fascicularis*) and Malay civets (*Viverra tangalunga*)—apparently have always been uncommon above about 1100 m elevation, which is the upper limit of floristic elements of lowland forest. The other three—palm civets (*Paradoxurus philippinensis*), Philippine wild pigs (*Sus philippensis*), and Philippine brown deer (*Cervus mariannus*)—ranged from near sea level to the highest peaks, although palm civets may be uncommon above about 2000 m (fig. 3.8). As discussed in chapters 4 and 6, macaques and both species of civets are probably not native to Luzon, but they have become fully naturalized in forest habitat. Unfortunately, because all of these large mammals are severely overhunted in most places (chapter 7), we are unable to assess their variation in abundance along elevational gradients and are limited in the ways we can investigate the role they play in mammal communities.

In summary, these elevational distribution data show that bats are most diverse at low elevations, and their diversity in general decreases with increasing elevation. All other groups show a different pattern: only a few are limited to lowland forest, some occur along the entire elevational gradient, and the great majority

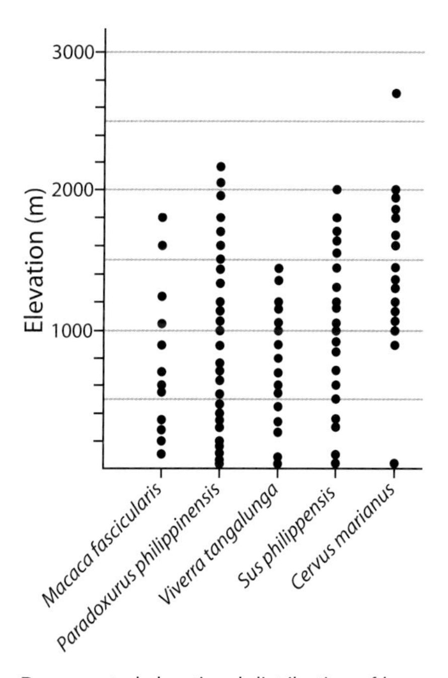

FIG. 3.8. Documented elevational distribution of large mammals on Luzon Island. Each dot corresponds to a given elevational record; thus one dot may represent many specimens from multiple localities. Based on museum specimens examined by the authors and on our published and unpublished field observations (Division of Mammals Archives, FMNH).

occur only in montane and mossy forest (or penetrate only a short distance downslope into the lowlands). Unambiguously, non-volant native mammals on Luzon, which form about half of the mammal fauna, are not most diverse in lowland rainforest (Heaney, 2001).

Changes in Species Richness of Small Mammals along Elevational Gradients

These observations lead immediately to three questions about the non-volant mammals. First, does the pattern of increasing diversity with increasing elevation occur on each of the many mountain ranges on Luzon? Second, do mountain ranges with higher elevations have more species overall, as a result of the increase in diversity with higher elevations? Third, is the increase in the number of endemic species with increasing elevation a consistent pattern? Because the large mammals have been so severely impacted by overhunting, we exclude them from this analysis and focus only on the generally diverse small mammals, including both native Luzon shrews (*Crocidura grayi*) and the many native rodents. Among the native rodents, we do not include the giant cloud rats unless stated otherwise, both because our sampling procedures allowed us to assess only their presence, not their abundance, and because they, too, have been impacted by overhunting.

We conducted our surveys along elevational gradients using consistent methods (including standardized types of traps and baits), with placement of the traps on the ground, in trees, and on vines (chapter 1). Our inspection of species accumulation curves indicates that our estimates of species richness (i.e., the number of species at a given locality) are likely to be reliable. We consider the measures of abundance presented in this volume to be less definitive; that is, they are accurate indices of abundance, but not precise measurements. In other words, readers should use the abundance figures here as good estimates and look to the individual publications mentioned in the references to fully understand the limitations of our numbers and more nuanced interpretations of them.

A clear and fairly simple example comes from our study on Mt. Anacuao, in the Northern Sierra Madre (fig. 1.8; Heaney et al., 2013a). In transitional lowland–montane forest at 940 m elevation and in lower montane forest at 1125 m elevation, five species of small mammals were present. At both places, *Bullimus luzonicus* were the most abundant species. Luzon shrews (*Crocidura grayi*) were represented by a single capture or were inferred to be present because they occurred both lower and higher along the elevational gradient (fig. 3.9). The same five species were either captured or inferred to be present up to 1500 m elevation, but the relative abundance of each species varied along the gradient: shrews remained uncommon, arboreal mice (*Apomys microdon*) increased greatly in abundance at 1300 m and above, the large Sierra Madre forest mice (*A. sierrae*) increased steadily in abundance with increasing elevation, and *B. luzonicus* and *Rattus everetti* reached their maximum abundance at about the middle of the transect. As a result of this variation, the number of species captured or inferred to be present up to 1500 m elevation remained constant at five, but the total abundance of small mammals increased steadily from 1.8 per 100 trap-nights (= "t-n") to about 4.2 per 100 t-n with increasing elevation.

At the highest sampling area (1760 m elevation), the first one located in prime mossy forest, the same patterns are evident within the five species, but three species have been added: *Chrotomys whiteheadi* and *Soricomys musseri* are earthworm mice, and *Musseromys anacuao* is a tiny cloud rat. This sampling area thus has the highest species richness (eight small mammals), with the three newly added species occurring at low abundance (fig. 3.9).

This example exhibits several noteworthy aspects: (1) each species varied in abundance along the elevational gradient, with considerable variation where any given species was most abundant; (2) maximum diversity occurred at the highest elevation sampled (near the peak); (3) two of the eight species at the peak (*Musseromys anacuao* and *Soricomys musseri*) are endemic to the Northern Sierra Madre, and the others are more geographically widespread; and (4) no non-native mammals were captured anywhere along the transect.

With an elevation of 1850 m at its peak, Mt. Anacuao

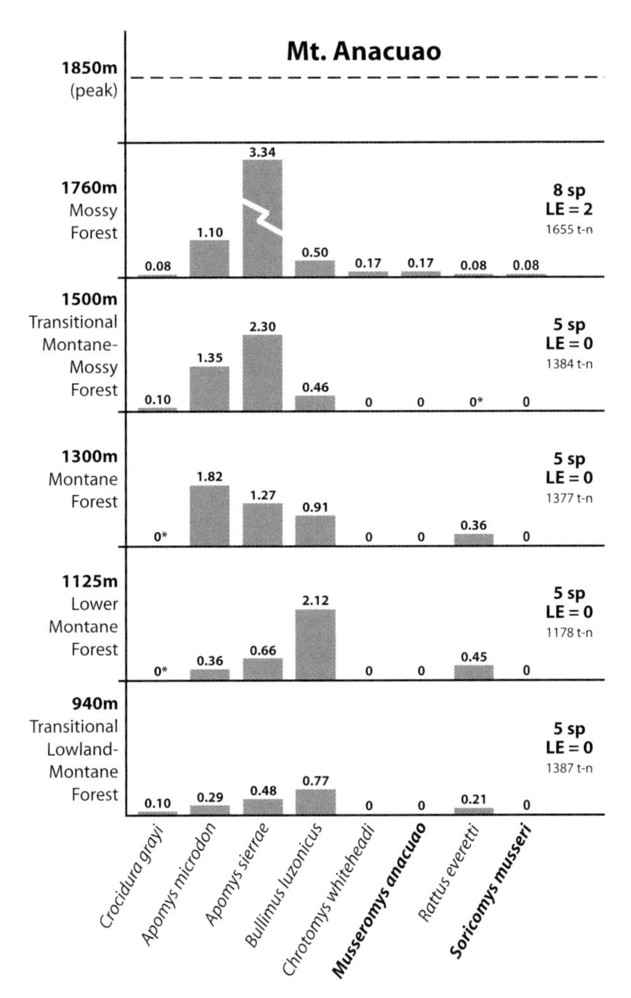

FIG. 3.9. Elevational patterns of presence and abundance in small mammals on Mt. Anacuao, Aurora Province. Numbers at the top of each column are the number of animals captured per 100 trap-nights; the height of each column is scaled to that value. The column at the far right gives the total number of species present and inferred, the number of locally endemic (LE) species (in this case, endemic to the Northern Sierra Madre), and total sampling effort (number of trap-nights = t-n). Asterisks indicate species that are inferred to be present at a given elevation because they occur both below and above the sampling area along the transect. Locally endemic species are shown in boldface. Data from Heaney et al. (2013a: table 1).

is one of the higher mountains in the Sierra Madre, but it is much lower than many peaks in the Central Cordillera. Luzon's highest mountains are in the Central Cordillera, so they present an opportunity to further investigate the patterns identified on Mt. Anacuao. Our transect on Mt. Bali-it, in Balbalasang-Balabalan National Park in the Central Cordillera, began at the lowest elevation for any of our transects (fig. 1.8). It started at 925 m elevation in mixed agricultural land, fallow fields, and regenerating secondary lowland forest and reached from there up to mossy forest at 2150 m elevation, just below the peak at 2238 m. Fifteen species of small mammals were present: 2 that are non-native and 13 that are native (fig. 3.10).

In the results from this transect, several features stand out (Rickart et al., 2011a, 2011b). The two non-native species (*Rattus exulans* and *R. tanezumi*) were abundant in the area heavily utilized by people at the lowest elevation, but three native species—the vermivorous *Chrotomys whiteheadi*, a generalist mouse (*Apomys abrae*), and a generalist forest rat (*R. everetti*)—were also fairly common (fig. 3.10). Only one individual of the non-native rats was captured at the next-higher elevation, and none at all above there. As on Mt. Anacuao, non-native species were not part of the small mammal community in mature forest, but native species were a common component of disturbed habitat.

Paralleling our findings on Mt. Anacuao, the most common species on Mt. Bali-it belong to the genus *Apomys* (subgenus *Megapomys*), in this case *A. abrae* and *A. datae*. They overlapped at only one sampling locality, where together they were roughly as common as either species was at localities where only one of the two species was present. This hints at the potential for competition between these two closely similar species (chapter 10) as a plausible explanation for their largely separate distributions. Aside from this one possible exception, each species apparently varied independently in its abundance along the elevational gradient, usually in some regular fashion; that is, each reached its highest abundance at some point along the gradient and was less abundant both above and below that elevation.

Seven of the species we found on Mt. Bali-it are endemic to the Central Cordillera. These included the two most abundant species (*Apomys abrae* and *A. datae*), the least abundant species (*Carpomys phaeurus*), and several that were intermediate in abundance; in this respect there was no shared pattern. There was a

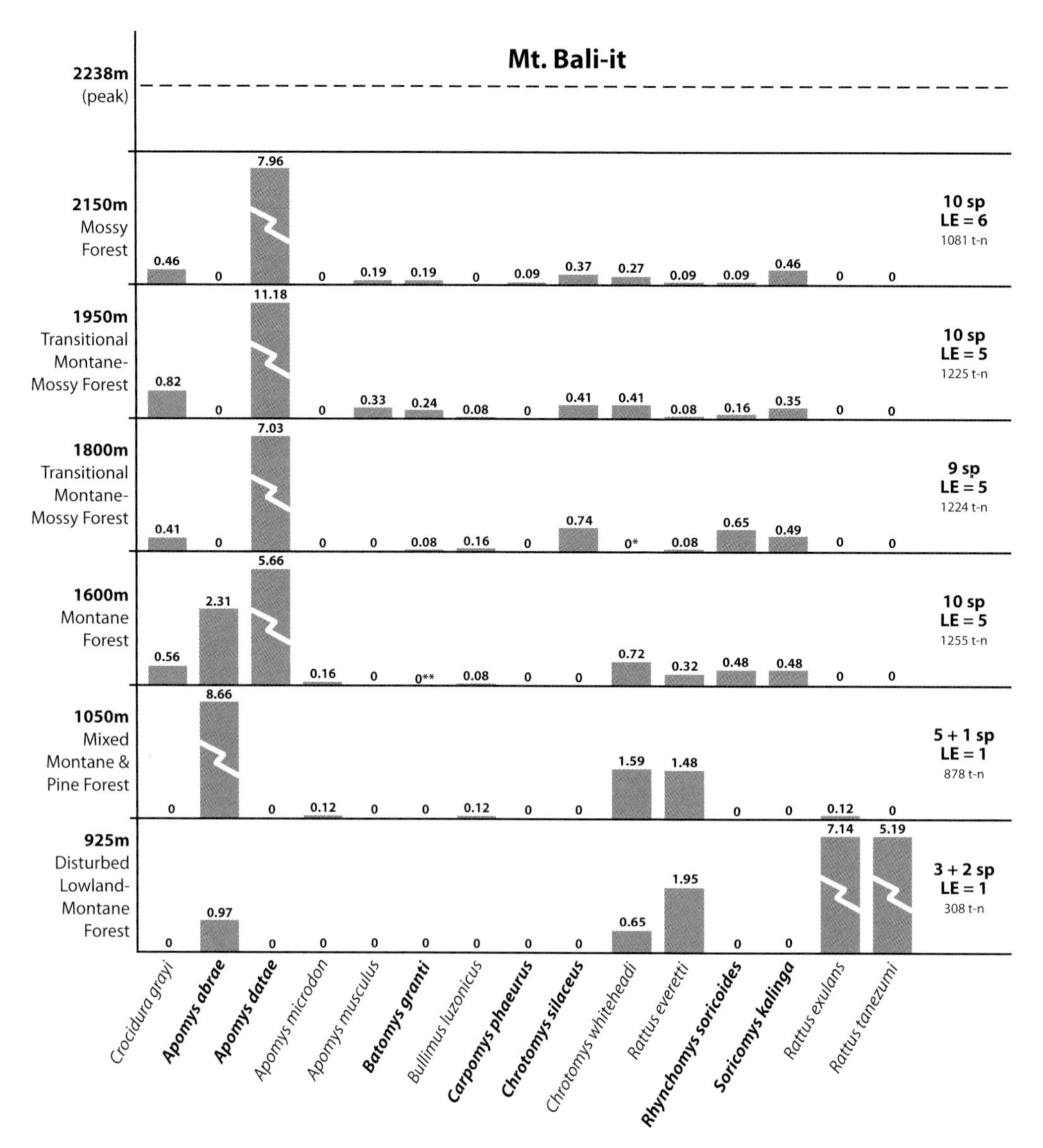

FIG. 3.10. Elevational patterns of presence and abundance in small mammals on Mt. Bali-it, Kalinga Province. The column at the far right gives the total number of native species present and inferred, the number of non-native species, the number of locally endemic (LE) species (in this case, endemic to the Central Cordillera), and total sampling effort (number of trap-nights = t–n). One asterisk indicates a species that is inferred to be present at a given elevation because it occurs both below and above the sampling area along the transect. Two asterisks indicate a species that was seen but not captured at a given elevation. Locally endemic species are shown in boldface. Data from Rickart et al. (2011a).

clear increase, however, in the *number* of species endemic to the Central Cordillera as the elevation rose: only one species in the lowest forested locality (1050 m elevation), increasing to five endemic species in montane forest at 1600 m, and six endemics in mossy forest at 2150 m. We note that although *A. musculus*, which occur only at high elevation, are currently considered to be widespread on Luzon, there is evidence that this taxon is a complex of multiple species, and that the form on Mt. Bali-it is endemic to the Central Cordillera; thus *A. musculus* fits this pattern as well. The addition of those endemic species beginning at 1600 m elevation raised the number of species on Mt. Bali-it from 6 native species (at 1050 m elevation) to 10, a figure that was sustained (or nearly so, with 9 species at 1800 m) up to 2150 m. In other words, the number of species increased with rising elevation, largely as a result of the increase in locally endemic species at the higher elevations.

The third and final transect that we present in this chapter comes from Mt. Amuyao, also in the Central Cordillera and one of the highest mountains on Luzon, with a peak at 2710 m elevation (fig. 3.11). Areas with lowland forest were distant from the town where we based our operations, so our transect went from 1650 m to 2690 m in mature montane (1650–2100 m) and mossy (2300–2690 m) forest. We captured 13 species, and all of them were native species—once again, nonnative species were absent from mature forest (chapter 6; Rickart et al., 2011b, submitted).

For elevations at which the Bali-it and Amuyao transects overlapped, species richness was similar between

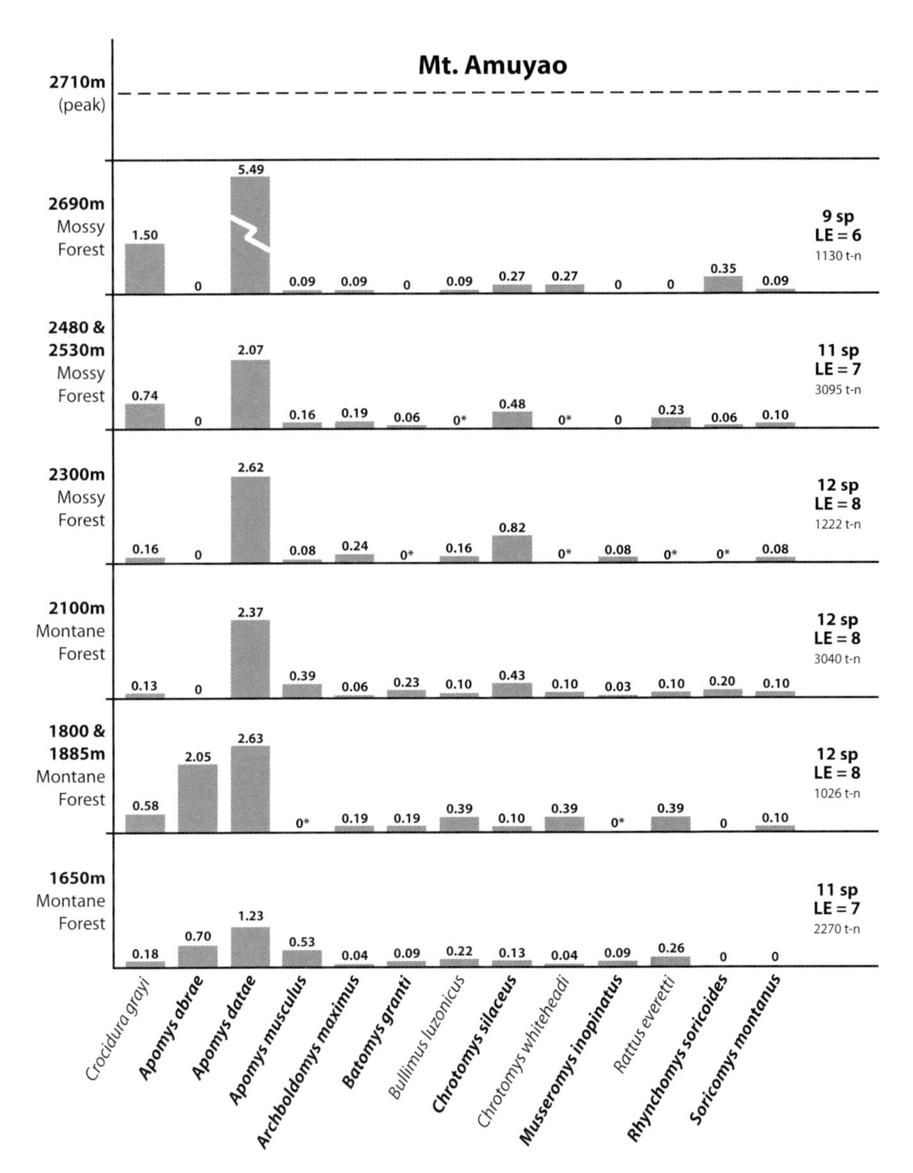

FIG. 3.11. Elevational patterns of presence and abundance in small mammals on Mt. Amuyao, Mountain Province. The column at the far right gives the total number of species present and inferred, the number of locally endemic (LE) species (in this case, endemic to the Central Cordillera), and total sampling effort (number of trap-nights = t–n). Asterisks indicate species that are inferred to be present at a given elevation because they occur both below and above the sampling area along the transect. Locally endemic species are shown in boldface. Data from Rickart et al. (submitted).

the two mountains but slightly greater on Mt. Amuyao: from 1600 m to 2150 m elevation on Bali-it, there were 9–10 species, and on Amuyao there were 11–12 (fig. 3.11). At the highest elevation on Amuyao, richness dropped from 12 species to 9, indicating a consistent decline. The two species of *Apomys* that were abundant on Bali-it were again abundant on Amuyao and showed overlap at two localities. The nine species endemic to the Central Cordillera included both these two most-abundant species and the two least-abundant species (*Musseromys inopinatus* and *Soricomys montanus*). Most of the Cordillera endemics were present at 1650 m elevation, but one occurred only at 1800 m and above, and one other just at 2100 m and above. Species richness thus ranged from 11 to 12 over most of the transect, falling to 9 only at the uppermost locality. The number of locally endemic (Cordillera) species also was high, ranging from 7 at 1650 m elevation to 8 over most of the transect, dropping to 6 only at the top of the mountain.

Summary of Elevational Patterns

Earlier in this chapter we posed several questions about the pattern of diversity along the elevational gradient. The first of these addresses the relationship between elevation and species richness. Figure 3.12 combines data from Mt. Anacuao, Mt. Bali-it, and Mt. Amuyao with similar data from five additional mountains (Heaney et al., 2013a, 2013b; Rickart et al., 1991, 2011a, submitted; unpublished data from Mt. Pulag in FMNH). On each of these mountains, maximum diversity of native small mammals occurs either *at* the top of the mountain, or *near* the top of the mountain with a subsequent decline at or near the peak (fig. 3.12). For Mt. Pulag, we have data only from three localities near the peak, at elevations of 2650 m, 2695 m, and 2780 m;

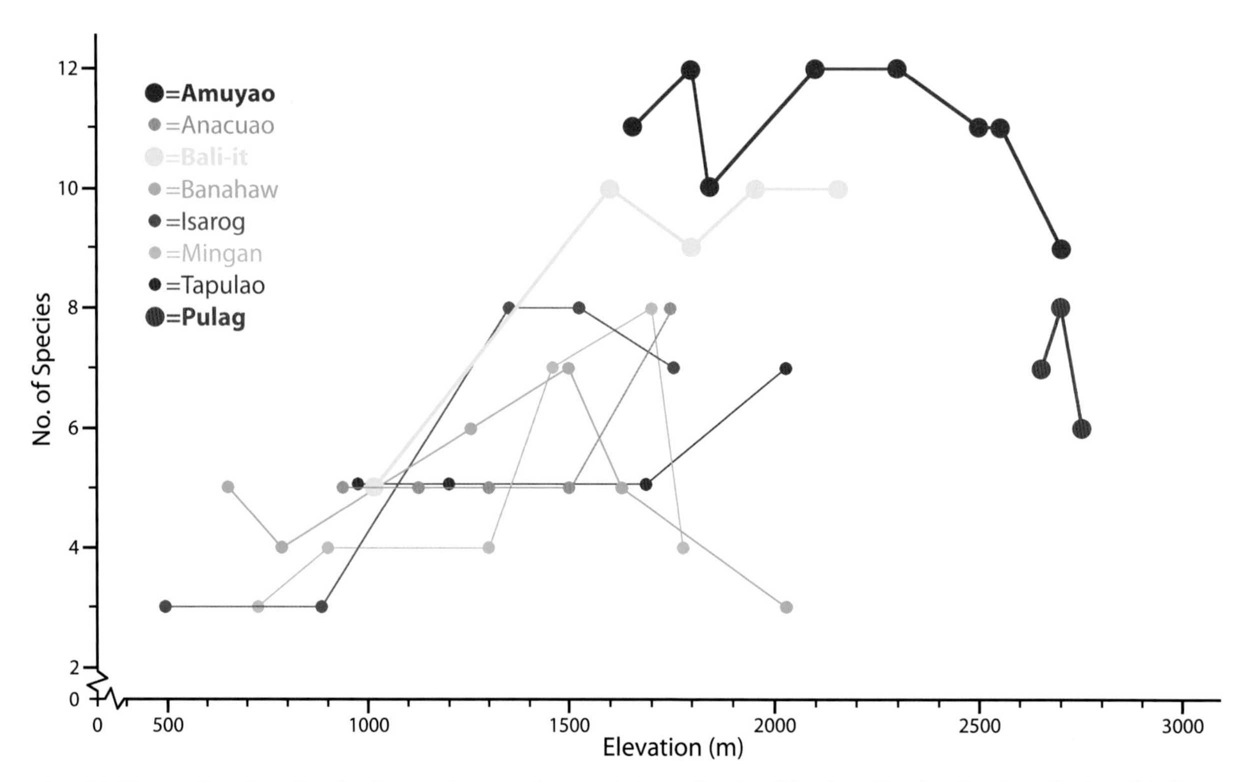

FIG. 3.12. The number of species of native small mammals present at sampling localities along the elevational gradient on eight intensively sampled mountains on Luzon. Circles represent the number of species at the given locality. Several low-elevation mountains that show little variation are not included. Data from Balete et al. (2009); Heaney et al. (2011a, 2013a); Rickart et al. (1991, submitted).

TABLE 3.1. Luzon mountains that have had extensive surveys of small mammals, with the maximum elevation and number of native species recorded. Data from Heaney et al., 2013a.

Name and Province	Elevation (m)	No. of species	Source
Mt. Natib, Bataan	1253	5	Rickart et al., 2013
Mt. Labo, Camarines Norte	1544	3	Rickart et al., 2011a
Mt. Palali, Nueva Vizcaya	1707	6	Alviola et al., 2011
Mt. Anacuao, Aurora	1850	8	Heaney et al., 2013a
Mt. Mingan, Aurora	1901	10	Balete et al., 2011
Mt. Isarog, Camarines Sur	1966	8	Rickart et al., 1991
Mt. Tapulao, Zambales	2037	9	Balete et al., 2009
Mt. Banahaw, Quezon	2158	10	Heaney et al., 2013b
Mt. Bali-it, Kalinga	2239	13	Rickart et al., 2011a
Mt. Amuyao, Mountain Province	2710	13	Rickart et al., 2011b

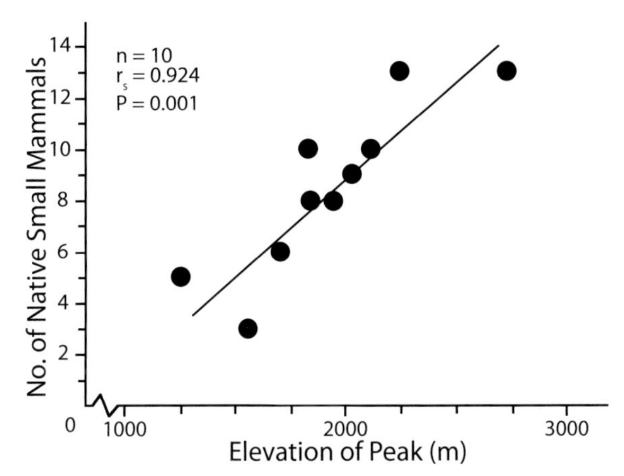

FIG. 3.13. The number of species of native small mammals on each of 10 intensively sampled mountains, plotted against the elevation of the mountain's peak. Data from table 3.1 and Heaney et al. (submitted).

taken together, these data appear to indicate a continuation of the decline in number of species that is evident above 2000 m on Mt. Amuyao.

Although there is much individual variation on Luzon, the overall pattern seems clear: diversity is low in the lowlands (three to five species), and begins to increase at about 1100 m elevation. Mountains with peaks at 2000 m or lower show a general pattern of increasing diversity with elevation, but they usually exhibit a decline near their peaks. In the Central Cordillera, species richness reaches a general maximum at about 2200 m elevation, at which point diversity begins to decline. The overall Cordillera pattern (fig. 3.12) is quite similar to that on Mt. Kitanglad, one of the tallest peaks (2950 m elevation) on Mindanao, where species richness in lowland forest at 1100 m is two species, rising to the maximum of nine species at about 2250 m, and declining to five at 2700 m (Heaney, 2001; Heaney et al., 2006b). On mountains with elevation reaching about 3000 m or more, this pattern of species richness increasing fairly steadily from the lowlands up to about 2200–2500 m and then declining at still higher elevations is rather common, and we accept it as the predominant pattern in the Philippines (Heaney, 2001; McCain, 2005; Rahbeck, 1995).

Our second question asked if mountain ranges with higher elevations have more species as a result of the increase in diversity with elevation. We conducted standardized surveys on 10 mountains on Luzon (table 3.1); as shown in figure 3.13, there is a strong correlation between the elevation of the peak of any given mountain and the number of species of native small mam-

TABLE 3.2. Mountain ranges (and isolated mountains) on Luzon, with the maximum elevation and number of locally endemic non-volant mammal species. Data from the listed sources.

Mountain range	Elevation (m)	No. of local endemics	Source
S. Sierra Madre	1530	1	Balete et al., 2013a
Mt. Labo	1544	1	Balete et al., 2013b
N. Sierra Madre	1850	5	Heaney et al., 2013b
Mingan Mts.	1901	5	Balete et al., 2011
Mt. Isarog	1966	5	Rickart et al., 1991
Zambales Mts.	2037	5	Balete et al., 2009
Mt. Banahaw	2158	5	Heaney et al., 2013a
Central Cordillera	2930	15	Rickart et al., 2011b

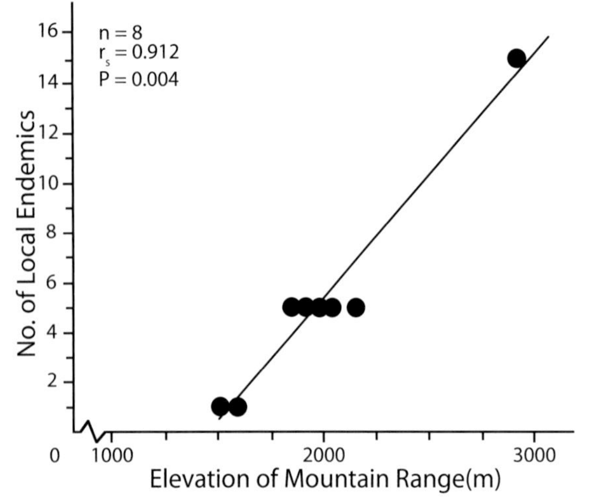

FIG. 3.14. The number of locally endemic species, plotted against the maximum elevation of the mountain range on which they are endemic. Data from table 3-2 and Heaney et al. (submitted).

mals that occur on the mountain (Heaney et al., submitted). The number of species in the lowlands appears to be fairly stable at three to five (e.g., Balete et al., 2009; Heaney, 2011, 2013b); it is only above about 1100 m elevation, where lowland forest vegetation gives way to montane forest, that the increase in species becomes apparent. From there, the rise is steady, with a roughly four-fold increase in species richness on the highest peaks. Clearly, the addition of species at progressively higher elevations is what drives this pattern.

This brings us back to the third question we asked above: is the increase in number of endemic species with increasing elevation a consistent pattern? We have sampled eight mountain ranges on Luzon that have locally endemic mammal species (table 3.2). A plot of the number of local endemics against the elevation of the mountain range (fig. 3.14) shows a strong correlation: the higher the mountain range, the more local endemic species are present (Heaney et al., submitted). Because of the physical constraints due to the number and heights of mountain ranges on Luzon, this analysis is less than ideal: it would be best if there were, for example, several ranges that reach a peak of 2200 m, 2400 m, and 2600 m elevation, so the pattern could be more fully documented. But within the constraints imposed by reality, the correlation is clear and strong: the higher a given mountain range, the more locally endemic species of native small mammals it supports.

It is worth noting that our assessment of this pattern is heavily based on the data that have resulted from the Mammals of Luzon Project described in chapter 1. In 1900, as a result of John Whitehead's studies, only two areas of endemism on Luzon were known: in the Central Cordillera and the Northern Sierra Madre (fig.

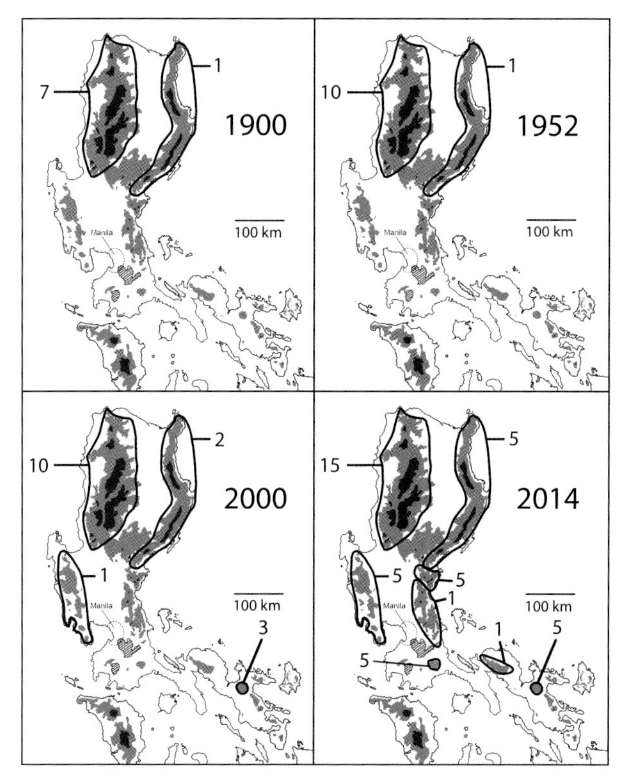

FIG. 3.15. The number of species known in each local area of endemism on Luzon, in 1900, 1952, 2000, and 2014. Data from Heaney et al. (submitted).

3.15; Heaney et al., submitted). Little changed by the middle of the twentieth century; at the time we began our Luzon studies in 2000, we could see no apparent pattern but also few relevant data—which raised many questions (chapter 1). After a decade of intensive field surveys, we now know of eight centers of mammalian endemism on Luzon, with the pattern shown in figure 3.15. We know of no other geographic region, on island or continent, that has so many endemic mammal species in such a small area. We can safely conclude that the topography of Luzon is intimately connected both to total diversity and to patterns of endemism among native small mammals. Those patterns are strongly associated with the present climate and habitat (chapter 2). We thus conclude that the current ecology of the species and their interactions with the environment play a prominent role in determining the patterns that we see.

But recognition of these patterns leads to another set of questions. How did this diversity arise, and when did it arise? Luzon is a land of active geological processes. Indeed, it is one of the most geologically active places on Earth, and the topography that currently exists, which has such demonstrable impacts on mammalian diversity, has its origins in relatively recent geological history. What is the connection (if any) between the evolutionary development of the highly distinctive and diverse Luzon mammal fauna and the geological development of Luzon, with its highly varied topography? To answer this question, we turn to those geological processes and their history in chapter 4, and then, in chapter 5, to the biological processes of speciation and diversification that have produced the diversity that we see today, examined in the context of that geological history.

4

Geological History and Fossil Mammals

There is an understandable tendency for people to think of the Earth as stable, mountains as fixed entities, and shorelines as permanent edges to islands. Understandable—but quite untrue. The Earth is dynamic and has changed constantly throughout its roughly 4.5 billion years of existence. In the case of the Philippines, we need not go back nearly that far—only to about 40 million years ago, far past the extinction of the dinosaurs and long after most major groups of mammals had evolved. Most of our story about the mammals of Luzon can be told during just the past 20 million years.

In some parts of the world, 20 million years would not be enough time for much geological change; Madagascar, for example, has undergone major climatic changes during that period, but the topography and geology of the island has not altered much (Wells, 2003). In stark contrast, the Philippine islands have been far more dynamic, in part because they lie in one of the most geologically active places on Earth: the western edge of the Pacific Ocean, along the circle of volcanoes that is often described as the "ring of fire." This nearly continuous ring of volcanoes stretches from the Philippines (fig. 4.1) to Japan, the Aleutian Islands in Alaska, along the west coast of Canada and the United States, and reaches south all the way to the southern tip of South America. This volcanic activity is perhaps the most prominent aspect of both the geological history and the current topography of Luzon (and the rest of the Philippines). The origin of that volcanism is the first—and crucial—part of the story of Luzon's geological history.

Plate Tectonics, Subduction, and the Origin of the Philippines

Research during the past 60 years has produced conclusive evidence that the surface of the Earth, long thought to be relatively static, is highly dynamic. It is composed of a large number of plates (i.e., segments of the Earth's crust) that move away from their points of origin and toward the places where they seem to disappear. This movement, once regarded as speculative, is properly known as plate tectonics but is more widely called continental drift.

The places of origin of these plates are the midocean ridges, where hot materials from deep within the Earth well up and form new ocean lithosphere roughly 100 km thick. None of these ridges are near to the Philippines; the one that most strongly influences the Philippines lies most of the way across the Pacific, toward South America. A map of the vicinity of the Philippines (fig. 4.2) shows the other end of the process—places where adjacent plates meet and form deep-sea trenches where water depths may exceed 10,000 m. The Philippines is actually bracketed by these trenches on its eastern and western sides, a key fact in explaining the history and development of the archipelago (Yumul et al., 2009).

FIG. 4.1. Volcanoes on the Bicol Peninsula of southeastern Luzon, viewed from ca. 1450 m elevation on the south slope of Mt. Isarog.

Given the immense size of the plates, their rate of movement is astounding. The Philippine Sea Plate, which is traveling in a north-northwesterly direction, moves at a rate of about 7–9 cm per year: that's 7–9 m per century. Over the lifetime of one long-lived person, 7 m of ocean floor that is many kilometers thick simply disappears beneath the islands extending from southern Mindanao to southern Luzon.

Of course, all of that ocean crust cannot simply disappear, and the presence of the deep trenches gives a clear clue as to where it goes. We can visualize this most easily in the simplified profile shown in figure 4.3. At the trench, the ocean floor descends steeply along the line of contact between two plates. The adjacent plates are likely to differ in density, due to age and composition, and where they meet, the denser of the two plates is forced down into the Earth's mantle. This process of plate subduction is especially active along the Philippine Trench, due to the speed of the Philippine Sea Plate's movement. At the beginning of the process, the ocean floor is made up of cold rocks that are saturated with water and capped by a layer of sediment that has settled from the water column above. The force of the plate's movement may shove some of the ocean plate and some of the top layers of sediment up onto the edge of the land mass (Luzon, in this case); we return to that topic later. But the bulk of the ocean plate is driven down deep into the Earth's mantle, at a depth where it is hot enough to melt rocks. As water is released from the descending plate into the overlying hot mantle, a portion of the thick layer of rock does just that—it melts.

If the subducting material were entirely identical in composition to the hot magma in the mantle, it would simply become part of a homogeneous mass and that would be the end of the story. But the top of the ocean plate contains sediment that may include layers of calcium-rich shells of marine invertebrate animals and

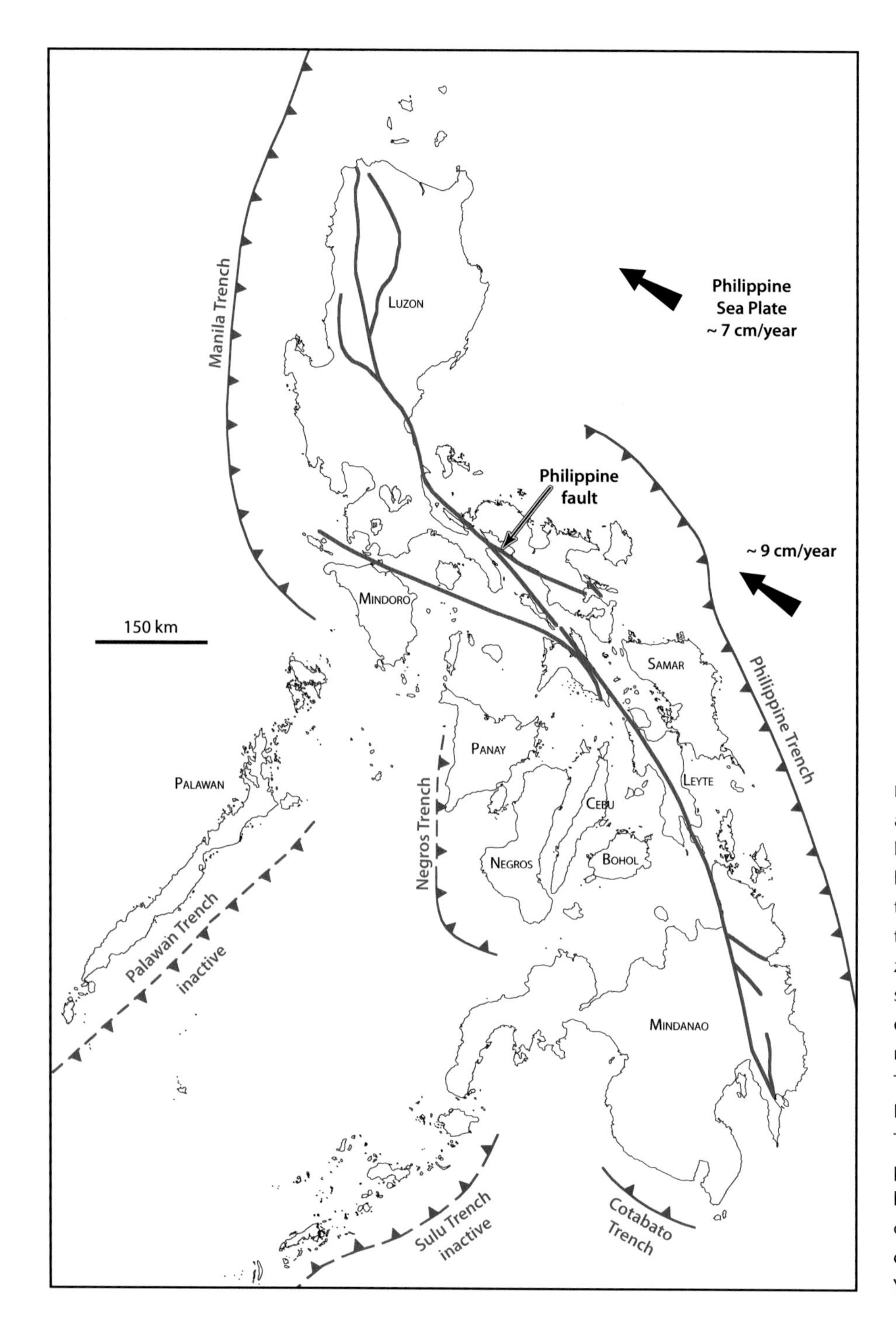

FIG. 4.2. The Philippine archipelago, with the shorelines of current islands (thin black lines), the locations of trenches associated with active and inactive subduction zones (red lines with triangles showing the direction of subduction), and the locations of major faults (thick blue lines). The extent and activity of the Negros Trench is uncertain. The direction and speed of plate movement is indicated by the black arrows. Based on Hall (2012, 2013, pers. comm.); Hamilton (1979); Yumul et al. (2008).

different amounts of the primary elements that make up rocks. Even more importantly, it always contains water. The water dissolves into the adjacent hot mantle, but because of the extreme pressure at such great depths, it does not turn to steam or remain as a distinct liquid—rather, it becomes part of a complex molten mixture. This mixture is less viscous and more buoyant than the surrounding mantle, and it very slowly but inevitably moves toward areas of less pressure—in this case, toward the surface of the Earth. When this buoy-

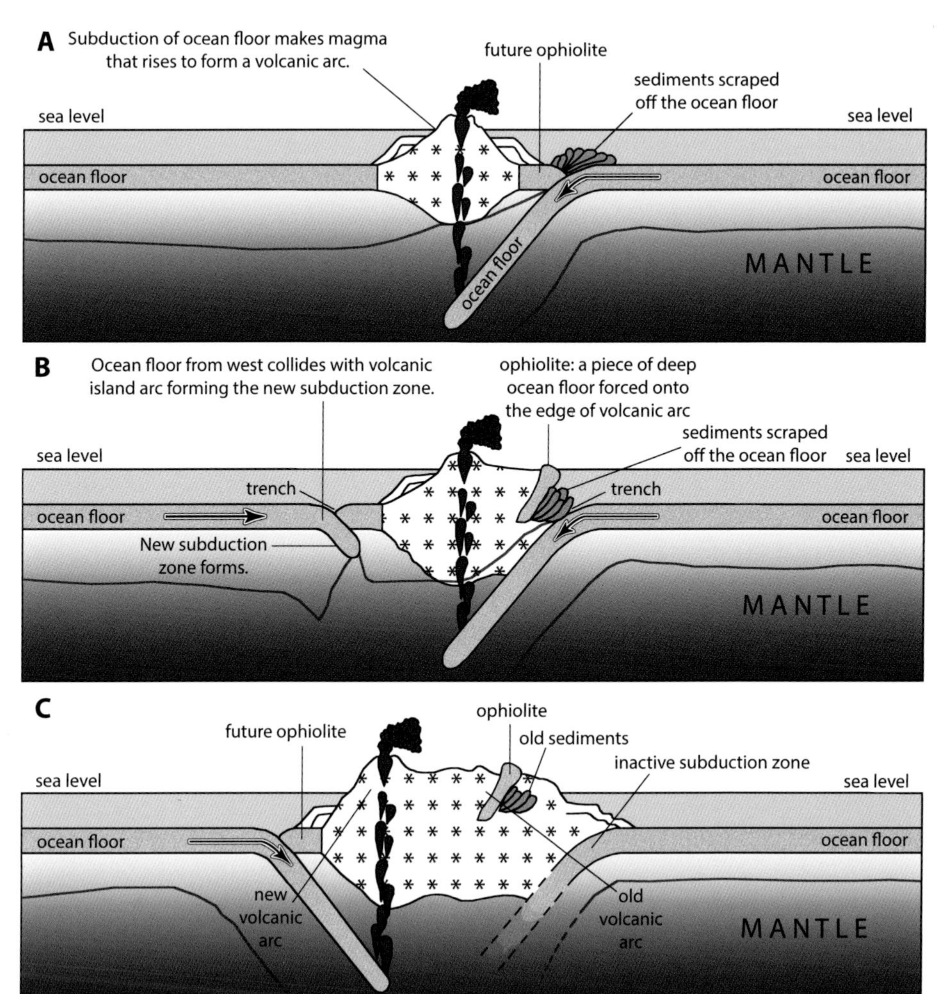

FIG. 4.3. Diagrammatic sketch of geological features and processes associated with the production of the Philippine island arc. Modified heavily from Meldahl (2011).

ant magma finally reaches areas of lower pressure near the Earth's surface, the water separates out and flashes to superheated steam (similar to the way bubbles form in soft drinks when the bottle cap is removed, reducing pressure), expanding with great speed and force, and carrying superheated bits of rock with it—an event we know as a volcanic eruption. The volcanoes that have produced the bulk of the Philippine islands are the result of the melting of the mantle above the subducting ocean lithosphere, typically about 100 km below the Earth's surface.

Over time, the eruption of volcanoes along a subduction zone (with its active trench) will produce a huge volume of material that builds mountains. These mountains, which begin as separate islands, may merge into single larger islands if the amount of volcanic debris is large enough to fill the space that intervenes between the volcanoes. This process is clearly visible in the photograph of the Bicol Peninsula (fig. 4.1): a string of young volcanoes has erupted, too recently to form a large area of continuous high elevation, but over a long enough period of time to fill the intervening spaces and create a single elongated peninsula that lies above a long edge of subducted ocean crust.

Over the course of time, changes in the rate and direction of movement of the tectonic plates may cause one subduction zone to shrink or become inactive and allow a new subduction zone to form on the opposite side of the emerging island arc. The volcanoes that are produced are now toward the side with the active subduction zone (the west, in this instance), so volcanic activity is most prominent to the west of the older line of volcanoes. In this way, the size of the island builds over time, with a ridge of volcanically produced mountains on either side of a relatively narrow valley. The longer there is movement of the oceanic plates that results in subduction zones and volcanic arcs, the larger and higher the island becomes.

But the movement of the plates does not always result in the development of trenches where ocean crust disappears and volcanoes form. In some cases, there is thrusting within or at the edges of the arc. That is happening currently along the north coast of Luzon, and the force of the moving plate is pushing Luzon to the west. Pressures such as this may force some former ocean floor that formed part of the forearc up onto the edge of the island. These portions of old ocean floor and underlying mantle are known as ophiolites (figs. 4.3 and 4.4); they are the rocks that often produce ultramafic (chapter 2) soil. These pressures may also result in deformation of the land mass, giving rise to movement of various parts of the land mass in different directions. These areas of movement and geological stress are commonly known as faults, where earthquakes often take place. The extreme pressures in the Philippines have produced the Philippine Fault, which spans virtually the entire length of the archipelago, with a great many smaller faults where the strain of deformation tears the land and moves it in different directions (fig. 4.2). These faults often mark the edges of major geological units within the archipelago, and even within parts of Luzon. The pace at which the land on opposite sides of the Philippine Fault move relative to one another is truly amazing: for the past 2–4 million years, the movement has been 1.9–2.5 cm per year, or 1.9–2.5 m per century (Ringenbach et al., 1993). It is no wonder that Luzon is known as one of the most earthquake-prone places on Earth.

Current Geology of Luzon

Figure 4.4, showing the major geological features of Luzon, demonstrates many aspects of its history. To the north of the Philippine Fault lie Luzon's two largest mountain chains, the Central Cordillera and the Northern Sierra Madre, each composed primarily of the remnants of ancient and recent volcanic eruptions. Between them lies the Cagayan Valley, containing layers of volcanic ash that either rained down directly from eruptions in the two mountain chains or were the products of slow erosion of the mountains caused by the heavy rains that fall each year. On the east side, there are several areas of ultramafic rocks—the portions of former ocean crust and mantle referred to as ophiolites that have been pushed up against the edge of the island (Encarnación, 2004). One of these ophiolites, the Mingan Mountains, is bounded by a small fault (and a narrow but deep valley) to its north, and by the Philippine Fault (and another narrow but deep valley) to the south. Areas of exposed limestone are scattered about, especially along the edges of the Cagayan Valley. These are formed from thick layers composed of the shells of marine invertebrates that lived for millennia in the shallow seas that once filled the valley (Vondra et al., 1981).

To the south of the Philippine Fault, along the eastern edge of Luzon, lies the Southern Sierra Madre, mountains made up of a mixture of limestone and volcanic materials, and to their west the Central Valley and the Zambales Mountains. Much of the northern Zambales range is composed of an ophiolite, thrust up from the west, and the range's more southern part is made up of recent and active volcanoes, including Mt. Pinatubo. South and east of Manila Bay and Laguna de Bay (a former part of Manila Bay, until volcanic sediment rather recently cut it off as a separate body of water) is a large area of scattered volcanic peaks, some inactive and heavily eroded, others temporarily quiescent, known as the Macolod Corridor (see below). Continuing south and east, Luzon becomes narrower and lower, with the extreme reached at the north end of Ragay Gulf, where the Philippine Fault crosses Luzon; the height of land along this narrow declivity is only about 4 m above sea level. A rise in sea level of just a few meters would cut Luzon in two, separating the Bicol Peninsula from the rest of Luzon.

From that low point above Ragay Gulf, the Bicol

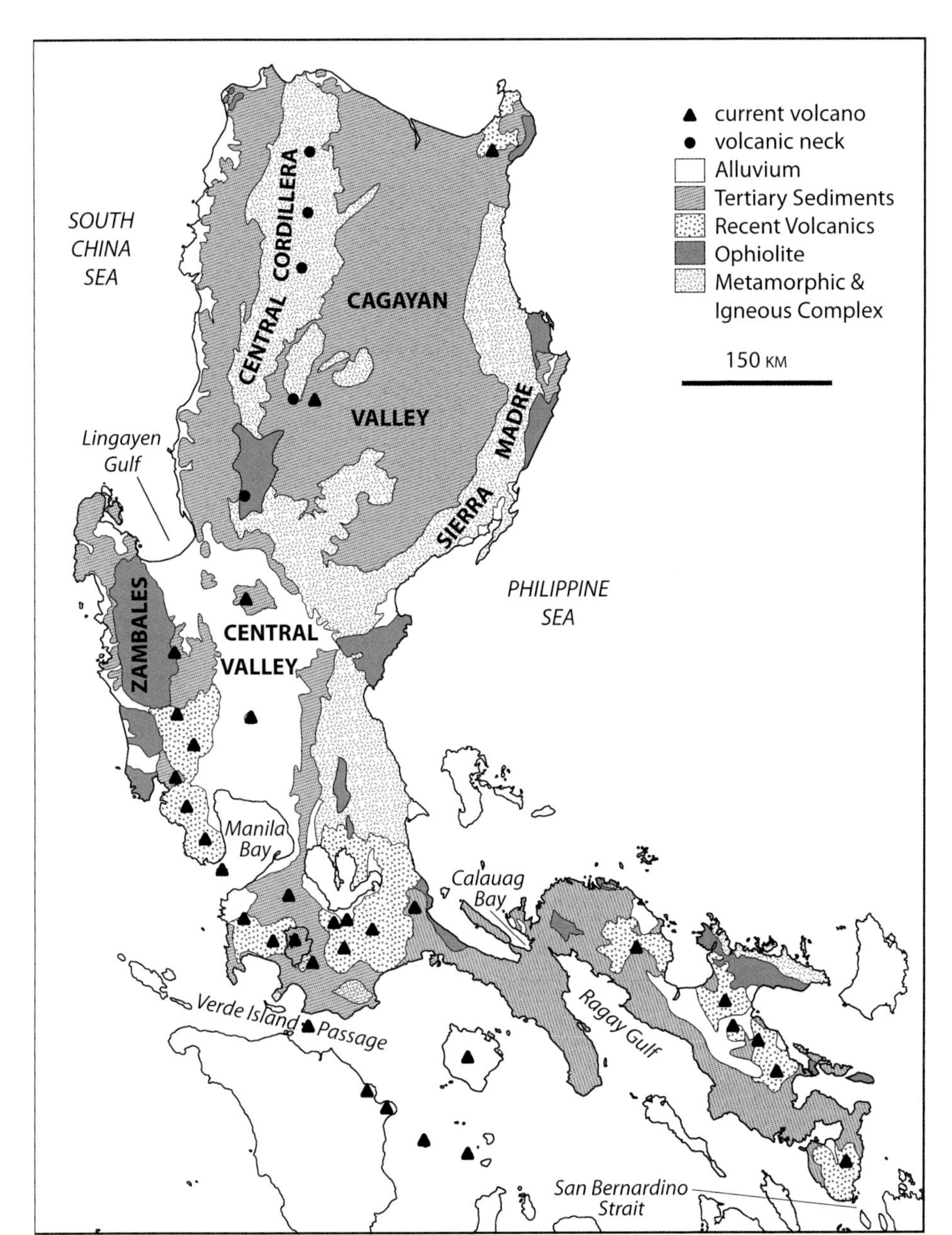

FIG. 4.4. Luzon Island, with the primary types of surface geological materials, the locations of major mountain chains and valleys, and the locations of current major volcanoes (triangles) and volcanic necks (circles). Modified from Bachman et al. (1983); Bureau of Mines and Geosciences (1982, 1986); Encarnación (2004); Philippine Bureau of Mines (1963).

Peninsula stretches southeast, a narrow band of lowlands surrounding a series of volcanic peaks, a few uplifted areas of exposed limestone (often little more than poorly consolidated coral reefs), and a few small ophiolites. The most conspicuous feature is a series of volcanoes that lie in a nearly straight line, extending from Mt. Labo in the north to Mt. Bulusan in the south (fig. 4.1). At the southern end of this chain of Bicol volcanoes lies the San Bernardino Strait, which separates Luzon from Samar Island and has a depth of just over 140 m.

The impact of tropical storms, including typhoons, has been discussed in chapter 2; with a rough average of 20 striking the Philippines each year, their impact on climate and rainfall and their damage to vegetation, is unmistakable. Volcanic eruptions occur less often—perhaps one large eruption every hundred to a thousand years or so—but their effect is even greater.

They erupt often enough that, if we accept one every hundred to a thousand years as a crude range, there are roughly 1000–10,000 major eruptions every million years. And the best estimate is that there have been volcanoes erupting within the area of the modern Philippines for over 30 million years. Volcanoes are, and have always been, an ever-present part of the Philippine environment.

Mt. Pinatubo serves as an instructive example of a major eruption (Newhall and Punongbayan, 1996). For centuries prior to 1991, the mountain was simply a heavily eroded, well-vegetated volcanic peak (1745 m tall before the eruption) that was thought to be extinct, like dozens of others on Luzon. Renewed activity in 1991 showed spectacularly that it was merely dormant. Earthquakes were noted in mid-March 1991, small eruptions began in early April, and a cataclysmic eruption took place on 15 June. It was the second-largest eruption of the twentieth century, exceeded only by Mt. Novarupta in Alaska; the 1980 eruption of Mt. St. Helens in Washington State was about one-tenth as large. The Mt. Pinatubo eruption produced about 10 km^3 of magma and reduced the height of the mountain by nearly 260 m. This was simply the latest in a series of eruptions by "Modern Pinatubo"; the first of these took place about 35,000 years BP, and others around 17,000 BP, 9000 BP, 5500 BP, 3000 BP, and 500 BP. The first of these produced an estimated 25 km^3 of magma, and each of the five produced more than the 1991 eruption. But the first volcano at the same location, known as "Ancestral Pinatubo," initially erupted about 1.1 million years ago, built up to about 2300 m tall, and continued to erupt for millennia. Together, Ancestral Pinatubo plus Modern Pinatubo may have produced well over 200 km^3 of volcanic materials. As massive as it was, Mt. Pinatubo's 1991 eruption was by no means unusual in the history of the Philippines.

Luzon, an Island of Islands

A full description of the history of Luzon would require a book longer than this one; as noted earlier, this is one of the most geologically active and complex parts of the world. But for our studies of mammals, there are certain aspects that are crucially important. When did each part of Luzon appear above water? What size did it attain? How high did each reach? And when did the individual pieces become connected to other parts of Luzon? We address these questions in a roughly chronological order. Our description is based largely on summaries presented by Robert Hall (1998, 2012, 2013), with additional details from the sources cited and from Hall (pers. comm.). As Hall (2012:27) has explicitly stated, far too little is known about the details of Philippine geology, and all of these reconstructions must be recognized as hypotheses that need to be refined, based on additional data; for now, it is best to view them as approximations.

About 40 million years ago, virtually nothing existed that would become the Philippines—at least nothing above sea level. But by about 35 million years ago, forces generated by plate tectonics began to cause the production of subduction zones north and south of the modern Philippines, as well as centers of oceanic spreading nearby. By 30 million years ago, two major processes had begun. Volcanic activity powered by the subduction zones produced the first volcanoes, in the area of the modern Central Cordillera, as well as in places that would eventually become parts of the central and southern Philippines. These were few and scattered, but the first islands appeared and remained above the waves (Bachman et al., 1983; Wolfe, 1988). The second process was far to the north, along what would come to be known as the southern coast of China; there, a spreading zone produced a rift that gradually widened, carrying south a piece of continental crust that would eventually contribute to the modern islands of Palawan, Mindoro, and a bit of Panay.

By about 25 million years ago, the continental crust that would eventually become the basis for Palawan and Mindoro had moved farther south; the open sea they left behind to their north later became known as the South China Sea (Hall, 2012). But Palawan and Mindoro still lay far north of what would become Luzon (fig. 4.5) and probably emerged intermittently above

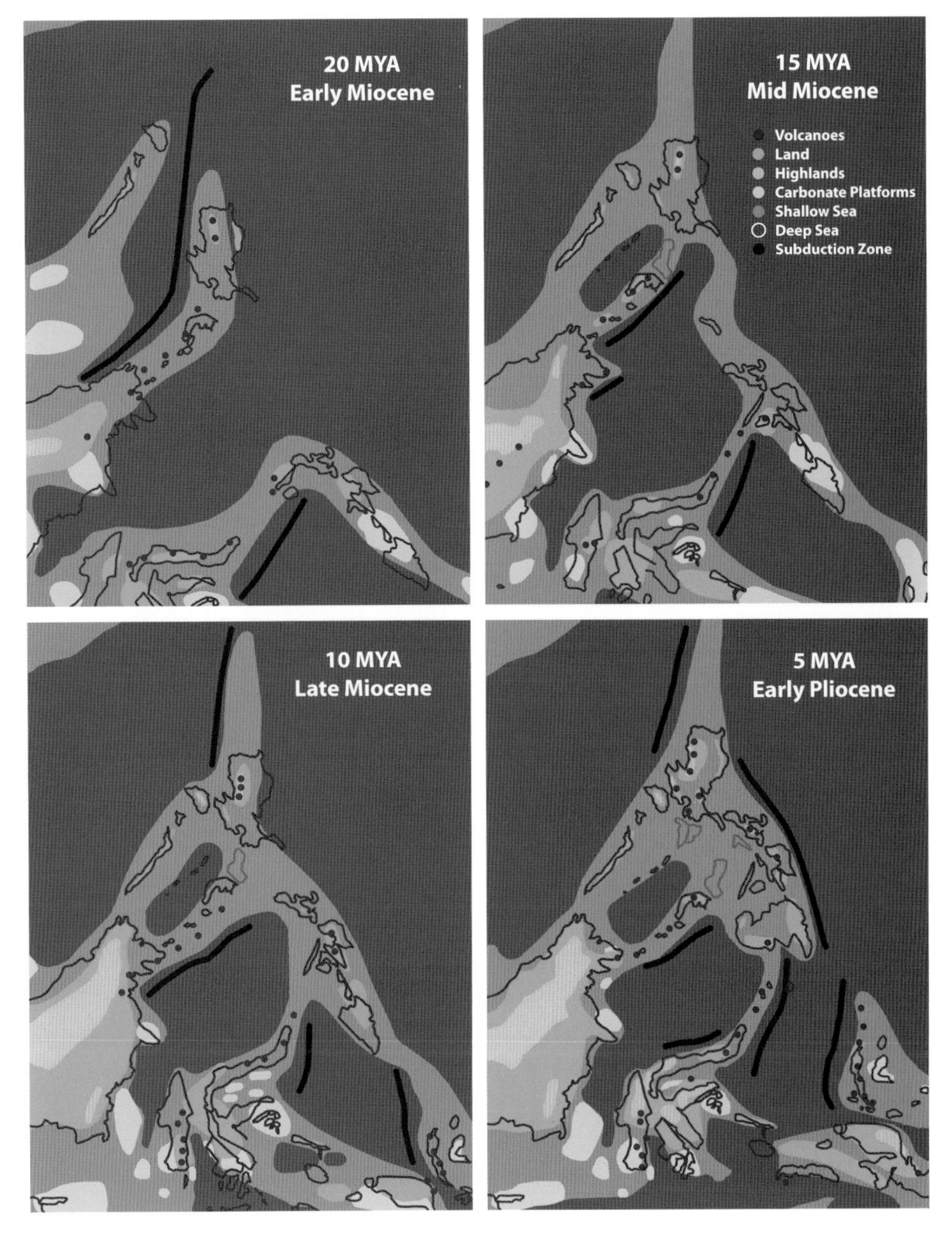

FIG. 4.5. Geological model of the development of the Philippines and adjacent islands. Maps show the approximate size and location of islands, areas of volcanic activity (not necessarily the location of specific volcanoes), shallow seas, and carbonate platforms, as well as the approximate locations of active subduction zones. These are rough estimates; more geological research is needed to make this precise. The outlines of modern islands are included only to show the locations of the rock units that would eventually have these shapes. Based on Hall (2012, 2013, pers. comm.) and modified from sources cited in the text.

sea level only as a few low-lying islands. A small amount of volcanic activity began 26–20 million years ago in what later became the Central Cordillera, probably as a set of scattered and perhaps impermanent islands. Another, somewhat smaller island (or set of islands) was formed during the same period in what would some day be known as the Northern Sierra Madre, but activity ceased and the islands eroded and disappeared (Bachman et al., 1983; Yang et al., 1995; Yumul et al., 2008). By 20 million years ago, the rock units that would become Palawan and Mindoro had moved farther south (Hall, 2013), initiating uplift in Mindoro and the Zambales Mountains (as an isolated island) soon thereafter. Volcanic activity continued in the area of the modern Central Cordillera, gradually increasing the area above water, though the extent and degree of dry-land connection is uncertain. A few other volcanic islands lay just to the south but probably were not permanent features—mostly serving as heralds of larger land masses that would develop later. Much farther to the south, the rock units that would some day make up the Bicol Peninsula and most of the Visayas and Mindanao were gradually moving northward.

By 15 million years ago, increased volcanic activity began to produce several rather large islands in the northern Central Cordillera (Bachman et al., 1983; Wolfe, 1988), probably with a gradually growing highland area above 1500 m elevation, though the details are unclear. The Zambales Mountains were thrust up beginning about 18 million years ago, largely as a result of forces associated with the southward movement of the Mindoro/Palawan mass of continental rocks, with Mindoro in its earliest phase of uplift (Bachman et al., 1983; Yumul et al., 2008).

By 10 million years ago, the Central Cordillera (still existing as an isolated island) had grown still further, expanding on the east side to fill the western edge of the shallow sea that was to become the Cagayan Valley (Vondra et al., 1981; Wolfe, 1988). Zambales had perhaps grown a bit but remained isolated, Palawan/Mindoro had moved farther south, Mindoro had uplifted further, and the rock units that later formed the central and southern Philippines moved ever closer to Luzon. Below sea level, some other changes were dramatic. The Central Valley, which had been a rather deep, open seaway, slowly filled with a mixture of limestone and volcanic materials, while simultaneously subsiding. The sediments eventually built to a depth of 8 km, with the seaway eventually becoming shallower and creating lowlands along the edges of both the southern Central Cordillera and the eastern Zambales Mountains. A similar process took place in the Cagayan Valley, filling it with sediments to a depth of about 10 km (Vondra et al., 1981).

By 5 million years ago, the rock units that make up the Luzon of today were nearly in their current positions, but the islands were far smaller than at present and still isolated from one another by shallow seas. Luzon was composed of a gradually enlarging Central Cordillera and the slowly growing Zambales Mountains, but new islands had emerged as well. The Northern Sierra Madre, which had disappeared below sea level for 15 million years, began to reemerge along the Pacific coast, though it may have remained as a series of small, isolated islands for several more million years (Yang et al., 1995). What was to become the Southern Sierra Madre appeared at about this time, in the same general area as some previous, short-lived volcanic islands (Bachman et al., 1983). For the first time, the rock unit that was to become the Bicol Peninsula was adjacent to the southern portion of Luzon, and an island had begun to develop about 6.6 million years ago in what is now the center of the peninsula (Knittel-Weber and Knittel, 1990; Ozawa et al., 2004).

Thus, by about 5 million years ago, it is likely that most of the mountainous regions of modern Luzon were present, but each existed as an isolated island. It is only within the past 5 million years that Luzon, as we know it today, began to take shape.

Luzon Rising: The Last 5 Million Years

On a geological timescale, things began to happen very quickly by the beginning of the Pliocene geological epoch, about 5 million years ago. The uplift of the Northern Sierra Madre and a burst of volcanic activity in the Central Cordillera at about this time continued to fill in the Cagayan Valley and completed much of the process by about 3 million years ago (Vondra et al., 1981). Coupled with the uplift of the Caraballo Mountains in the late Pliocene or early Pleistocene (about 3–2.5 million years ago), this relatively quickly created the modern shape of northern Luzon (fig. 4.6). At some unknown point during the past 2–5 million years, the Central Sierra Madre (the ultramafic ophiolite also known as the Mingan Mountains) was uplifted, creating a land-bridge to the Southern Sierra Madre, which had emerged slightly earlier. Expanded volcanic activity in the Zambales Mountains combined with volcanic activity in the Southern Sierra Madre to complete the process of filling the Central Valley, start-

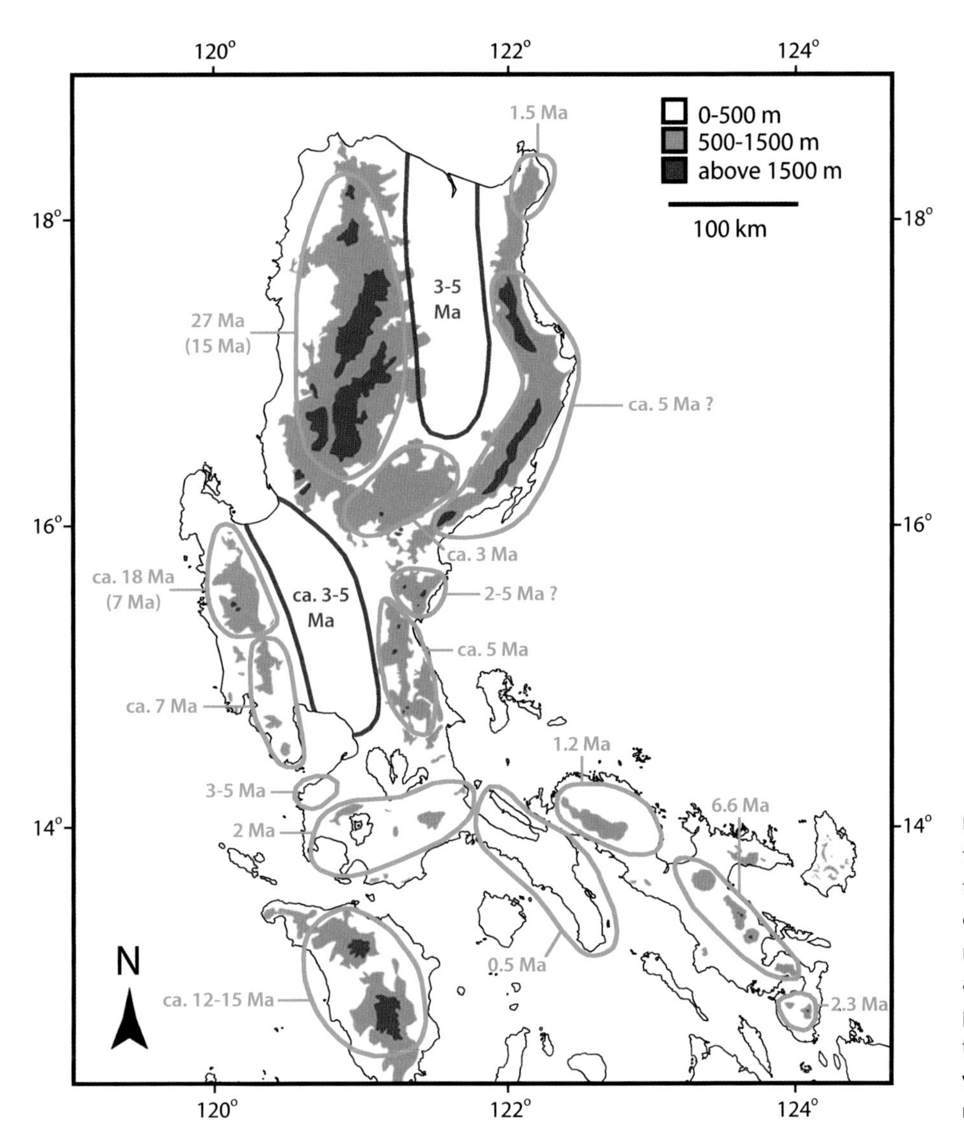

FIG. 4.6. Modern Luzon, with the estimated ages of the portions of the island that developed independently. Some movement has taken place among these areas during the past 5 million years, so positions and distances should be viewed as having been dynamic. See text for sources.

ing near its current center and expanding outward to the north and south, with the first dry-land connection probably occurring 4–5 million years ago. Thus by about 3 million years ago, both Central and Northern Luzon probably bore a great resemblance to their configuration today.

But south of about the latitude of Manila Bay, the growth and connection of dry-land areas happened even later. At 3 million years ago, a small island existed in Cavite, south of modern Manila Bay around the volcanic peak we know as Mataas na Gulod, and the island in what is now the central Bicol Peninsula had grown still further. By 2 million years ago, the large area south of Laguna de Bay, known as the Macolod Corridor, began a series of massive eruptions that slowly raised the entire area above sea level; the island of Mataas na Gulod was connected to the mainland only about 1 million years ago (Ku et al., 2009; Vogel et al., 2006; Wolfe and Self, 1983). Volcanic activity at the southernmost tip of modern Luzon, in Sorsogon, began only about 2.3 million years ago and remained as a separate island for at least a million years. Eruptions in the northern

Bicol Peninsula created the first island about 1.2 million years ago (around modern Mt. Labo), merging with the central Bicol island some time later (Ozawa et al., 2004). The last piece of the puzzle to appear, the Bondoc Peninsula, is poorly known, in part because it is made up predominantly of limestone that is difficult to date, but it appears to have uplifted during the late Pleistocene, perhaps less than a half-million years ago (Aurelio et al., 1991).

Thus up to a dozen areas in today's island of Luzon each most likely existed as a separate island for extended periods before merging into the gradually growing modern island. These mergers took place principally during the past 5 million years—much of it within the past 3 million years—and some crucial parts, including the formation of a dry-land connection between the Bicol Peninsula and central Luzon, within the past million years. That history, and that timescale, becomes a crucial matter in chapter 5, in which we examine the timing of speciation and diversification of the native mammals of Luzon. But before moving to that topic, we have one more issue to consider that involves the reshaping of the Earth—the great ice ages of the Pleistocene and the modification they brought to Luzon.

Changing Sea Level and Climate

To the best of our knowledge, no glaciers have ever formed in the Philippines, even at the tops of the highest peaks during the extremes of the coldest of the ice ages, when massive ice sheets covered much of northern North America, northern Europe, and many other areas. But even in the tropical Philippines, the periods of time when those glaciers formed had a distinctly different climate than what we experience today.

The primary cause of the ice ages—the period of geological time about 2.6 million years long, encompassed by the Pleistocene epoch—is now fairly well understood (for further details, see Bintanja et al., 2005; Brigham-Grett et al., 2013; Zachos et al., 2001). In brief, the Earth's orbit around the sun is nearly but not perfectly circular, and the orbit changes from almost circular to slightly elliptical. This astrophysical cycle is quite precise—the length of the fluctuation is about 98,000 years.

When the orbit is elliptical, winters in the Northern Hemisphere are cold and the summers are hot; the result is that the winter snows melt entirely each summer. When the orbit is nearly circular, both the northern winters and the summers are mild; snow still falls during the milder winters, but the summers are not hot enough to melt it all. As a result, during the ice ages the snow gradually built up each year and eventually packed down into masses of ice that were thousands of meters thick, covering huge areas of land.

As a result of this climatic fluctuation, the Earth cycles, on a 98,000-year repeated pattern, from periods of time with extensive continental glaciers to periods with almost no glaciers. This cycling has taken place four times, most recently reaching a peak of glacial development about 20,000 years ago—and previously at about 120,000 years ago, 220,000 years ago, and so on. Prior to that, other variations (such as changes in the angle of the Earth's axis relative to its orbital plane) had caused a similar but somewhat less-dramatic variation in climate in cycles of about 40,000 and 22,000 years, extending back to about 2.6 million years ago (Zachos et al., 2001). The only prior major glacial periods occurred long before the origin of the Philippines.

There were two major impacts from these glacial cycles that were felt in the Philippines. The first was a fluctuation in temperature that happened worldwide. During the times when continental glaciers were present, they formed huge areas of white snow that covered much of the Earth. Sunlight that fell on these white areas was largely reflected back into space, so energy was lost into space and did not warm the ground or air. During the interglacial periods (when only relatively small areas of land were covered by glaciers, such as currently), sunlight falling in the same areas of the Northern Hemisphere was absorbed by the ground, heating it and then warming the adjacent air. Thus, with so much energy from sunlight being lost during the glacial periods, on average the Earth's temperature was about 5°C cooler than it is today. The Philippines did not have glaciers during the ice ages, but it was cooler than at present (though we do not have any directly relevant hard data at this time). Based on changes elsewhere in the tropics, the temperature in the Philippines was perhaps as little as 2°C or as much as 8°C cooler (Lomolino et al., 2010:323). This decrease in temperature may have been greater at higher elevations than at low elevations, as it was in New Guinea (Morley and Flenley, 1987).

We return to the issue of cooler temperatures later, but first we need to consider the second major impact in the Philippines of the development of continental glaciers elsewhere. The moisture that falls as rain or snow over North America and Europe originates mostly from evaporation from the oceans of the world. When this water falls as rain, it either evaporates (a little of it) or runs downhill (most of it) until it returns to the ocean. When it falls as snow, it stays where it fell until it melts (a little bit of snow "evaporates" directly, but only very little). Thus, as continental glaciers formed during the ice ages, water that evaporated from the oceans did not return. Over the years, the snow built up, and sea level in the oceans correspondingly began to drop. At the peak of the last three or four ice ages, sea level worldwide fell to about 120 m below the present level because of all of the snow that had remained on land (Bintanja et al., 2005).

The impact of a 120 m drop in sea level in the Philippines is enormous. Figure 4.7 shows the modern islands of the Philippines in dark green, and the islands as they existed about 12,000–20,000 years ago (and during each of the three prior ice ages) in light green. Luzon expanded outward to become about 25% larger than it is today, encompassing Catanduanes, Marinduque, and a host of smaller islands; this bigger ice-age island has been referred to as Greater Luzon. Mindanao merged with Basilan, Bohol, Leyte, Samar, and a great many other islands to form Greater Mindanao, with nearly twice the area of modern Mindanao. Cebu, Masbate, Negros, Panay, and nearby small islands merged into Greater Negros-Panay, Palawan expanded out to about triple its current area, and Jolo and Sulu merged to form Greater Sulu. Mindoro is surrounded by deep water, and although it was not far from the shorelines of Greater Luzon and Greater Palawan, it remained isolated from the other, larger islands. A number of small islands (such as Camiguin, Lubang, and Sibuyan) also remained isolated, as did the islands that make up the Babuyan and Batanes Islands that lie north of Luzon (Heaney, 1985; for earlier perspectives, see also Dickerson, 1928, and Inger, 1954; for additional details, see Brown et al., 2013).

These Pleistocene ice-age islands hold special significance for studies of biodiversity. All (or nearly all) of the mammal species that occur on Leyte and Samar, for example, also occur on Mindanao; they are functionally part of a single area of endemism, often referred to as the Mindanao Faunal Region. All of the species of mammals that occur on Catanduanes and Marinduque also occur on Luzon—again, they are part of the same area of endemism, the Luzon Faunal Region. But although the San Bernardino Strait between Greater Luzon and northernmost Greater Mindanao (i.e., the northwestern corner of Samar) is only 15 km wide, there is an astoundingly great difference between the mammal faunas of these two regions (Heaney, 1986; Heaney et al., 2010). Four orders of mammals—gymnures (Erinaceomorpha), tarsiers (Primates), flying lemurs (Dermoptera), and tree shrews (Scandentia)—that are present on Greater Mindanao are absent in Greater Luzon, and a diverse and widespread family of rodents (squirrels, Sciuridae) is also absent on Luzon (aside from one species of squirrel introduced from Thailand in the 1960s). Instead, the Luzon side has two endemic, highly diverse, species-rich adaptive radiations of small mammals (chapter 5). Moreover, 93% of the currently recognized non-flying native mammals are endemic to Greater Luzon, and the percentage on Greater Mindanao (which is less well known) is quite

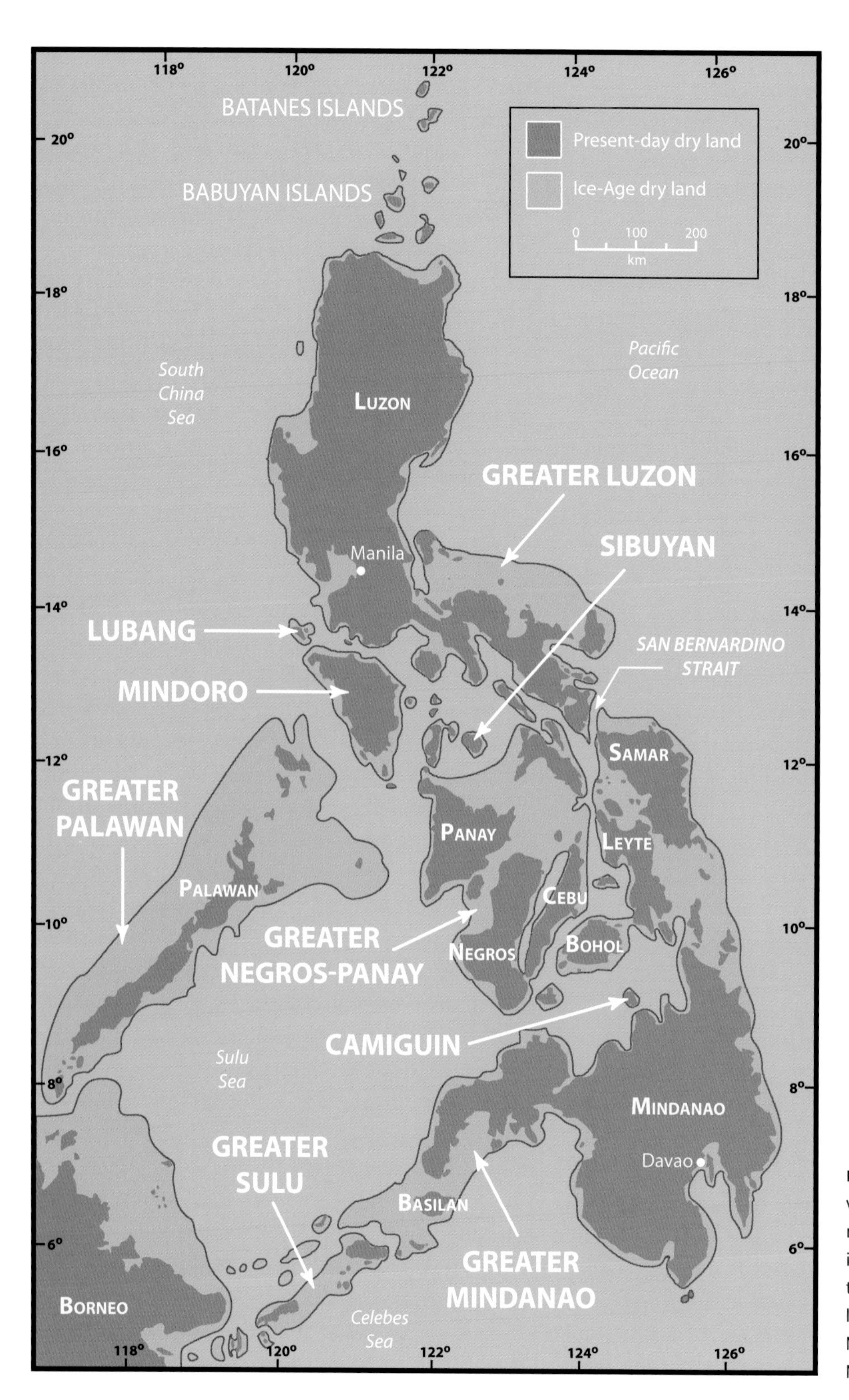

FIG. 4.7. The Philippines, with the boundaries of both modern islands and the islands that existed during the maximum lowering of sea level during the Last Glacial Maximum of the Pleistocene. Modified from Heaney (1986).

similar (Balete et al., 2008; Heaney, 1986; Heaney et al., 2014a). Similar (though not identical) patterns of endemism have been documented for birds, amphibians, and reptiles (Brown et al., 2013); clearly, the impact of this historical factor has been profound.

To put this into a broader perspective, the change in mammal fauna from Madagascar to Africa is perhaps even greater than the one between Luzon and Samar, but the channel between the former areas is over 400 km wide. The only comparable faunal break over a similar distance that comes to mind is the one between Bali and Lombok, made famous by Alfred Russel Wallace (1859); at about 30 km wide, it marks Wallace's Line, one of the most dramatic and widely recognized faunal boundaries anywhere. By any definition, the change in mammal fauna between Greater Luzon and Greater Mindanao is world class.

As variation in sea level took place during the ice ages, there was also fluctuation in global temperature. There are virtually no hard data from the Philippines that are directly relevant, but since the global average decline in temperature was about 5°C (Lomolino et al., 2010:323), we can take that as a rough estimate for beginning our discussion of the change in the Philippines. Be forewarned: this discussion will be rather murky at the outset and will get murkier as we go along.

If temperatures across the Philippines had declined by about 5°C during the height of the Last Glacial Maximum (about 20,000 years ago, perhaps continuing until about 12,000 years ago), the Philippines might have felt like a much less tropical place. Where today we might expect daily highs of about 30°C in the lowlands (chapter 2), temperatures around 25°C might have been common. Because temperatures currently decline by about 6°C with every 1000 m increase in elevation, all of the temperatures shown in figure 2.7 may have been an average of about 5°C cooler.

Because temperature is one of the major factors that influences the environment, it is possible that the lower boundaries of the montane and mossy forest might have shifted downslope in response. If this took place in a direct fashion, a decline of 5°C equates to an elevational decline of about 830 m. Since the lower limit of montane vegetation on Luzon is about 900 m elevation, this suggests that montane forest might have occurred down to perhaps 200 m above sea level during the coolest portions of the Pleistocene (bear in mind that sea level then was about 120 m lower than at present). There are two complicating factors, however. When seawater cools, less water evaporates into the atmosphere, and there is less moisture available to fall as rain. While some parts of Southeast Asia remained covered by rainforest, there is substantial evidence that rainfall was indeed lower during the periods of cooler temperature (and lower sea level) in parts of Borneo and the adjacent exposed parts of the continental shelf of Southeast Asia, causing Palawan to develop areas of seasonal forest and perhaps savanna where only lowland rainforest grows today (Bird et al., 2005; Heaney, 1984; Piper et al., 2011). Thus climatic conditions on Luzon during the ice ages were perhaps both cooler and drier.

The second complication has to do with the fact that when drier air rises along an elevational gradient, it cools at a faster rate than that for more humid air (the rate of cooling is referred to as the lapse rate). The humid air that typifies the Philippines today cools at a rate of about 0.6°C per 100 m elevation (chapter 2). But less-humid air cools at a rate of 0.8°C per 100 m elevation, and very dry air (such as in a desert) may cool at 1.0°C per 100 m (Lomolino et al., 2010:55). We have no relevant hard data from the Philippines, but a change like this took place on New Guinea during the height of the Last Glacial Maximum, resulting in a temperature change of about 2°C cooler at sea level that gradually increased to a change of about 8°C cooler at 3000 m elevation, relative to current conditions (Morley and Flenley, 1987). In other words, the compression of vegetation zones potentially was greatest at high elevations and was much less at sea level. Based on this, lowland forest on Luzon might have been compressed downslope by only a few hundred meters (from the current maximum elevation of ca. 900 m down to 600 or 700 m elevation), even during the coolest phases of the Pleistocene.

To further complicate the picture by combining these possible (and generally hypothetical) effects, while cooler temperatures might have been a factor in causing species distributions to shift downslope, for taxa that require moist conditions (such as earthworms and the species of mammals that feed on them, as well as many plants), the drier conditions in the lowlands might have inhibited that trend. In other words, perhaps the availability of moisture is a more important factor than temperature.

Unfortunately, for now all of this is speculative, because we have no hard data from the Philippines about Pleistocene rainfall patterns and, still more unfortunately, no directly relevant information about the distribution of plant communities on Luzon during the periods of glacial development. As a result, the impact of Pleistocene climatic fluctuations on Luzon is highly uncertain. We can be sure that conditions were different—probably somewhat cooler and drier, with elevational zones that may have shifted lower—but a great deal of additional research is needed to provide clarity.

There is one last place, however, that we can turn to for information about ice-age climates: the fossil record of mammals on Luzon. This is an indirect approach, and the data are scanty indeed, but there is enough to be tantalizing.

The Fossil Mammals of Luzon

The various types of rainforest that covered the Philippines until recently provided wonderful habitat for many living species of mammals and a great many other taxa (chapter 2), but rainforest is a notably difficult place to find fossils. But by 100 years ago, some extensive areas had been cleared of nearly all natural vegetation and converted to open grassland, including the Cagayan River Valley (chapter 7). This valley is a sedimentary basin that has gradually filled with volcanic ash and other debris, much of which provides a very good environment for preserving fossils (Vondra et al., 1981). Although there are also scattered fossils of mammals from Pangasinan and Rizal Provinces, most Luzon mammal fossils (and most from the country as a whole) have been found in the Cagayan Valley. With one exception, details were not well documented when the fossils were discovered decades ago; most were found on the surface of the ground, so their age could not be determined more precisely than to some time in the Pleistocene (i.e., at some point during the past 2.6 million years or so).

Some of the finds are not too surprising: fossil deer (*Cervus*) and pigs (either *Sus* or *Celebochoerus*) may simply be the ancestors of the living species, or perhaps their counterparts from an earlier colonization from the mainland (Van der Geer et al., 2010). Another species is illuminating: a fossil dwarf water buffalo (*Bubalus*), apparently related to the extant tamaraw of Mindoro and the extinct, even smaller dwarf water buffalo of Cebu (Croft et al., 2006). Other finds have been more startling: at some point during the Pleistocene, a rhinoceros (*Rhinoceros*) and at least two different lineages of elephants (*Elephas* and *Stegodon*) were also present on Luzon (Beyer, 1957; de Vos and Bautista, 2001, 2003; Van der Geer et al., 2010; Vondra et al., 1981). These latter animals are associated elsewhere in Indo-Australia with relatively dry, at least partially open habitat, not with closed-canopy rainforest. The thought of the Cagayan Valley supporting herds of elephants and rhinos, along with dwarf water buffalo, deer, and native wild pigs, all moving across an open savanna-like habitat, probably with forest along the river margins, puts an entirely different perspective on what the fauna of Luzon might once have been like. Unfortunately, as things stand currently, that is about the full amount of what we can glean from the fossil record. Research focused on fossils from the Cagayan Valley (and probably elsewhere) should surely open doors to a very new and different view of the history of Philippine mammals and their habitats.

That door has also recently opened a crack further, in a way that may be equally revealing about the Luzon fauna. Beginning in 1976–77, archeologists from the

National Museum of the Philippines began a series of excavations of caves near Peñablanca, at elevations near 100 m; further excavations were conducted in 2003–9 by researchers from the University of the Philippines (Mijares, 2008). Dating of charcoal and other materials from one of them (Callao Cave) show that it was inhabited periodically by humans back to about 25,000 BP, and, quite astoundingly, at least briefly around 67,000 BP (Mijares, 2008; Mijares et al., 2010). Little can be gleaned at this time about those earliest humans in the Philippines—even older than those from Palawan, dating from roughly 16,000 BP to possibly as far back as 47,000 BP (Détroit et al., 2004; Dizon, 2003). But more can be inferred about the mammalian fauna of the area, both from what is present in the cave sediments and what is not.

The layers of sediment from Callao Cave older than 25,000 BP yielded only limited bones, so interpretations must be made cautiously. A deer tibia with what appear to be cut marks from butchering was dated to ca. 60,000 BP, and two teeth from deer were dated at about 52,000 and 54,000 BP. In the layers that are 25,000 years old and younger, the archeologists found hundreds of fragments of mammal bones and teeth, some of which showed clear evidence of butchering by humans. Most of the bones came from deer that are not distinguishable from the living species on Luzon (*Cervus mariannus*); most of the rest were wild pigs. Rhinoceros and elephants were absent; although it is possible that they simply were not hunted, this might also mean that they had disappeared from the Cagayan Valley by that time. Notably, bones from monkeys (*Macaca fascicularis*) and the two species of civets now on Luzon (*Paradoxurus philippinensis* and *Viverra tangalunga*) were not recovered. Since civets and monkeys are frequent prey of human hunters on Luzon today, their absence from the cave sediments implies that they also were not present on Luzon during that period of time. Archeological deposits in caves elsewhere in northern Luzon dating from ca. 7000 BP to ca. 3600 BP also yielded bones and/or teeth of deer and pigs, but no other mammals (Mijares, 2008).

More recently, the door into the past has opened another crack, with much smaller animals. Three rodent mandibles were in the same series of older sediments from Callao Cave as the human remains, dated from about 50,000 and 68,000 BP (Heaney et al., 2011b). One of the mandibles was identified as *Apomys microdon*, a species of small mouse that lives in the area today at about the same elevation as the cave, as well as much higher (at least to 1400 m elevation) on the adjacent mountains (Duya et al., 2011). The other two mandibles represent a species of *Batomys*; today this genus is found only above about 1300 m elevation in montane and mossy forest. The fragmentary nature of the *Batomys* fossils left their species designation uncertain (they were a bit larger than *B. granti*, the common species in the Central Cordillera), but since *Batomys* occur today only above 1300 m elevation and only in the Central Cordillera (with a closely related species on Mt. Isarog in far southern Luzon), we can conclude that the native rodent fauna was certainly different in some ways from that of today, with at least one species occurring both much lower and in a very different geographic region than where any member of the genus is found today.

This glimpse of the fossil history of the highly diverse mammal fauna on Luzon is tantalizing, with hints of a large mammal fauna that has mostly disappeared; wild pigs, deer, and the tamaraw on Mindoro are perhaps the last traces. But we also have some limited evidence that the native small mammal fauna differed in both its elevational distribution and geographic extent, along with the possible absence of monkeys and Luzon's two current species of civets.

On the latter point, we can add a bit more. The oldest records of monkeys and civets from Luzon are about 3500 years old, after the arrival of Malayo-Polynesian-speaking people, who had cultures that remained in place until the arrival of the Chinese and Spanish (Bellwood, 1997, 2013; Diamond, 2014). At this time, given the high genetic similarity of Philippine populations of macaques and civets to those of Borneo, we tentatively conclude that these species were introduced to Luzon by humans in relatively recent times (chapters 6, 7, and 9). This leads to one of the more striking conclusions about the mammal fauna of Luzon: this ecologically rich fauna included no native mammalian predators. Thus the extant mammalian fauna of Luzon is distinctive both in what is present (the highly diverse rodent and bat fauna) and what is absent (native mammalian carnivores). In chapter 5, we approach the first topic—the diversity of the small mammal fauna—on the basis of what we can deduce from living species. We return to the implications of the second topic later (chapters 6 and 8) but emphasize now that it most likely had a profound impact on the evolution of the small mammal fauna.

5

Producing Diversity: Speciation and Diversification

With 110 native species (and the number continuing to rise) on an island of just 103,000 km^2, Luzon has an exceptionally great concentration of mammalian diversity—perhaps the highest in the world. Chapters 2 and 3 have documented their distribution patterns over the island and along elevational gradients but have not dealt with one utterly fundamental question: how did this diversity arise? To answer that question we also must ask, when did the ancestors of the current species arrive, and when did the current species themselves arise? Of the original colonizing species, how many have only a sole living descendant, and how many gave rise to multiple species? As we attempt to seek answers, we need to consider the geological history of Luzon and the rest of the Philippines (chapter 4), since we know that Luzon is one of the most geologically active places on earth. Depending on the timing of the processes that have produced the current mammalian diversity, what we think of today as Luzon might have been a quite different place when the ancestral lineages first arrived on Luzon and subsequently during the diversification processes.

These questions about the geographical and temporal origins of Luzon mammals are brought into high relief by considering a simple comparison of the species that are able to fly with those that cannot (table 5.1). Among the 57 species of bats, 38 (67%) are widespread in Indo-Australia. A further 17 (30%) are restricted to the Philippines but are widespread within the archipelago. Only 2 species (4%) are restricted to Luzon. Although some of the currently recognized species of bats are poorly studied and may actually consist of species-groups that may contain some Philippine endemics (chapter 11), the general pattern is clear and unlikely to change dramatically: most bats on Luzon are members of quite widespread species and show little evidence of developing endemism on Luzon.

In contrast, among the 47 species of native murid rodents, which includes all but 1 of the native small mammals (the other is the shrew *Crocidura grayi*), none of the species are widespread in Indo-Australia, and only 2 are widespread within the Philippine archipelago. Instead, 45 species (96%) live only on Luzon. Further, 37 of these species (79%) do not even occur throughout all of Luzon; rather, they occur only within a limited area on Luzon (i.e., they are locally endemic), as shown in figure 3.15.

Obviously, the ability of bats to fly has shaped their diversity patterns in ways that differ greatly from the roughly equally diverse non-flying small mammals. For a species to be widespread in Indo-Australia, there must be enough gene flow among all populations, now or in the recent past, to allow the spread and incorporation of new mutations across the entire population. Otherwise, the mutations remain localized and, as they accumulate, the isolated populations become different

TABLE 5.1. The number of all bat species, fruit bat species, and murid rodent species within the defined geographic areas. Updated from Heaney and Rickart, 1990.

	All bats	Fruit bats	Murid rodents
Native species on Luzon	57	15	47
Species widespread in Indo-Australia	38 (67%)	6 (40%)	0
Species endemic to oceanic Philippines	17 (30%)	7 (47%)	2 (4%)
Species endemic to Luzon only	2 (4%)	2 (13%)	45 (96%)
Species locally endemic within Luzon	0	0	37 (79%)

and eventually are unable to interbreed. The same is true for species that are widespread within the Philippines: to maintain genetic cohesion among populations, rates of gene flow must now be high or must have been so in the recent past. On the other hand, when we say that a species is endemic to some given area (e.g., to the single island of Luzon or a single mountain range on Luzon), this means that gene flow was interrupted at some point in the past, and the populations began to accumulate new genetic mutations and eventually became consistently different from other populations. Bats can fly, and therefore they more readily maintain gene flow among populations on different islands (or even among different archipelagoes). In contrast, among the 47 native murid rodents on Luzon, only 2 have had high enough gene flow between islands, or became isolated sufficiently recently, to remain a single widespread species. Small non-flying mammals apparently have great difficultly moving between islands. The observation that 79% of these species occur only in a single mountain range on Luzon suggests that many even have difficulty moving between mountain ranges on a single island.

Investigation of the processes that produce these patterns of species distributions requires quite a lot of information. It would be ideal if we had enough detailed data to look into the basis of the differences between all of the bats and all of the non-flying mammals on Luzon. Unfortunately, that is not the case. The studies conducted up to this time are sufficient only for two groups: the fruit bats (Family Pteropodidae) and the native mice and rats (Family Muridae). Shrews (Family Soricidae) are well known, but only a single native species is present on Luzon, making the story comparatively simple. We initially consider these first two groups separately, starting with the one well-studied family of bats, and then turn to the murids. Not surprisingly, given the data shown in table 5.1, the stories unfold rather differently in these two families, but they do so in ways that provide deep insights into the impact of geographic circumstances on the diversification of mammals in the Philippines.

Fruit Bats

The 15 species of fruit bats known from Luzon include some of the most striking species of mammals in the Philippines. For example, golden-crowned flying foxes (*Acerodon jubatus*) are one of the largest species of bats in the world, at times reaching about 1.4 kg (chapter 11). Among the 15 species, 6 (40%) are widespread in Indo-Australia, 7 (47%) are widespread within the oceanic Philippines (i.e., excluding Palawan), 2 (13%) are endemic to Luzon, and none are confined to a single mountain range or mountain on Luzon (table 5.1). In other words, nearly half of the species occur widely in Southeast Asia; apparently, members of each of these species fly between islands frequently enough that each species maintains itself as a genetically integrated unit (Heaney and Roberts, 2009; Heaney et al., 2005). This includes many of the species that live in disturbed habitat around humans, such as *Eonycteris spelaea*, *Macroglossus minimus*, *Pteropus hypomelanus*, and

Rousettus amplexicaudatus (chapter 11). These species often forage in open pastureland habitat, flying among the scattered fruit trees (often wild figs or shrubs such as *Melastoma*) that have been left by humans, and sometimes appearing in agricultural areas among planted fruit crops. It seems that flying across a narrow sea channel is not difficult for a species that typically flies across open agricultural landscapes (Heaney and Roberts, 2009).

Another set of fruit bat species—again, nearly half—are widespread within the Philippines but not elsewhere: this includes the giant *Acerodon jubatus*, as well as the medium-sized *Ptenochirus jagori* and rather small *Haplonycteris fischeri*. This group also probably includes *Cynopterus brachyotis*, which functions as an endemic species within the Philippines but is part of a species-group that is in need of further taxonomic study (chapter 11). Apparently these can fly between islands within the Philippines but are unable to colonize other, more distant islands. Detailed studies have shown that these species often show evidence of great genetic similarity among islands that were joined together during the Pleistocene periods of low sea level (fig. 4.7), and of lower genetic similarity among islands that were always isolated by sea channels that remained as seawater, even when sea level dropped to its lowest levels (Roberts, 2006a, 2006b). This implies that the sea channels reduce the rate of gene flow for these taxa: not enough to allow speciation, but enough to permit some local differences to emerge. It also shows that these patterns of genetic similarity and difference that exist today within single species have been influenced by changes in sea level that took place over 10,000 years ago (and perhaps much further in the past). Thus these are patterns of genetic geographic similarity that develop over long periods of time.

Finally, two species of fruit bats (*Desmalopex leucopterus* and *Otopteropus cartilagonodus*) occur only on Luzon. The genus *Desmalopex* is endemic to the Philippines: there is one species on Luzon, one (*D. microleucopterus*) on Mindoro, and an unnamed species on Dinagat that is currently under study by Kristofer Helgen and colleagues (pers. comm.). The genetic differences between the Luzon and Mindoro species (Esselstyn et al., 2008) demonstrate not only that there is no gene flow presently taking place, but also that there has been none for a long time. Because their closest relatives live on the Solomon and Bismarck Islands, east of New Guinea, it seems that a population colonized the Philippines from that area long ago. Subsequently, what was once a single ancestral species has become three species, associated with colonization within the Philippines—involving (with unknown directionality) Luzon, Mindoro, and Dinagat—that happened too rarely to maintain gene flow. Thus speciation has produced three species on two adjacent islands and one more-distant island, none of which have ever been connected by dry land.

Otopteropus cartilagonodus (Luzon dwarf fruit bats) present a somewhat different picture. The genus contains only one species, but it is closely related to two other small fruit bats endemic to the Philippines: *Alionycteris paucidentata*, which occur only on Mindanao, and *Haplonycteris fischeri*, which are widespread within the oceanic Philippines (chapter 11). These three genera, each with a single species recognized currently, are members of a clade that occurs only in the oceanic Philippines (Heaney and Roberts, 2009). Apparently their common ancestor arrived in the Philippines long ago and underwent speciation within the archipelago, but never was able to colonize outside it. The restriction of one of the species to Luzon and one to Mindanao, as well as evidence that there is geographic structure to the genetic variation within *Haplonycteris* (Roberts, 2006b), implies that these bats have a limited ability to fly between islands.

Overall, these data on fruit bats indicate that most of the 15 species now on Luzon are the direct descendants of bats that arrived from outside the Philippines, and nearly half have maintained high enough gene flow to remain as single, widespread Asian species. Some others have not maintained gene flow with populations outside the archipelago but have had enough gene flow not to have speciated within the Philippines. Only in two cases among Luzon fruit bats—*Desmalopex* and the three genera of dwarf fruit bats—do we see clear evidence that there has been speciation within the archipelago, and that has always been between (not within) Pleistocene islands (such as Greater Luzon and Greater Mindoro). In other words, the current species diversity of fruit bats on Luzon has been driven mostly by colonization from outside the Philippines, with only a small amount produced by speciation within the archipelago. While the other families of bats are too poorly studied to address these matters in detail, there is nothing currently known that would lead us to suspect that they differ substantively (e.g., Esselstyn et al., 2012b). Bats can fly between islands, at least occasionally, and it seems that this has made all the difference for their patterns of speciation and diversification.

Murid Rodents

One of the most distinctive aspects of the mammal fauna of Luzon is the preponderance of one family, the Muridae. Because these animals are related to household rats and mice, the response to the name is sometimes strongly negative. This is terribly unfortunate, because the species that cause so many problems for humans are not native to the Philippines; they were brought in from the Asian mainland, probably entirely by accident, within the past few thousand years and are only distantly related to most of the Luzon murids (chapters 6 and 10). The native species typically are quite attractive animals that are highly diverse, looking much like squirrels, shrews, gophers, and the like from other parts of the world. Native murids are rarely associated with humans, and when they are, some of them perform valuable functions, such as the lowland chrotomys (*Chrotomys mindorensis*), which feed on earthworms that dig holes in ricefield walls and snails that can damage crops and carry disease. Most of the native species live in forest in the high mountains, where they perform a diverse set of roles in the web of nature (chapter 3). The fact that these murid rodents occupy so many distinctive niches is one of the most striking parts of the story of the mammals of Luzon.

Interpreting the great diversity among these small mammals requires us to develop as much understanding as possible about how they are related, so we can attempt to answer the questions posed at the start of this chapter. When did the ancestors of the current species arrive, and when did the current species arise? How many of the species are the sole living descendant of an original colonizing species, and how many of the original colonizing species are the ancestors of more than one of the living species?

Traditionally, biologists learned about relationships among mammals by conducting detailed studies of their anatomy and seeking similarities that have been inherited from common ancestors, especially novel features that show a close relationship among some of the species. Great progress has been made in that fashion, and this remains a critically important aspect of evolutionary studies today, especially those involving extinct species represented by fossils.

In recent decades, however, some new techniques have been developed that allow for more precision and expanded investigations. These involve the use of DNA, the molecule that holds the genetic information essential to the functioning of nearly all aspects of live mammals. In brief, it is now possible to determine the precise chemical composition of either portions of the DNA or the entire genome of an individual, and to compare this DNA with that of other individuals. When identical portions of DNA are found in a set of individuals, these indicate that those individuals descended from a common ancestor. And the greater the number of identical portions of DNA, the more closely related the individuals are. The actual analysis is more complex than this, but we won't go into such details; suffice it to say that these techniques can define relationships of species of mammals (and other organisms) quite conclusively. The approach is essentially the same as the one employing anatomical similarities, but examining DNA allows the use of vast amounts of information, and it is that vastness that allows great precision.

One standard result from either type of analysis

(i.e., based on anatomy or on DNA) is a diagram that shows the relationships of a set of species (or genera, or families, etc.). These are shown in a format that looks similar to the branches of a tree and leads to the frequently used term "tree of life." A branching diagram of this type is called a phylogeny. In the rest of this chapter, we talk quite a lot about phylogenies, since they are crucial in understanding the history of mammalian diversity on Luzon.

A second aspect of these DNA-based phylogenies is also essential to our knowledge about the fauna of Luzon. By developing phylogenies based on anatomy that include some fossils of known age as well as living species, and by also including information from the DNA of the living species, we can estimate when the various species arose and when they last shared a common ancestor. The more fossils that are included, the more precise the estimates are of the dates when the various speciation events took place.

Among Philippine mammals, we have virtually no fossil record that helps us in this regard. Elsewhere in Asia, however, many fossil rodents are known, and they have been incorporated into phylogenies that include the Philippine murids (e.g., Jansa et al., 2006; Rowe et al., 2008; Schenk et al., 2013). As a result, we can estimate when the ancestors of the living species in the Philippines arrived and when the many diversification events took place. These estimates are not precise; it is best to think of them as being within a range of plus or minus about 20%.

As an example, an analysis of this type for the cloud rats (fig. 5.1) shows that the five living genera had their most recent common ancestor about 10 million years ago; that ancestor diverged from murids on the Asian mainland about 14 million years ago. All of these dates (and those used hereafter) are estimates, and these estimates are shown on the timelines in the figures. For example, the last common ancestor of cloud rats is estimated to have lived 10 million years ago, but it might have been anywhere from 8 to 12 million years ago. That is a rough estimate, but it still places that event at a fairly specific time in the geological history of Luzon.

The first split in the phylogeny of the cloud rats gave rise, on the one hand, to the ancestor of four of the genera and, on the other, to the ancestor of *Phloeomys*, which is a highly distinctive genus with two species. These two species live in the lowlands, and one of them (*P. pallidus*) also occurs high in the Central Cordillera (chapters 3 and 10). About 7 million years ago, the other branch gave rise to a lineage that eventually

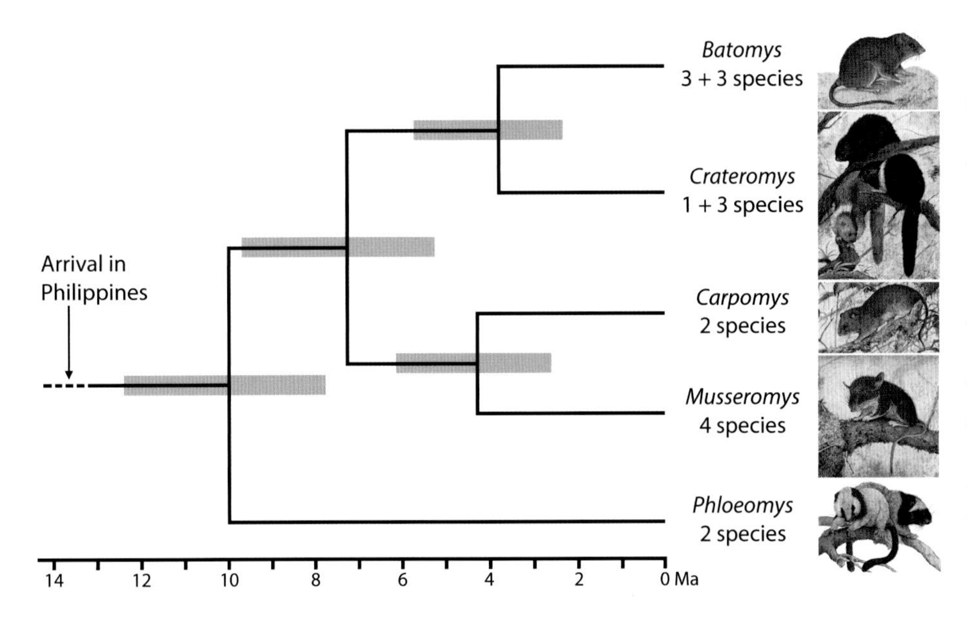

FIG. 5.1. Phylogeny of cloud rats. The phylogenetic relationships are strongly supported (i.e., the branching pattern itself is highly likely), but the timing of the individual events are estimates, with the range of possible dates indicated by the gray bars. The number of species in each genus is shown as the number on Luzon plus the number elsewhere in the Philippines (when these exist). Based on Heaney et al. (2014b); Schenk et al. (2013).

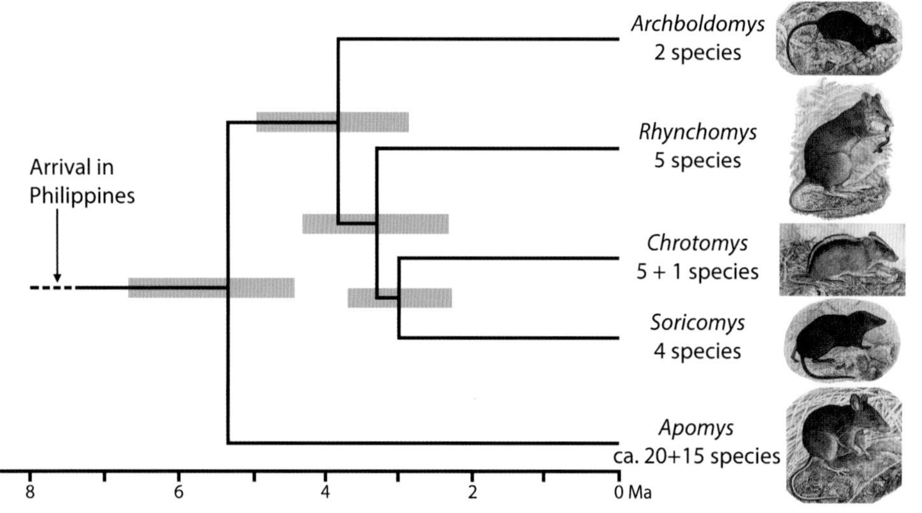

FIG. 5.2. Phylogeny of earthworm mice. The phylogenetic relationships are strongly supported (i.e., the branching pattern itself is highly likely), but the timing of the individual events are estimates, with the range of possible dates indicated by the gray bars. The number of species in each genus is shown as the number on Luzon plus the number elsewhere in the Philippines (when these exist). Based on Balete et al. (2012); Schenk et al. (2013).

split into what we now recognize as *Musseromys* and *Carpomys* (at about 4 million years ago), and to *Batomys* and *Crateromys* (at a little less than 4 million years ago). Thus all five of the extant genera had come into being about 4 million years ago (always recognizing that these are estimates of the timing, as shown by the gray bars in fig. 5.1). Twelve species of clouds rats now living on Luzon are descended from this ancestral cloud rat, as well as six species (in the genera *Batomys* and *Crateromys*) found elsewhere in the Philippines (mostly on Greater Mindanao). On Luzon, all of the genera currently occur in the Central Cordillera, two species (the species of *Phloeomys*) occur widely in the lowlands, and only one species in the highlands anywhere other than the Central Cordillera (i.e., the one species of *Batomys* on Mt. Isarog; chapter 10). The descendent species differ hugely in body size, ranging from about 22 g for the smallest *Musseromys* to 2.7 kg for the largest *Phloeomys*; thus there has been extensive morphological and ecological radiation, produced at an average rate of about 1.4 species per million years. All of these species are primarily arboreal and feed only on plant material; the largest (*Phloeomys*) eat bamboo shoots and tender young leaves, and (to the best of our very limited knowledge) the smallest (*Musseromys*) consume oil-rich seeds (Heaney et al., 2014b).

A phylogeny of the earthworm mice shows many similarities, but some differences (fig. 5.2). Five genera currently exist, including even more species (about 36) on Luzon and at least 16 species elsewhere in the Philippines. This total diversity from the original common ancestor has developed more rapidly than that for cloud rats: the common ancestor of the earthworm mice is estimated to have arrived in the Philippines about 7 million years ago, with an overall average rate of about 7.4 species produced per million years. The five genera of earthworm mice do not vary greatly in size, from about 20 g to 160 g (chapter 10), but their body shapes differs markedly: *Chrotomys* are stout burrowers; *Soricomys* and *Archboldomys* are small, lightly built, shrew-like animals; *Rhynchomys* hop about on their sturdy hind legs much of the time; and *Apomys* look much like everyone's idea of a (rather cute) mouse. Differentiation of the five living genera occurred between about 5.5 and 3 million years ago. All of the genera are represented in the Central Cordillera, but earthworm mice are fairly widespread over Luzon, including at least one *Chrotomys* and several *Apomys* species that occur in the lowlands (chapter 10). All of the species preferentially consume earthworms and other soft-bodied soil invertebrates, but *Apomys* also eat seeds. Given these feeding habits, it comes as no surprise that most of the species live on the ground; only the members of the subgenus *Apomys* are primarily arboreal.

In addition to these two large endemic clades, four additional colonizations of Luzon by murids have taken place. All of them arrived in the Philippines relatively recently—during the past 5 million years or so—and they are thus often referred to as the New Endemics, in contrast to the cloud rats and earthworm mice, which are considered Old Endemics (chapter 10). One of these new groups includes two enigmatic genera, *Abditomys* and *Tryphomys*; during our entire intensive survey on Luzon, we did not encounter them. Because there are so few specimens of these two taxa and so little information to work with, we do not even know with certainty that they are each other's closest relatives (Musser and Heaney, 1992). We think that they are related to a large, poorly understood group of murid genera that occur widely in Indo-Australia, the Tribe Rattini (Fabre et al., 2013; Schenk et al., 2013). We very roughly guesstimate that their ancestor arrived in the Philippines about 4 or 5 million years ago, simply because this is when many of the other genera in that group first evolved. Assuming that these enigmatic taxa are indeed a distinct clade, that clade has produced two species in roughly 4 million years.

The genus *Bullimus*, which is only found in the Philippines, is much better known. These are fairly common, ground-living animals (three species, one of which is widespread on Luzon) that are related to the genus *Sundamys*, which occurs on the Sunda Shelf and the Southeast Asian mainland (Fabre et al., 2013; Schenk et al., 2013). *Bullimus* diverged from *Sundamys* about 3–4 million years ago; incomplete studies that we are conducting currently (Heaney and Kyriazis, unpublished data) suggest some divergence among populations on Luzon, but for now we treat all of the Luzon individuals as one species, *B. luzonicus*. It occurs all along the elevational gradient in many places but is absent from most of the Bicol Peninsula.

One of the most widespread and abundant of the New Endemics species is *Rattus everetti*. This species provides an example of the limitations of our current knowledge about the phylogenetic relationships of Luzon mammals: current genetic data (e.g., Fabre et al., 2013; Schenk et al., 2013) indicate that they are most closely related to *Tarsomys* and *Limnomys*, which are genera endemic to Mindanao, not to other members of the genus *Rattus*. Thus it is clear that the scientific name of this species needs to be changed to that of some other genus, but we do not yet know to *which* genus, so we operationally continue to use *Rattus*. At any rate, it seems as though what we call *R. everetti* diverged from its relatives about 2 million years ago and spread over much of the Philippine archipelago. On Luzon, the species occurs over the entire elevational gradient but tends to be most abundant in the upper reaches of lowland forest, and it is more abundant in disturbed forest than in old growth.

The final clade of native murids on Luzon is represented by *Crunomys fallax*. This, too, is a highly enigmatic genus. In addition to the one species on Luzon, there are two others on Greater Mindanao, and another on Sulawesi (Achmadi et al., 2013; Rickart et al., 1998); they are most closely related to the genus *Maxomys*. DNA data suggest that the genus diverged from *Maxomys* only about 1 million years ago, and since all of the relatives of *C. fallax* occur well to the south, this implies that this species arrived on Luzon even more recently. But because *C. fallax* is known only from a single specimen obtained by John Whitehead in 1894, everything about this taxon remains murky.

Shrews

In addition to the 47 species of murid rodents, which clearly have evolved into a wide variety of ecological and morphological niches, there is one additional small mammal that is native to Luzon: the shrew *Crocidura grayi*. This species has been included in the most intensive study of speciation and biogeography of any single Luzon mammal (Esselstyn et al., 2009, 2010, 2011). The story of their evolutionary biogeography is told on pages 64–65.

It is remarkable that there is only one native small mammal on Luzon that is not a murid rodent. We know of no other large island with a diverse mammalian fauna for which a single mammalian family so

Speciation in Shrews (*Crocidura*) in the Philippine Archipelago

Jacob A. Esselstyn, Museum of Natural Science, Louisiana State University, Baton Rouge, Louisiana, USA 70803

Shrews are small, ground-surface-living mammals with small eyes, a long flexible nose, and a fast metabolism. In the Philippines their diversity is modest, with about 10 native and 1 non-native species currently recognized. *Suncus murinus*, the non-native species, is widespread in the archipelago and usually common in urban areas; it probably arrived as a stowaway on boats. The native species, all in the genus *Crocidura*, are widespread as a group, but individual species are typically endemic to a single island or group of islands. Among the species of *Crocidura* that are known to occur on multiple islands, some are spread among islands that were connected by dry land during Pleistocene periods of low sea level, but others occur on permanently isolated islands. Most of the large islands in the country support only one species of *Crocidura*.

In phylogenetic analyses of regional shrew diversity, 8 of the 10 species of Philippine *Crocidura* form a clade (i.e., a natural grouping of closest relatives). Two species, however, are more closely related to shrews that live outside the Philippines. The first, *C. batakorum*, a Palawan endemic, is more closely related to shrew species from Sulawesi Island, Indonesia, than to any known species in the Philippines. The second, *C. tanakae*, which occurs in the Batanes group of islands at the far northern extremity of the Philippines, is closely related to populations of the same species living in Taiwan and on the Asian mainland. Mindanao, Mindoro, and Palawan are exceptions to the one-shrew-species-per-island rule, with each of these islands supporting two species of *Crocidura*. Phylogenetic research suggests that the species occurring on Mindoro (*C. grayi* and *C. mindorus*) and Palawan (*C. batakorum* and *C. palawanensis*) are not each other's closest relatives. This tells us that these species were probably not generated by processes that occur within an island, but rather by a process that occurs between islands. In other words, the modern species are the products of inter-island colonization. In fact, when we take the wide distribution of shrews across the archipelago, together with (a) individual species having localized distributions, (b) most islands holding a single species, and (c) some islands holding two species that are not closely related, we can infer from these patterns that inter-island colonization is the primary process that has generated shrew diversity in the Philippines.

How would a shrew get from one island to another? It seems highly improbable that a tiny animal, weighing about 10 g, could cross an oceanic barrier without help. While extremely unlikely over short time scales, such transfers must happen occasionally over long evolutionary periods. These shrews have been in the Philippines for several million years, and we need infer only 10 over-water colonization events to explain their geographic distributions and phylogenetic relationships. It only takes one successful inter-island colonization event every few hundred thousand years to generate the patterns of shrew diversity we see today. The Philippine Islands are regularly hit by strong typhoons, which bring heavy rains, strong winds, and flooding. These events can wash large chunks of soil and vegetation out to sea, providing a raft on which plants and small animals, such as shrews, can ride. Some of these floating islands are large enough that they support standing trees and hollow logs. If these floating islands eventually make landfall in a suitable place, they provide a vehicle for small mammals to expand their geographic distribution and lead to the long-term isolation necessary to generate different species. But computer simulations of this process suggest that an entirely random island-colonization scenario is somewhat unlikely to have generated the current distributions of Philippine shrews. Rather, these inter-island dispersal events appear to have been somewhat more common than we would infer simply from examining their current distributions. Thus some other process must affect colonization. One plausible explanation is that competition and incumbency play a role. Perhaps a few individual shrews riding a raft to a neighboring island are at the mercy of history—if another, similar shrew species is already there, the new arrivals will fail to establish a population, due to a lack of available resources. On the other hand, if the island is shrew free when the flotsam-riding shrews arrive,

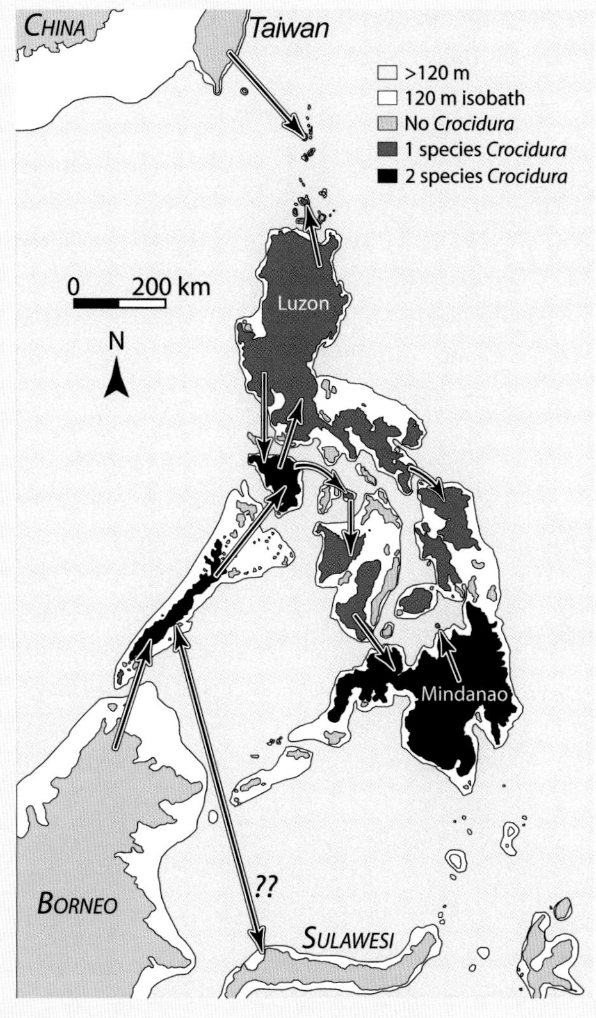

FIG. 5.3. The Philippines, with the likely routes of colonization of the archipelago by shrews of the genus *Crocidura*.

they have a fair chance to establish a new population and survive over long time scales, generating a new species in the process.

References: Esselstyn and Brown, 2009; Esselstyn et al., 2009, 2011.

thoroughly constitutes virtually the entire non-volant mammal fauna.

Geological History and the Diversification of Murids on Luzon

Having laid out the outlines of the historical diversification of the small mammal fauna of Luzon—including the 47 murid rodents plus 1 shrew—we can now place that diversification into the geological context developed in chapter 4. Before doing that, we note that no information on past extinction rates is currently available for these animals, and the following discussion proceeds as if extinction has been a minor (or absent) component of the overall diversification process. The degree to which extinction has occurred has an impact on the calculations that follow: the more extinction that took place, the more speciation must have taken place to produce the current richness of the mammal fauna. In other words, our estimates of diversification rates are minimum estimates. Data from fossils that document actual rates of extinction are badly needed to fully understand the diversification process.

Figure 5.4 combines the phylogenies of the cloud rats (from fig. 5.1) and the earthworm mice (from fig. 5.2), along with the four New Endemic groups. The 47 murids on Luzon are the result of six successful colonization events that began about 14 million years ago. By comparing these phylogenies with the timeline that is shown (and remembering always that these dates are estimates), we can see that the number of genera (or of genus-level lineages) did not rise to two until about 9 million years ago, when the ancestor of *Phloeomys* separated out from the ancestor of the rest of the cloud rats. From that point on, splitting within the existing lineages, coupled with the arrival of newly colonizing lineages, resulted in an increasingly rich murid fauna. The rise in the number of lineages appears to have accelerated, starting about 4 million years ago, as both the cloud rats and earthworm mice began to diversify rapidly and the New Endemics began to arrive. The most recent split, producing two genera within Luzon, occurred between *Soricomys* and *Chrotomys* about 3 million years ago. We also note that the two subgenera of *Apomys* (*Apomys* proper and *Megapomys*) split about 3.2 million years ago. Currently there are 15 genus-level murid lineages on Luzon: 11 of them developed as a result of diversification within Luzon, and only 4 by recent colonization. *Bullimus luzonicus* actually diverged from its closest relative, *B. bagobus* (which is native to Greater Mindanao), only about 1.2 million years ago, so the former species may only have arrived on Luzon at that time, rather than about 3.5 million years ago, which is when the genus *Bullimus* split from *Sundamys*, its closest relative (Schenk et al., 2013).

For the earthworm mice, we currently have enough data on four of the five genera (excluding *Rhynchomys*, for which we have no DNA data) to consider diversification at the level of species as well. In every case, speciation within the genera began at about 1.6–1.8 million years ago (including within each of the two subgenera of *Apomys*). We come back to this topic in more detail below.

We may now place these phylogenetic data into the context of the history of what we now call Luzon Island. As summarized in chapter 4, Luzon is made up of many areas that began as separate islands and merged into the current island over the past 5 million years. This history of development and merging is shown in figure 5.4 (below the timeline). The first to emerge above sea level, over 20 million years ago, was the Central Cordillera, originally formed as a set of small volcanic islands that gradually grew and joined together. Beginning about 15 million years ago, increased volcanic activity caused it to grow fairly rapidly and develop extensive upland areas. The Northern Sierra Madre had emerged roughly 26 million years ago (somewhat earlier than the Central Cordillera), but volcanic activity there ceased by about 20 million years ago, and it disappeared below the waves until much later, reappearing due to renewed volcanic activity. The Zambales Mountains first emerged about 18 million years ago. They gradually grew by uplift (not by volcanic activity, at least until about 7 million years ago) but did not merge with the Central Cordillera until between 4 and 5 million years

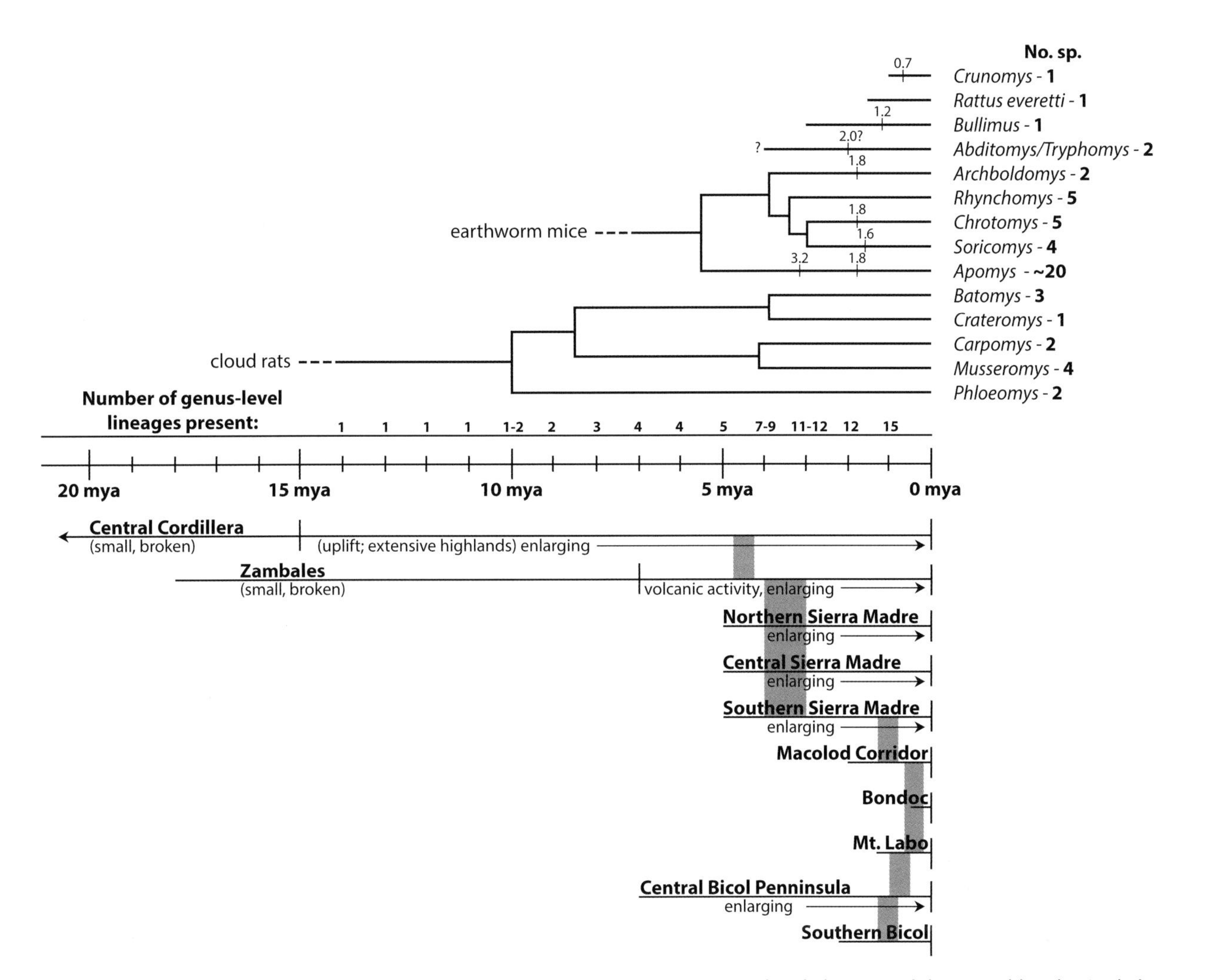

FIG. 5.4. Above the central timeline (from 20 million years ago to the present) are the phylogenies of the two Old Endemic clades (from figs. 5.1 and 5.2), plus the estimated arrival time of the New Endemics. The number of species on Luzon within each genus is shown in boldface numbers to the right of the phylogenies. Below the timeline is a summary of the history of each island that makes up modern Luzon, including a gray bar that shows when each become connected to its neighbors (chapter 4). The timing of all events is approximate, as discussed in the text.

ago, when the Central Valley first filled up enough to make a land-bridge connection. The three parts of the Sierra Madre all emerged about 5 million years ago (with the Northern Sierra Madre reappearing after its much earlier appearance and disappearance), and they combined with each other and with the Central Cordillera / Zambales Mountains between 3 and 4 million years ago. The center of the Bicol Peninsula emerged as an island about 7 million years ago and gradually grew, but it did not merge with the island comprising northern and central Luzon until after the Macolod Corridor, Mt. Labo, and Bondoc Peninsula emerged. The last of these occurred only within the past half-million years.

Several more facts need to be kept in mind to place all of this into context. First, it is important to note that all of the genera of cloud rats and earthworm mice on Luzon currently occur in the Central Cordillera, and no other part of Luzon comes close to that level of generic diversity. Further, none of the species that occur outside of the Central Cordillera diverged from their closest relatives more than about 3 million years ago. Second, we need to remember that the Central Cordillera was the only island with a substantial upland area (i.e., high enough to support montane or mossy forest) from about 15 million years ago until roughly 7 million years ago, when Zambales began to grow substantially. The conclusion based on these observations seems inescapable. The ancestral species of the cloud rats and earthworm mice most likely first arrived on what is now the Central Cordillera, simply because there was nothing else in existence above sea level at the times they arrived except the Zambales Mountains, where few genera are currently present (only *Apomys*, *Chrotomys*, *Rhynchomys*, and *Phloeomys*) and all of those appear to be closely related to species in the Central Cordillera (i.e., every murid that lives in Zambales arrived there within the past few million years). The Central Cordillera is where it all began.

We must next ask whether the speciation took place entirely within dry-land portions of Luzon as it merged into the current large island, or if any of the species presently on Luzon arrived where they now occur by movement over water before the individual islands coalesced. Our current data indicate that speciation within the genera began about 1.8 million years ago, after northern and central Luzon had emerged and combined; this helps to eliminate some possibilities and focus our attention on others. But at this point, the limitations of our current data become apparent. While our studies have come a long way, a great deal of research remains to be done to conclusively answer this question. Fortunately, we can address the issue at least partially with data we have at hand.

As things stand currently, our best estimate indicates that *Archboldomys luzonensis* split from *A. maximus* about 1.8 million years ago. This implies to us that the former species most likely reached the Bicol area over water, since central Bicol was an island separate from the Central Cordillera at that time. This is certainly worthy of note but perhaps should not be considered surprising: *Chrotomys* got to both Mindoro (recently) and Sibuyan (longer ago; Jansa et al., 2006; Rickart et al., 2005; Schenk et al., 2013), so they are capable of this type of short-distance overwater colonization. We also know that *Apomys* (subgenus *Apomys*) have spread throughout much of the central and southern Philippines by dispersing over the many water channels (Steppan et al., 2003). *Apomys* (subgenus *Megapomys*) also colonized Mindoro and Lubang over water from Luzon (Heaney et al., 2014a; Justiniano et al., 2015). The details are murky at this point, but certainly we can conclude that there are at least a few cases in which overwater colonization probably took place to parts of Luzon when these areas were separate islands, as well as to islands that are nearby but still isolated from Luzon.

The other side of the question—whether speciation took place over dry land as parts of Luzon merged—thus far has been investigated in detail only for mice of the subgenus *Megapomys*. A new study using DNA phylogenetic techniques yielded the results shown in figure 5.5 (Justiniano et al., 2015). It is a single example, but one that included quite a few species and produced very informative results. All of these mice are among the most common species of small mammals in the areas where they occur; most of them occur only above about 1000 m elevation, and very few occur below 500 m (chapter 3).

Of the 13 species of *Megapomys* currently known, 11 occur on Luzon. Apparently, about 2.8 million years ago a population colonized across the Verde Island Passage to Mindoro, where it evolved into the species we know today as *Apomys gracilirostris*. Far more recently—perhaps a quarter-million years ago—another population dispersed from Luzon (possibly from Zambales, where its closest relative, *A. sacobianus*, lives today) to Lubang Island (just north of Mindoro) and there differentiated into *A. lubangensis*. Cross-channel colonization clearly has taken place within this clade, but not very often. The other 11 species of *Megapomys* appear to have undergone speciation entirely within Luzon during the past 2 million years, well after northern and central Luzon merged into one island.

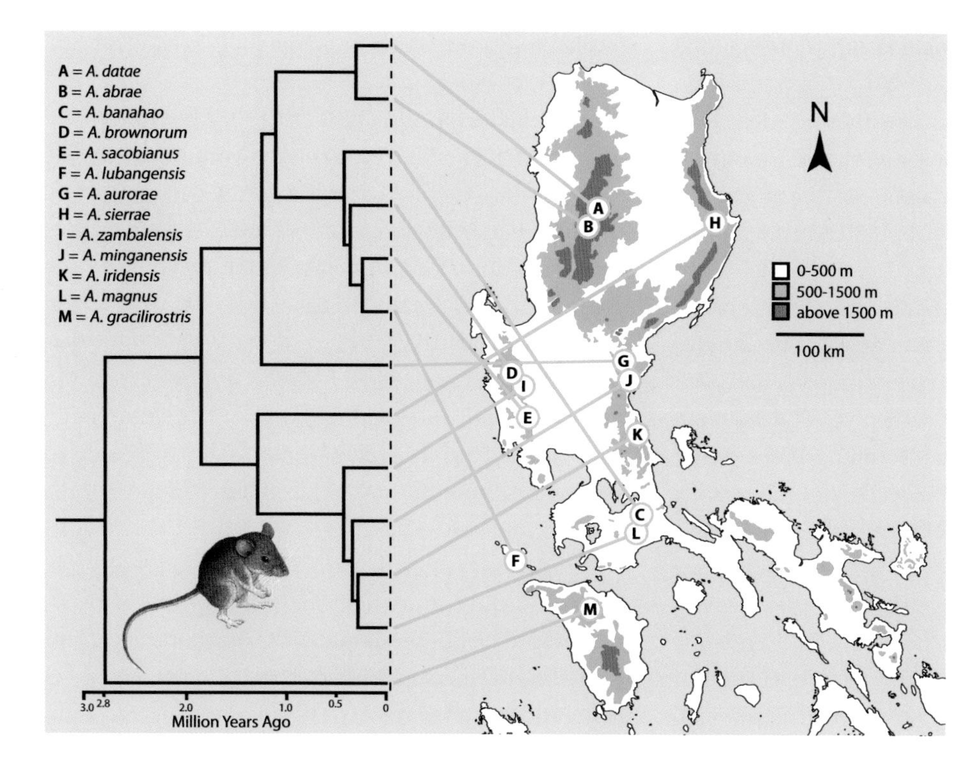

FIG. 5.5. Phylogeny and geographic patterns of speciation within the subgenus *Megapomys* of the genus *Apomys*. Adapted from Justiniano et al. (2015).

We can distinguish three species-groups within *Megapomys*. The first includes only *Apomys gracilirostris*, the species on Mindoro that was discussed above. The second group includes the two species from the Central Cordillera (*A. abrae* and *A. datae*), two from Zambales (*A. brownorum* and *A. sacobianus*), one from the Mingan Mountains (*A. aurorae*), and one from Mt. Banahaw (*A. banahao*). The third group includes one from the Northern Sierra Madre (*A. sierrae*), one from Zambales (*A. zambalensis*), one from the Mingan Mountains (*A. minganensis*), one from the Southern Sierra Madre (*A. iridensis*), and one from Mt. Banahaw (*A. magnus*).

Several patterns are immediately apparent. First, there are no species of *Megapomys* south of Mt. Banahaw, even though the Bicol Peninsula has many areas of mountain forest with habitat that looks identical to the habitat in central and northern Luzon. The half-million years or so during which the Bicol Peninsula has been connected to central Luzon apparently has not been long enough for these mice to cross the lowlands and arrive successfully in Bicol. Dispersal across both water and land that is not suitable habitat does happen, but the process is apparently slow, and sometimes dispersal is notable for *not* having happened (yet).

Second, over most of the mountains of central and northern Luzon, there are two species of *Megapomys*; in a few places there is only one species, and in one area (the Zambales Mountains), there are three. In the Central Cordillera the two species are each other's closest relative (i.e., sister-species). But everywhere else, the species within a single mountain range are more distantly related. It seems that species can coexist on a single mountain only when they are fairly distantly related (with rare exceptions).

Third, the diversification of this group of mice has happened entirely after the joining of the originally isolated islands that now form northern and central Luzon. None of these data imply, for example, that the mice colonized the Zambales Mountains before this landform merged with the Central Cordillera; instead, we see evidence of one early colonization of Mindoro, and then a burst of colonization (and speciation) as the mice spread throughout a single island that included central and northern Luzon. Perhaps this was associated with the climatic fluctuations that occurred during the Pleistocene ice ages; even with the murkiness caused by our very limited knowledge of any details, we know that there was a change that involved at least some cooling, and perhaps that helped the mice move between mountainous areas. But the pattern of speciation events does not show any temporal clumping of the type that one might expect if lots of speciation took place at the specific times when the glacial maxima reached their peaks; instead, the timing looks rather steady, and not clumped into specific points in time. So, while the Pleistocene climatic variations may have helped, they have not left a pulsed "signal" that we can detect in our data (Justiniano et al., 2015).

Our studies of other taxa among the Old Endemic genera yield generally similar patterns. We have never documented more than one species of *Archboldomys*, *Rhynchomys*, or *Soricomys* within a single mountain range, and when two species of *Chrotomys* or *Musseromys* occur within the same range, they are only distantly related. As best we can judge, at this time the evidence points toward a pattern of speciation among murids in which there is colonization of a mountain range; followed by isolation of the two populations; and then followed by divergence over lengthy periods of time, probably typically more than 1 million years (fig. 5.5). Subsequently, these mice become sufficiently different that they are able to live on the same mountain but at different elevations; and, as our morphological studies have shown (Heaney et al., 2011a), they have different tooth and cranial morphologies that may be related to how they obtain and process food.

Thus, over time, the number of species can build up by repeating this cycle of colonization, isolation, morphological and genetic divergence, and secondary colonization and overlap. In the Central Cordillera, this has led to the situation in places such as Mt. Amuyao (fig. 3.11 and Rickart et al., submitted), where we have documented eight members of the earthworm mouse clade that occur together (three *Apomys*; two *Chrotomys*; and one each of *Archboldomys*, *Rhynchomys*, and *Soricomys*), and four genera of cloud rats (one each of *Batomys*, *Crateromys*, *Musseromys*, and *Phloeomys*). The communities of small mammals that exist on Luzon have resulted partly from recent colonization by New Endemics (*Crocidura grayi*, *Bullimus luzonicus*, and *Rattus everetti*), but the great majority of the small mammals at nearly every locale where we have conducted surveys in the mountains of Luzon are members of the two Old Endemic clades.

The diversity of small mammals on Luzon is predominantly—nearly overwhelmingly—the story of diversification within the island by the descendants of two early, highly successful species that happened to get lucky and ride a log across the sea from the Asian mainland. Once there, isolation on mountaintops on a geologically evolving landscape determined most of the rest of the story. Subsequent colonization across water barriers within the Philippines also took place and generated some of the extant diversity, but the bulk of the diversity on Luzon has resulted from speciation within the island. The diversity we see today has evolved over a period of about 14 million years, gradually producing morphologically and ecologically diverse communities that are composed almost entirely of members of a single family, the Muridae, that have evolved to fill a remarkably wide range of niches, a process that is often referred to as "adaptive radiation." More recently, colonizing members of this family from outside Luzon have added modestly to its species richness and ecological and morphological diversity. The story of non-volant mammal diversity on Luzon is thus primarily one of highly successful diversification by early colonizers, gradually evolving over millions of years.

6

Habitat Disturbance and Invasive Species

The native mammals of Luzon evolved over the course of millions of years as members of complex biological communities within forested habitats. During the past century, the once-vast areas of mature rainforest that covered much of the island have been greatly reduced. In 1900, about 70% of the total land area of Luzon was forested. By recent estimate, only about 20% of the island is now forested, and much of this consists of highly disturbed, regenerating second-growth forest (chapter 7). The degradation and loss of natural habitat poses the single greatest threat to biodiversity on Luzon, and protecting the remaining forest is the single most important conservation objective (though we note that protection of special habitats, such as caves where bats roost, are important as well). But regardless of what the future holds, the forests of Luzon are hardly pristine. There is no "virgin" forest; to a greater or lesser degree, habitat everywhere on the island has been altered. Frequent major disturbances from typhoons, earthquakes, and volcanic eruptions, which are occasionally catastrophic, have been part of the natural scheme on Luzon since well before the arrival of humans, and humans have been present and active as hunters and as causes of habitat disturbance for many thousands of years (chapters 2 and 4). In this chapter, we discuss two areas of disturbance ecology that have emerged from our research: how native mammals cope with habitat disturbance, and how they interact with invasive, non-native species.

Forest Disturbance and Regeneration

Forest communities are subject to varying levels of disturbance, both natural and human caused. Natural disturbances range in scale from local events, such as the fall of a single ancient tree, to widespread damage caused by severe storms and volcanic eruptions. Likewise, human disturbances also vary in scale and intensity, from the selective and sustainable harvest of medicinal forest plants to complete deforestation due to commercial logging, mining, agriculture, and urban expansion.

Forest disturbance is counterbalanced by the process of regeneration. Under natural circumstances, disturbance and regeneration constitute a normal cycle, the length of which depends on the scale and intensity of the disturbance and the intrinsic rate of recovery. A highly localized disturbance may be "repaired" relatively quickly. In the most extreme situation, where a large deforested area is allowed to regenerate, the process involves a complex series of stages. The cleared section is first colonized by pioneer species: grasses and other plants (some of which are non-native) that have easily dispersed seeds and are fast growing. These pioneers, in turn, are followed by light-loving shrubs and trees that form an early-stage second-growth forest. As the forest canopy continues to develop, it limits the amount of light that reaches the ground surface, creating conditions that favor slower-growing, shade-tolerant plants. The regeneration of forest is facilitated by animals, particularly fruit-eating birds and bats that disperse the seeds of plants such as figs (*Ficus*), pandans (*Freycinetia*), shrubs (e.g., *Melastoma*), and peppers (*Piper*; Ingle 2003; Utzurrum 1995).

Over time, the structure of the forest becomes increasingly complex. The trees themselves provide habitat for a variety of epiphytes, including mosses, ferns, orchids, and pitcher-plants (*Nepenthes*). Vines—including pandans, climbing bamboo (*Dinochloa*), and commercially valuable rattan (*Calamus*)—climb up the trunks of trees and spread through the forest canopy. The variety of shade-tolerant understory and ground-cover plants increases, and at higher elevations a thick layer of humus develops on the ground surface. Finally, the forest community reaches a stage where the plant diversity and structural complexity are relatively constant, at levels that are determined by local physical conditions (including climate, elevation, underlying geology, and soil chemistry). This final, comparatively stable phase defines the climax community, and it is represented by the various types of old-growth (or primary) forest formations of Luzon (Fernando et al., 2008). The principal forest types seen along elevational gradients on Luzon include lowland (dipterocarp), montane (sometimes called "lower montane"), and mossy (sometimes called "upper montane") forests (chapter 2).

The natural process of forest succession can be (and generally is) interrupted by disturbance events that may delay or even prevent achievement of the final climax community. If disturbance is repeated at regular intervals, it results in a persistent intermediate stage referred to as a subclimax. Such is the case for storm forests that develop in areas frequently hit by typhoons, causing tree-falls, limiting the height of standing trees, reducing the number of epiphytes, and creating gaps in the forest canopy (fig. 2.2). Another type of subclimax community that results from human activity is the pine forest that dominates large midelevation areas in the Central Cordillera (fig. 6.1B). Benguet pine (*Pinus kesiya*) is a pioneer species, one of the first trees to colonize during the early stages of forest regeneration. Under natural plant succession, pines are eventually replaced by broadleaf trees that ultimately dominate the climax stage of the old-growth forest. But repeated fires, either deliberately or accidentally set by people, remove any colonizing saplings of broadleaf trees and result in a subclimax consisting of nearly pure stands of pines, along with understory grasses and bracken ferns that are adapted to fire (Kowal, 1966).

Some human-caused disturbances—such as traditional, small-scale swidden agriculture (*kaingin*)—may be readily accommodated in the natural cycle of forest regeneration (Kowal, 1966). In areas that have been entirely deforested, however, the establishment of second-growth forest requires many years, even if it is encouraged through managed natural-reforestation efforts. Furthermore, decades or even centuries must pass before the forest matures to a point where it begins to approach the levels of diversity and complexity seen in old-growth forest. Because of the length of time involved and the fact that, under current conditions, most secondary forests are continually disturbed, the process of regeneration rarely achieves its final state. Nevertheless, we have found that secondary forests, particularly where they are well developed, are extremely important for biodiversity conservation (Rickart et al., 2011b).

Non-Native Mammals on Luzon

Habitat disturbance is also a principal factor in the spread and establishment of alien plants and animals, including non-native mammals. Not counting domesticated species, there are eleven non-native mammals that occur on Luzon. Some are known to have been present prehistorically, from fossil evidence discovered at archaeological sites (chapter 4). Most non-native species probably arrived through accidental transport, along with early human migration and trade. Some species, however, may have been introduced deliberately. This is certainly the case for variable squirrels (*Callosciurus finlaysonii*), a species that was released in Metro Manila in the 1960s and has since become firmly established (chapter 10).

Long-tailed macaques (*Macaca fascicularis*), palm civets (*Paradoxurus philippinensis*), and Malay civets (*Viverra tangalunga*) are found throughout the Philippines in a wide range of natural and human-modified habitats. In the past, these three species were considered to be non-endemic natives, that is, widespread Asian species that reached the Philippines through natural colonization fairly recently (i.e., during a recent Pleistocene glacial period). Evidence from the archaeological record and from comparative genetics, however, suggests that they arrived much more recently, probably through human agency, and perhaps deliberately (as mouse catchers, pets, or food sources). We therefore treat them as introduced, non-native species (chapters 4 and 9).

The non-native small mammals that occur on Luzon include several rodents and one shrew species. House shrews (*Suncus murinus*) are common throughout the island. They generally are confined to buildings and agricultural areas, and at worst are considered only a minor nuisance, since they eat cockroaches and other household bugs. In contrast, all the non-native rodents are major pests, destroying at least 10% of the annual rice crop in the Philippines. They also pose a significant health threat, as vectors of serious human diseases (Singleton et al., 2008). Two species in particular, Polynesian or spiny ricefield rats (*Rattus exulans*) and Asian house rats (*R. tanezumi*), are widespread on Luzon. Some of these non-native species can occur in forested areas, but their potential impact on forest communities in the Philippines is still poorly understood. In other tropical regions, the actual (or potential) threat to native biodiversity from non-native species is a growing concern, especially in conjunction with increasing levels of habitat disturbance, which promote their spread. Island faunas in particular are often considered to be highly vulnerable to ecological disruption from non-native mammals (Berglund et al., 2009; Corlett, 2010; Sax et al., 2005), and non-natives may cause the extinction of native mammals, either through direct competition or the spread of exotic wildlife diseases (Duplantier and Duchamin, 2003; Harris, 2009; Wyatt et al., 2008). We have learned, though, that the situation involving these species on Luzon is more complex and appears to be less dire than we initially expected.

Habitat Disturbance and Small Mammals: Insights from Gradient Studies

The native mammals of Luzon, along with countless other organisms, have evolved as integral components of natural forest communities. As such, it is clear that they require healthy forest communities for their continued survival. When a forest is clear cut, the entire community is lost, including all of the forest-dependent native mammals. But short of wholesale destruction, what are the effects of habitat disturbance on these complex forest communities? Specifically, how do native species vary in their responses to different levels of disturbance? How does habitat disturbance influence the interactions of potentially competing native and non-native species? Finally, how do species and communities respond to habitat regeneration following disturbance?

During the course of our research on Luzon, we gathered information to address these questions and other topics by studying the distribution of mammals across local gradients of habitat disturbance (fig. 6.1). This approach involved standardized trapping surveys of small mammals in areas with different levels of current or previous human-caused habitat disturbance. The gradients included all or most of the following types of habitat, ranging from least to greatest disturbance: areas of intact or lightly disturbed old-growth forest; old growth forests that had been significantly disturbed through logging or other activities; secondary forests of varying ages that were naturally regenerating in areas previously deforested or severely disturbed; secondary pine forests that were maintained through frequent burning (fig. 1.1); and entirely deforested areas, ranging from the earliest stages of second growth to actively

FIG. 6.1. Habitats represented along the Barlig–Mt. Amuyao survey gradient. A: Rice terraces near Barlig town, ca. 1550 m elevation, illustrating the disturbed agricultural and early second-growth habitats. B: Secondary pine (*Pinus kesiya*) forest, with understory grasses and ferns, along a ridgeline at 1760 m elevation. C: Lower montane forest, ca. 1650 m elevation. D: Mossy forest, ca. 2550 m elevation.

cultivated agricultural land. Data on the species richness and relative abundance of mammals across these gradients was used to assess the effects of disturbance on the general structure of mammal communities, as well as the disturbance tolerances of individual species and native and non-native species-groups. Additionally, information from secondary (regenerating) forests was used to infer the responses of species to habitat recovery (Rickart et al., 2007, 2011b).

We surveyed four areas in the highlands of the Central Cordillera: Mt. Amuyao, Mountain Province; Balbalasang-Balbalan National Park, Kalinga Province; Mt. Data National Park, Mountain Province; and Mt. Pulag National Park, Benguet Province (fig. 1.8). In each, we did comparative inventories at multiple localities. We categorized each locality based on the level of habitat disturbance, ranging from mature forest with little disturbance to deforested cropland. Results of these studies reveal strong and consistent patterns across all four gradients (fig. 6.2). The total number of native species is greatest in intact and moderately disturbed old growth forest, where 12 native species were typically recorded. Likewise, the overall relative abundance of native species, as reflected by the average trap success for the group as a whole, also is greatest in old-growth forest. Both species richness and relative abundance of native species decline with increasing levels of disturbance. Although the relative abundance of native mammals is lowest in deforested areas with the most severe disturbance, it is remarkable that one-third of the native species on these gradients were recorded in highly disturbed habitat outside of forest.

Based on their distribution patterns across these gradients, it is clear that native species differ in their tolerance of habitat disturbance. For example, *Rhynchomys soricoides* may be common in areas of intact old-growth forest, but they are much less common where forest is moderately disturbed, and they do not occur in habitat that is more severely disturbed. It is very important to note that the relative abundance of a species does not, in itself, reflect its ability to tolerate disturbance. This is quite clear in a comparison of the two most abundant species in the Cordillera, the native forest mice *Apomys datae* and *A. abrae* (fig. 6.2). These closely related sister-species have quite different distributions across disturbance gradients. *Apomys datae* is by far the most abundant species found in old-growth forest, but it is much less common in regenerating secondary forest, and it does not occur at all in more disturbed areas. By contrast, *A. abrae* is an uncommon species in intact old-

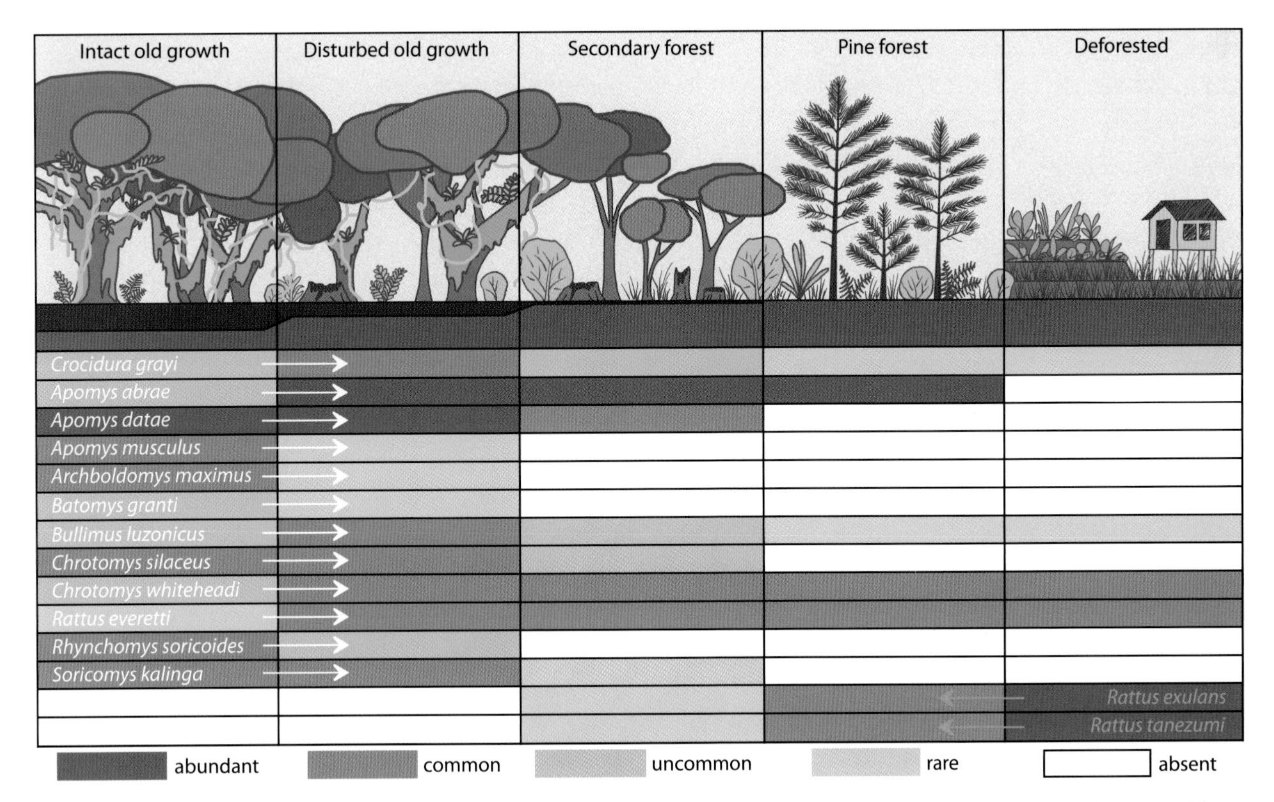

FIG. 6.2. Generalized schematic diagram of the response of small mammals to disturbance gradients at high elevations (above 1000 m) in the Central Cordillera of northern Luzon. The relative abundance of native species (names in white on the left-hand side of the diagram) along the gradient is shown in shades of green; note that some are most abundant in intact old growth, and others in more disturbed habitat. The non-native rats (names in red along the right-hand side of the diagram) are most abundant in cropland and around buildings. Based on Rickart et al. (2011b) and references cited therein.

growth forest, but it is very common in disturbed forest and is often the most abundant native species found in highly disturbed pine forest (Rickart et al., 2011b; see also Reginaldo and de Guia, 2014; Reginaldo et al., 2013).

Species that are ecological specialists have very specific evolutionary adaptations that often place them in narrow ecological niches. In comparison with ecological generalists that are capable of adapting to a wider range of habitat conditions, the more narrow requirements of many specialists may make them particularly vulnerable to the ecological disruptions that occur when habitat is greatly disturbed. This does, in fact, appear to be the case for some of the earthworm mice, such as *Archboldomys* and *Rhynchomys*, that have quite selective diets. These specialists are largely restricted to intact or lightly disturbed old growth. Indeed, our survey of Mt. Data failed to rediscover some of these ecological specialists that were known to be present there in the past (chapter 1). Expanding commercial agriculture on Mt. Data has led to rampant deforestation, leaving only isolated patches of highly disturbed forest. Apparently, what remains of this habitat is simply not sufficient to support those species that are least tolerant of disturbance, and they are presumed to be locally extinct.

Although there does appear to be a strong relationship between ecological specialization and sensitivity to habitat disturbance, there is a remarkable exception to this pattern. Members of the genus *Chrotomys*, strongly built animals adapted for burrowing, are among the most highly specialized rodents on Luzon. Their diet consists entirely of earthworms and other soft-bodied invertebrates, such as insect larvae and snails. One might expect them, like the other earthworm mice that have a similar diet, to be restricted primarily to pristine natural habitat, yet *Chrotomys* are among the most adaptable of native species. In particular, *Chrotomys whiteheadi* occur across the full spectrum of disturbance and are one of the few native species that are common in agricultural settings outside of forest. The extreme tolerance of *Chrotomys* for disturbed conditions is, in fact, a product of ecological specialization. As powerful burrowers that spend much of their lives underground, they create their own microhabitat that is protected from the daily temperature and humidity fluctuations that occur in deforested areas. They are active both day and night, providing even more flexibility in both foraging behavior and predator avoidance. Finally, their specialized feeding habits offer another advantage, allowing them to exploit a major food resource: the diversity of earthworms and other invertebrates that abound in disturbed agricultural settings. In fact, *Chrotomys* may be economically beneficial in controlling important crop pests, such as the highly destructive golden apple snail (*Pomacea canaliculata*; Stuart et al., 2007).

Among the native species most tolerant of human disturbance in the Cordillera are *Crocidura grayi*, *Bullimus luzonicus*, and *Rattus everetti*. Indeed, they tend to be more abundant where habitat is at least moderately disturbed (naturally or by humans) than they are in areas of intact old-growth forest. They occur in a wider variety of natural habitats than other native species, as reflected by their broad elevational ranges (chapter 3) and—at least in the case of the rodents—more generalized diets. These species have another distinction: in contrast to the other native species that belong to the very ancient, endemic cloud rat and earthworm mouse groups, they are members of groups that arrived in the Philippines much more recently (chapter 5).

In stark contrast to what we have observed for the native small mammals of the Cordillera region, non-native rodents show the opposite pattern of distribution across habitat-disturbance gradients. Two species of non-native rodents, *Rattus exulans* and *R. tanezumi*, are widespread in the Cordillera region, where they are serious agricultural pests. Both are abundant in the most heavily disturbed habitats, which include areas under active cultivation for rice or vegetables, fallow cropland, early-stage brushy second growth, and regularly burned pine forest. We found these non-native species to be much less common in well-developed secondary forest, and they were not recorded in areas of old-growth forest, even where that forest was significantly disturbed (see also R. Miller et al., 2008; Stuart et al., 2007).

Similar disturbance responses are seen for mammals inhabiting lowland areas of Luzon (fig. 6.3). Because species richness of non-flying mammals increases with increasing elevation (chapter 3), communities in the lowlands of Luzon contain fewer native species than those in the uplands of the Cordillera. The natives that occur in the lowlands are principally generalist species that have a high tolerance for habitat disturbance. As in the Cordillera, striped rats (in this case, *Chrotomys mindorensis*) are unique as ecological specialists that have adapted well to human-disturbed habitats in the lowlands. Non-native species make inroads farther into lowland forests than they do at higher elevations, specifically *Rattus exulans*, which penetrate areas of disturbed old-growth forest. In addition, house shrews (*Suncus murinus*) are occasionally found in forest, particularly where native Luzon shrews (*Crocidura grayi*) are absent. The situation with *Suncus* is particularly revealing and reflects a general pattern: non-native species can become established in forest only in places where native species have low diversity, and this process of invasion is augmented through habitat disturbance.

Some of the most compelling insights from our surveys across disturbance gradients came from secondary forests: areas that had previously been severely disturbed or even entirely deforested, but where forest has been allowed to regenerate. Secondary forests are, in fact, natural experiments that allow us to measure the potential recovery of forest communities. The disturbance gradients reveal that non-native species can only become established in areas where habitat disturbance is severe enough to eliminate most of the native species. They also demonstrate that the abundance of non-native species is a function of the degree of distur-

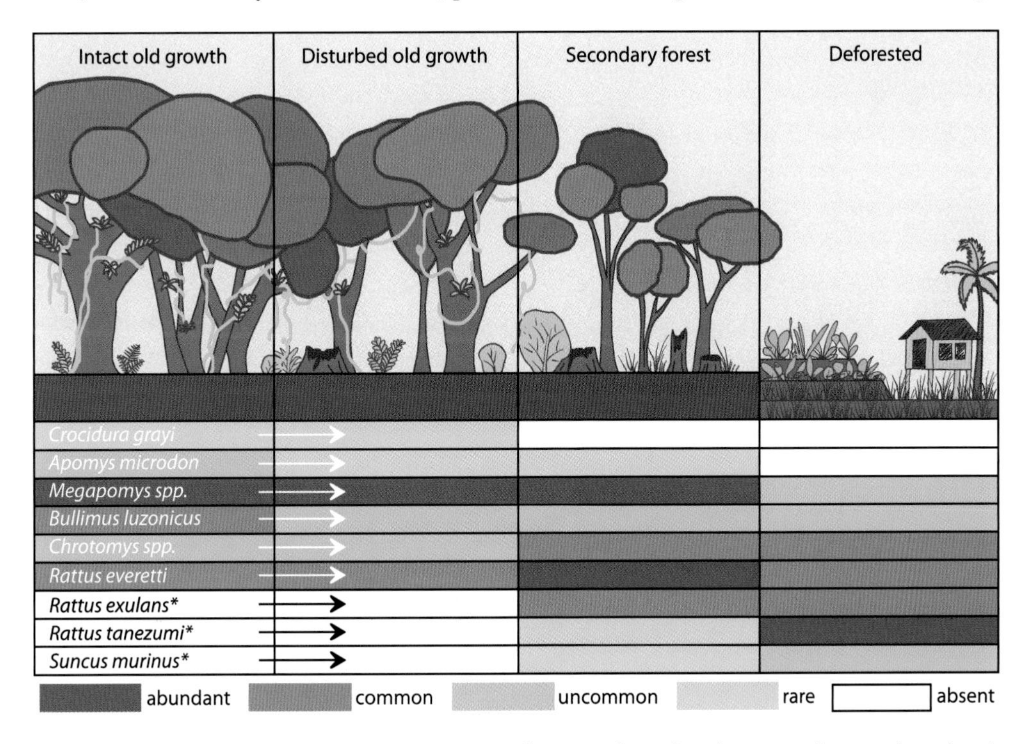

FIG. 6.3. Generalized schematic diagram of the response of small mammals to disturbance gradients at low elevations (from ca. 400 m to 600 m) in central and northern Luzon. The relative abundance of native species (names in white on the left-hand side of the diagram) along the gradient is shown in shades of green. The non-native shrew and rats (asterisked names in black) are most abundant in cropland and around buildings. Based on Balete et al., 2009, 2011, 2013a, 2013b; Heaney et al., 2013a, 2013b; Rickart et al., 1991, 2011b; Stuart et al., 2008.

bance, and that such species are most plentiful in the most heavily disturbed deforested areas. Yet in places where we have surveyed areas of well-established secondary forest (i.e., forest that has recovered for 20 years or more), we find mammal communities that are dominated by a subset of native species—those that are most tolerant of disturbance—and in which non-natives are much less common. This strongly suggests that during the process of forest regeneration, the hardiest native mammals recolonize and eventually displace the non-natives through competition. It further suggests that if the process of regeneration is allowed to continue, the eventual outcome would be the reestablishment of the old-growth forest community of native small mammals. Of course, this would require many decades. Furthermore, it could only occur where there are areas of protected old-growth forest sufficient to provide refuge for the source populations of native species that would eventually recolonize regenerating forest.

Bats and Disturbance

Although our discussion of disturbance ecology has focused on small, non-flying mammals, we have also gathered information on bats. Our most compelling data are for small fruit bats (in the Family Pteropodidae) that are diverse on Luzon and relatively easy to document through standardized surveys using mist nets (Balete et al., 2011; Heaney et al., 1999; Rickart, 1993). Due to their greater vagility, bats are able to make use of a broader area than non-flying small mammals, and their disturbance relationships are therefore more difficult to assess. For example, a particular bat species may require relatively undisturbed habitat for roosting yet utilize disturbed habitat for foraging. Nevertheless, bats show some general patterns with respect to habitat disturbance that mirror those seen for non-flying species (Heaney and Roberts, 2009; Ingle, 1992, 1993; Sedlock et al., 2008, 2011; Utzurrum, 1995, 1998). Some species, such as *Otopteropus cartilagonodus* and *Haplonycteris fischeri*, are relatively intolerant of habitat disturbance and rarely occur outside of old-growth forest or well-developed secondary forest. At the opposite extreme, there are other species, such as *Cynopterus brachyotis* and *Rousettus amplexicaudatus*, that are common in highly disturbed agricultural areas and rarely found within intact forest. Finally, there are species like *Macroglossus minimus* and *Ptenochirus jagori* that occur across a broad spectrum of disturbance conditions. As with non-volant mammals, the bat species that are least tolerant of disturbance are Philippine endemics (and may be ecological specialists), whereas those that are common in disturbed habitats are geographically widespread species. Some very widespread bat species, such as *Taphozous melanopogon* and *Scotophilus kuhlii*, are largely commensal with humans, and it may be the case that these bats colonized the Philippines relatively recently.

Summary of Disturbance Patterns

Several important generalizations have emerged from our disturbance-gradient studies of small mammals on Luzon. First, there is great variation in how native species respond to disturbance. Those that are least tolerant of disturbance are often ecologically specialized. They include locally endemic species that are restricted to particular areas of highland habitat. Nevertheless, a few of these endemic specialists have traits that allow them to persist, and often thrive, in highly disturbed habitat. Most of the native species that persist in disturbed habitats are ecological generalists that can live in a variety of habitats—from the lowlands to the highest peaks—and occur throughout, and in some cases beyond, Luzon. Some of these generalists are New Endemics, species descended from a relatively recent natural colonization of the Philippines. Widespread, non-native species of small mammals predominate in the most severely disturbed habitats, are much less common in moderately disturbed forest, and rarely occur in intact old-growth forest. The dominance of native small mammals in areas with regenerating secondary forest suggests that habitat restoration involves a reversal of the community changes associated with disturbance. The inverse patterns seen for native and non-native species are a strong indication that the community changes occurring during habitat disturbance and regeneration are driven by active competition between these groups. In the Central Cordillera, the diversity of native mammals is apparently the most important factor limiting the spread of non-native species and preventing their invasion of old-growth habitat. This contention is supported by the fact that in areas where native communities are less diverse, such as in the lowlands, non-native mammals make inroads farther into forest.

Prevailing opinion has it that island faunas are ecologically fragile and highly susceptible to disruption from both direct habitat disturbance and competitively superior invasive species (Sax et al., 2005; Whittaker and Fernandez-Palacios, 2007). From this perspective, the results of our gradient studies are surprising, to say the least. They reveal a fauna that can tolerate significant habitat disturbance, is highly resistant to invasion by non-native species, and can recover in response to habitat regeneration. These are not the hallmarks of an ecologically fragile island fauna; in fact, they are more characteristic of a continental fauna with the ecological complexity and diversity sufficient to withstand disturbance and resist invasion by non-native species.

The hardiness of the Luzon fauna is understandable only from the perspective of regional natural history. Luzon is an ancient island with a highly dynamic geography (chapters 4 and 5). The island has been shaped by powerful geological forces and is regularly subjected to strong earthquakes, volcanic eruptions, and devastating storms. There are countless places on Luzon that are no less than heavenly in their splendor, but potential disaster lurks everywhere. The native mammals evolved for millions of years in the context of periodic—often cataclysmic—disturbance; perhaps we should not be surprised that many of them are adept at tolerating substantial habitat disturbance. In this respect, they appear to be "preadapted" to cope with much of the disruption wrought by humans. Nonetheless, humans are capable of disturbance that far exceeds those associated with even the most severe natural phenomena. Unfortunately, the impact of humans continues to increase with their expanding population and ever-greater demand for natural resources, ultimately producing such extreme pressure that even the hardiest of native species can be placed at risk (chapter 7).

We leave this topic with an audacious prediction. At this time, Luzon is (to the best of our knowledge) the only island on which its native small mammals are thought to be highly tolerant of moderate levels of habitat disturbance and apparently are competitively superior to the non-native rats that travel widely with humans. We predict, however, that Luzon's uniqueness in this respect will be shown to be illusory, based mainly on the absence of data from similar islands, just as this pattern was unknown until very recently on Luzon itself (Rickart et al., 2011b). We think that any island in the Philippines with a diverse native small mammal fauna will exhibit the same patterns. Further, since a great many other islands in Indo-Australia also have widely varied faunas of native mammals, we predict that all such islands will prove to be similar. Specific examples include Sulawesi, which has a highly diverse known fauna (Musser, 1987) and is where discoveries of new species and genera are being made that are fully comparable with those we have made on Luzon (Achmadi et al., 2013; Esselstyn et al., 2012a; Rowe et al., 2014); Halmahera (Fabre et al., 2013); and the many large and small islands on the Sunda Shelf—such as Borneo, Java, and Sumatra—where the mammal fauna is relatively well documented (e.g., Heaney, 1984; Md. Nor, 1996). Testing this prediction would be relatively simple, and it might dramatically change the current prevailing view of the vulnerability of island communities of native small mammals from one of invariant, rather extreme fragility to a pattern of robust resistance that is overcome only by massive habitat alteration.

Colonization and the Cycle of Faunal Change

All of the native mammals that now occur on Luzon are descended from species that successfully colonized the island (chapter 5). The members of the endemic cloud rat clade evolved on the island, but all of them descended from just one species that arrived about 14 million years ago. The same is true of the earthworm mice, which are descended from another successful ancestral colonist that arrived several million years later. The New Endemic rodents are derived from several more-recent, but still ancient, colonization events (fig. 5.4). As an isolated oceanic island, Luzon was a difficult place to reach. Bats were certainly more successful than non-flying mammals, as evidenced by the many families of bats that arrived independently. Of course, given the very limited fossil record, we only know of those colonization events that were successful and left living descendants. There may have been others that failed, where arriving species may have died off either without becoming established or after a brief period of establishment. Nevertheless, rates of natural colonization were undoubtedly low. This changed dramatically with the advent of humans, who arrived on Luzon in multiple waves of colonization that began over 60,000 years ago, before the Last Glacial Maximum (chapter 4). Eventually, many other species of mammals arrived with these people, including domesticated species as well as non-native pests.

Non-native mammals offer important insights into the natural processes of colonization and establishment that were essential for the eventual evolution of the native mammal fauna of Luzon. These most recent colonists are models for understanding what may have occurred in the ancient past, when newly arrived species faced the challenge of becoming established. From our disturbance-gradient studies, it is evident that competition must have been very important in determining the likelihood of successful establishment. On Luzon, non-native pest rodents only occur in heavily disturbed habitats, but elsewhere in the Philippines this is not the case. On islands that are either small or geologically young, and that have few native mammals as a result, non-native species have become firmly established in old-growth forest (Heaney et al., 1989, 2006c; Rickart, 1993; Rickart et al., 1993). Where there are no ecologically equivalent native species, the probability of successful establishment is much greater, as illustrated by the successful naturalization of monkeys and civets throughout the Philippines. Such may be the case with the newest invader, variable squirrels (*Callosciurus finlaysonii*), should this species eventually find its way to areas of old-growth forest.

From a broader perspective, the ecological distribution of both native and non-native mammals on Luzon is consistent with some aspects of the taxon-cycle model of island colonization (Ricklefs and Cox, 1972; Wilson, 1961). The Old Endemics of Luzon, descended from the most ancient colonization events, dominate the highly diverse mammal communities of the interior highlands and include species that are locally endemic and often are quite specialized. The New Endemic species, derived from more-recent colonization, are ecological generalists, have wider geographic ranges, and occur at lower elevations. The most recent arrivals are the non-native species, some of which have become naturalized on Luzon, but most of which are restricted to highly disturbed habitats. But, unlike the predictions from the taxon-cycle model, the Old Endemic small mammals appear to be competitively superior to the more recent arrivals (except in the most heavily disturbed habitats) and have not only survived but also show evidence of continued speciation following the arrival of the New Endemics (fig. 5.4). Continuing colonization seems to have added to diversity, rather than replacing existing older species and clades—a topic we return to in chapter 8.

Monkeys and Carnivores

Current evidence indicates that macaques and both of the civet species on Luzon were brought there recently by humans, perhaps within the past 4000 years (chapter 4). We find it striking that, unlike non-volant small mammals, these taxa have been highly successful in becoming naturalized—that is, they now occur widely in both disturbed forest and old-growth forest, as well as in more heavily disturbed habitat (chapters 3 and 9).

Why should monkeys and small carnivores succeed in becoming invasive in natural forest on Luzon when non-native rats, mice, and shrews do not? The most likely answer may come from our studies of small mammals on small and geologically recent islands elsewhere in the Philippines, as cited above. On those islands, there are very few native small mammals, and the non-native mammals have invaded the forest successfully. In other words, in the absence of competition from native species with similar ecologies, the non-natives are able to become established once they arrive. This is relevant for a simple and obvious reason: to the best of our knowledge, no monkeys or small carnivores were present on Luzon when these species were introduced. Resources were available that were similar to those in the places where they originated (probably Borneo; chapter 4), and there were no close competitors. That, it appears, is a recipe for success.

Future Directions

Regardless of their impressive resiliency, the native mammals of Luzon evolved in the context of forest and certainly require forest for their continued existence. The preservation of remaining old-growth forest on the island is of paramount importance. The value of well-developed secondary forest as wildlife habitat is also evident from our studies, and efforts toward restoration of natural forest and the careful management of secondary forests are likewise extremely crucial. Regrowth of natural forest in upland areas adjacent to agricultural lands would have many benefits, not only in providing habitat for native wildlife, but also for water conservation, flood control, and a reduction in populations of agricultural pests that thrive in deforested areas but cannot persist in forest (chapter 7).

Our studies on the ecology of disturbance and invasive species have only scratched the surface. There is much yet to be learned, and we acknowledge that many of our conclusions are simply hypotheses that need to be tested and refined through further research. Some of the most critical areas for future work involve understanding the ecological and behavioral interactions of native and non-native species. How do they compete for living space and essential resources? What specific aspects of disturbance favor non-natives, and what aspects of regeneration favor native species? How might complex biotic interactions, such as the transfer of both native and exotic parasites and diseases, shape competition? Likewise, we know very little about the restoration of natural habitat. How long does the process take? How can it be enhanced through management activities? Clearly, there is much to learn and many important topics remain for future research.

7

Conservation of Luzon's Mammalian Diversity

We have documented that there are at least 110 native species of mammals on Luzon. These vary greatly in their habitat requirements, and thus in the extent of their natural distributions: nearly all of the bats are widespread on the island, but many of the small mammals occur only within areas that range from more than 10,000 km^2 to as small as a few dozen square kilometers. Fortunately, our evidence indicates that many of these species occur at moderate to high density and are tolerant of moderate levels of habitat disturbance (chapter 6), perhaps because of their evolutionary history in a land buffeted by frequent typhoons and volcanic eruptions. Further, our studies have shown that invasive species, especially non-native rats, fare poorly in areas of natural habitat; instead, the native species predominate (Rickart et al., 2011b). From these observations, one might think that the conservation of mammals on Luzon is not a subject for much concern.

That, however, would represent a serious misunderstanding of the current reality. Although Luzon is the fifteenth largest island in the world (nearly identical in area to Cuba), it has about 44 million people, at a density of about 440/km^2, with virtually all parts of the island inhabited and utilized by humans in some way. Among the world's large islands, only Java, Honshu, and Great Britain are more densely populated. Associated with that heavy human population density, the tropical forests that once clothed the island have been vastly reduced, and hunting pressure on many species is intense.

Vanishing Treasure: The Loss of Forest Cover

There are no meaningful hard data about the extent of forest cover on Luzon prior to about 1875, when the first comprehensive forest surveys and maps were prepared by the Spanish colonial government. Their figures were rough estimates, but maps from that era (fig. 7.1) show 65%–70% of the Philippines covered by forest, and probably a bit less for Luzon individually (Bankoff, 2006; Kummer, 1991; Suarez and Sajise, 2010). Most of the Central and Cagayan Valleys had been cleared, as had the Ilocos and Zambales coasts, the vicinity of Manila Bay and Laguna de Bay, and parts of the central and southern Bicol Peninsula. Mountainous areas, and even significant foothill areas, remained largely forested. These figures and maps undoubtedly did not account for the impact of shifting agriculture and rice culture along rivers in upland areas, which were inhabited by many distinctive ethnic groups; we know that burning was widely practiced, which promoted the spread of pine forest, but little else is certain.

By 1950–55, estimates for the country as a whole showed a reduction to about 50%–55% forest cover (Bankoff, 2006; Kummer, 1991). Maps based on data from that period are nearly as crude as those from the 1870s, but they do show the continuing reduction in

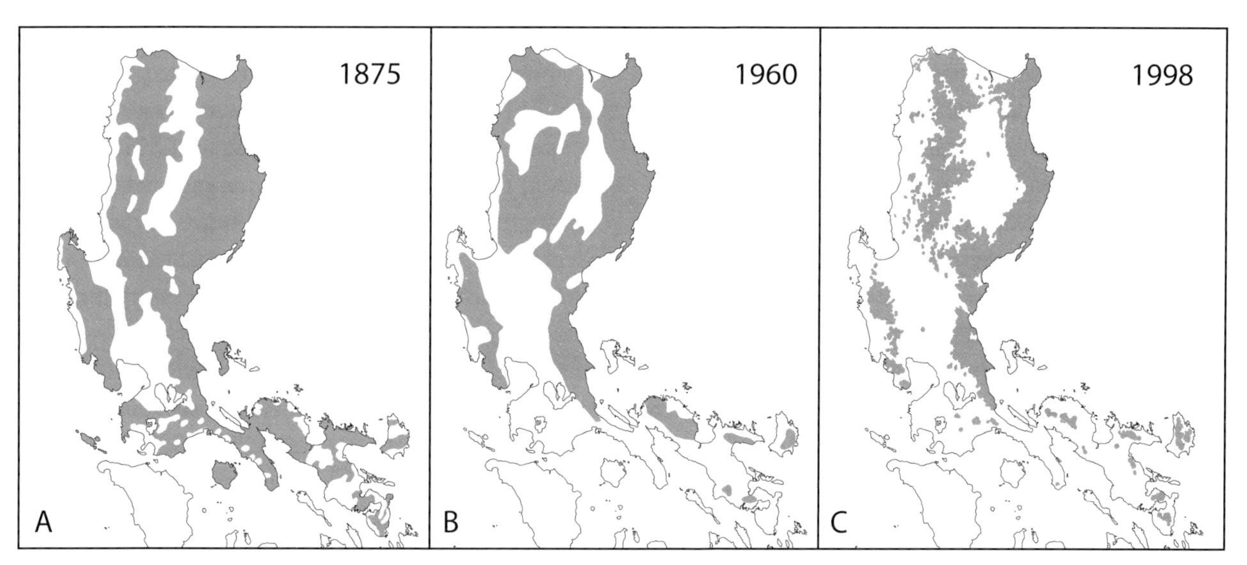

FIG. 7.1. The extent of forest cover on Luzon. These maps include all types of forest and should be taken as estimates (see text). A: About 1875. From Anonymous (1876). B: 1960. From Huke (1963). C: 1998. From Environmental Science for Social Change (2000).

forest cover and the retreat of forest into more isolated and higher mountainous areas. As with much of the mapping that followed, no distinction was made between old-growth and secondary forest.

With the advent of aerial surveys and, eventually, satellite imagery, it would be reasonable to expect more detailed and precise figures for forest cover. Unfortunately, differing and imprecise definitions of what constitutes forest has prevented a clear picture from emerging; this is due in part to the mandate for governmental agencies to provide information on the extent of commercial forest, generally defined as forest that could be commercially harvested for lumber. In these efforts, mossy forest, forest over limestone, and forest over ultrabasic soil (i.e., on ophiolites; chapter 4) often were categorized as having open canopy and were lumped together with previously logged forest. As a result, considerable uncertainty remains about the extent of the types of forest on Luzon (Environmental Science for Social Change, 1999; Walpole, 2010), but a reasonable figure for the total amount of forest on Luzon in 2002 is 2,074,960 ha out of 10,608,660 ha in total land area, or 19.6% forest cover (Walpole, 2010:table 3). This is similar to the 22.2% calculated for 1998 (Environmental Science for Social Change, 1999). Thus, from about 1955 to around 2000, forest cover declined from roughly 55% to 20%, including both old-growth and secondary forest.

Deforestation in the decades since World War II has been extensive. There are a few provinces on Luzon with more than 50% forest cover, including all forest types (e.g., Aurora and Apayao); many with less than 10% (Albay, Batangas, Camarines Sur, Cavite, Ilocos Sur, Laguna, Pangasinan, Sorsogon, and Tarlac); and several with less than 1% (La Union, Pampanga, and Metro Manila; Walpole, 2010). This deforestation has created massive economic problems because of the increased floods, landslides, and drought that follow; protection of watersheds surely must rank high on the nation's list of needs for economic and social stability. Because so little commercially valuable timber remains on Luzon, the rate of forest loss has slowed, but both legal and illegal cutting of what little remains persists (e.g., Broad and Cavanaugh, 1993; Catibog-Sinha and Heaney, 2006; FAO, 2008; van der Ploeg et al., 2003; Vitug, 1993).

Loss of forest cover has had many causes, often operating simultaneously. Logging has been the single greatest factor in the initial clearing of forest (Bankoff, 2006; Kummer, 1991), but others often followed quickly. Traditionally, fire has been used to remove forest and create grazing land for domestic livestock as well as for native species that are hunted (mostly deer on Luzon), and this practice remains in common use. Such burning is often indiscriminant, in that fires are allowed to sweep up steep mountainsides where grazing is unlikely. This not only removes habitat for the native small mammals and bats, but also creates habitat for the non-native pest rats that cause much damage to agricultural crops (Reginaldo and de Guia, 2014; Rickart et al., 2011b; Stuart et al., 2008). We have seen many places on Luzon where more carefully controlled and restricted burning would produce nearly as much grazing land, yet would also allow regeneration of forest that would reduce populations of destructive non-native rats and simultaneously improve watershed conditions, decreasing erosion and flooding.

Clearing land for agriculture has been another of the primary causes of deforestation; some of the land has been used for growing food for domestic consumption, and other areas for producing crops intended for export, such as abaca, tobacco, coffee, and sugar. Most of the lowland areas of Luzon that formerly supported forest have been converted to agricultural land. This practice allows some bats to persist (especially those that roost in buildings), but diversity is generally low and most of the species are those that are widespread in Southeast Asia (i.e., they are not Philippine endemic species). Among the small non-volant mammals, very few native species survive in cropland; complete removal of forest is unsupportable for all but a very few of them, even given their tolerance for disturbed forest (chapter 6).

Large areas of Luzon have been converted to plantations of trees or other large, tree-like plants that sometimes are mapped as "forest" (Stibig and Malingreau, 2003). These plantations include those that produce food (e.g., coconuts, fruits, coffee, and bananas), those grown for specific products (e.g., abaca fiber), and those grown for wood (gmelina and eucalyptus). We do not know the extent of these plantations, but we estimate that roughly a quarter of Luzon may support these crops. Very little is known about the utility of plantations as habitat for native mammals on Luzon. Our limited observations indicate that gmelina and eucalyptus plantations are virtually sterile, because these species are allelopathic: they produce chemicals that retard the growth of native vegetation and leave little in the way of cover or food for native mammals. On the other hand, coconut plantations that have been allowed to regenerate with small native trees, shrubs, and other plants provide habitat for at least certain species of bats, especially those most tolerant of disturbance, particularly when forest and/or caves are nearby to provide roosting sites. Likewise, other bats are closely associated with bananas and abaca and may be abundant where these crops are prevalent. Given the extent of the plantations, there is a great need to document the extent of biodiversity that exists in these areas under the varied conditions imposed by soil type, elevation, and other simultaneous land use (e.g., grazing). None of them should be considered forest; they are simply a form of cropland.

Lowland Forest Habitat

Lowland tropical forest probably covered the greatest area of Luzon prior to significant human impact, but today it has been vastly reduced by logging, agriculture, and urban development. Firm figures seem to be lacking, but certainly no more than roughly 2% of the island is covered by old-growth lowland forest today. Given that about 90% of Luzon would once have been covered by lowland forest (from sea level to about 1000 m elevation), not more than about 4% of its original extent remains (Walpole, 2010), though perhaps an equal or greater amount of second growth is present in the lowlands. In the past, the native non-volant mammal fauna in lowland forest probably consisted mostly of wild pigs, deer, two species of giant cloud rats (*Phloeomys cumingi* and *P. pallidus*), and low densities of several widespread small mammals (*Crocidura grayi*,

TABLE 7.1. Luzon bats that current data indicate are dependent on lowland forest to maintain populations. Data from chapter 11 and references cited therein.

Acerodon jubatus	*Emballonura alecto*	*Rhinolophus inops*	*Kerivoula whiteheadi*
Desmalopex leucopterus	*Coelops hirsutus*	*Rhinolophus macrotis*	*Murina cyclotis*
Dyacopterus rickarti	*Hipposideros antricola*	*Rhinolophus philippinensis*	*Murina suilla*
Eonycteris robusta	*Hipposideros bicolor*	*Rhinolophus rufus*	*Myotis ater*
Haplonycteris fischeri	*Hipposideros lekaguli*	*Rhinolophus subrufus*	*Myotis macrotarsus*
Harpyionycteris whiteheadi	*Hipposideros obscurus*	*Harpiocephalus harpia*	
Pteropus pumilus	*Rhinolophus arcuatus*	*Kerivoula papillosa*	

Apomys microdon, *Bullimus luzonicus*, *Chrotomys mindorensis*, and *Rattus everetti*), plus the non-native monkeys and civets (chapter 3). Only one non-flying mammal species that we know of (*Musseromys gulantang*) is apparently restricted to lowland forest (chapter 10). As such, loss of lowland forest habitat was probably not a serious blow for the non-flying small mammals overall, because virtually all are more abundant at higher elevations, but it probably represented a substantial loss for the deer and pigs that formerly had been plentiful in the lowlands.

The loss of lowland forest has undoubtedly had the greatest impact on bats, the most diverse group of mammals on Luzon. Only a few species of bats are abundant above about 500 m elevation, and most occur only below about 1000 m (chapters 3 and 11; see also Ingle, 1992, 1993; Sedlock et al., 2008, 2011; Utzurrum, 1992, 1995, 1998). Current evidence leads us to estimate that at least 29 (51%) of the 57 species of bats cannot maintain populations outside of forest. Most in are in lowland forest, as shown in table 7.1, plus two at higher elevations (chapter 11). Well-developed secondary lowland forest may provide adequate habitat for some (and perhaps many) of these bats, but with the combination of both secondary and old-growth lowland forest now reduced to about 4% of their original extent, there can be little doubt that populations of most of those species have been greatly reduced. Much additional research is needed to define more precisely the habitat needs of nearly all of these species, but the implication is clear: loss of lowland forest habitat since the 1800s—and even since the 1960s—has vastly reduced the population sizes of many Luzon bat species.

Highland Forest Habitat

Above about 1000 m elevation on Luzon, lowland tropical forest is replaced by montane and mossy tropical forest (chapter 3). Relatively few bats live in this forest, but the majority of the highly diverse small mammal fauna occur there. The area of Luzon above 1000 m elevation is about 971,728 ha, which is a little over 9% of the total land area. Prior to human disturbance, nearly all of this would have supported montane and mossy forest (Environmental Science for Social Change, 2000), but only about 216,047 ha (22.2%) remained forested in 2002 (Walpole, 2010). Retention of this forest has been highest in the Sierra Madre, where Aurora, Cagayan, Isabela, Quezon, and Quirino Provinces all had 50% or more of their original coverage remaining in 2002. In the Central Cordillera in 2002, Apayao and Kalinga Provinces had more than 25% of their area above 1000 m elevation still covered by this forest, but Abra, Ifugao, and Mountain Provinces had 10%–20%, and Benguet retained only 1%. Tarlac and Zambales Provinces had about 30% of their original montane and mossy forest remaining, but Pampanga had virtually none. Camarines Sur Province had nearly 50% of its small area of this type of forest remaining, but Albay and Sorsogon had virtually none.

The higher retention of montane and mossy forest along the Pacific coast of Luzon than in the Central Cordillera may be associated with the greater amount of rainfall along the Pacific coast (chapter 2), which makes forest clearing at high elevations especially difficult. In the generally drier and more seasonal Cordillera, pine forest is much more extensive, covering about 85,154 ha, which is roughly 8.8% of the area above 1000 m elevation (Walpole, 2010). Pines occurred naturally in the Cordillera before substantial human disturbance (Stevenson et al., 2010), but widespread burning has caused a great expansion of pine forest (Kowal, 1966). Where pine forest is frequently burned and grasses and bracken ferns predominate as ground cover, we have found low density and few species of native small mammals; it is quite poor habitat. When burning ceases and broadleaf montane or mossy forest regenerates beneath mature pines, however, the native small mammals move back in, and both their abundance and diversity increase dramatically (e.g., Rickart, 2011a, 2011b).

The combination of montane, mossy, and pine forest on Luzon thus currently occupies about 31% of the area above 1000 m elevation. This represents a severe loss of natural forest habitat, but it is certainly less drastic than the loss in the lowlands, where perhaps as little as 4% remains. Loss of highland forest in the Central Cordillera and the Bicol Peninsula has been extensive, but the loss of lowland forest has been much greater.

In recent years, the growing popularity of some vegetables and fruits (e.g., potatoes, green beans, radishes, cabbage, broccoli, carrots, and strawberries) that require a cool, moist climate has led to large-scale clearing of mossy forest in the Central Cordillera for commercial farming of those crops (Heaney et al., 2006b). Much of the reduction of highland forest is attributable to this cause, including the virtual denudation of Mt. Data National Park (chapter 1) and extensive clearing in Mt. Pulag National Park (fig. 7.2; Lapitan et al., 2010; Rickart et al., 2011b), as well as the reduction of highland forest cover in Benguet Province to less than 1%. This practice is increasing rapidly and is one of the most active sources of forest loss on Luzon today.

FIG. 7.2. Babadak Ranger Station, Mt. Pulag National Park, April 2008, from across one of the many vegetable fields that have been cleared and planted along the road that leads into the park.

Forest over Ophiolites and Limestone

Information on the extent and status of forest over ophiolites and limestone is very limited; we know of no hard data on the original extent or current status of forest in these habitats. In some respects, this lack of sure knowledge also extends to the mammals there. Both the Zambales and Mingan Mountains are primarily ophiolitic, and each has five known endemic species of mammals (chapter 3), but it is not clear if any of them are restricted to the ophiolitic areas; much further research is needed.

Mining has been practiced on Luzon since prehistoric times, with gold and copper being especially sought-after metals. Prior to 1900, mining probably occurred on such a small scale that it had little impact on biodiversity, but—especially since World War II—it has expanded greatly. Currently, extraction continues to include gold as a major focus, but the increased global value of nickel, chromite, and other rare metals has driven a substantial expansion of mining on Luzon (Environmental Science for Social Change, 1999; Stinus-Remonde and Vertucci, 1999). This is taking place at a time when we still know little about the mammals (or other organisms) that utilize forest over ophiolites, a specialized, distinctive habitat where much of the mining is concentrated at present.

With the expansion of the Philippine economy in recent decades, the demand for concrete has grown a great deal. This has created more of a need for a primary component of concrete, lime, which is produced by quarrying limestone on a large scale. We know of no scientific research that has documented the impact of limestone mining on biodiversity on Luzon, but it is quite clear from our own observations and those of others (Mould, 2012; Sedlock et al., 2014b; Struebig et al., 2009) that this quarrying frequently destroys caves that have previously been used as roosting sites by large populations of bats. Additionally, we have seen clear evidence that frequent disturbances from ecotourism in caves can have a severe negative impact on populations of cave-dwelling bats. We have abundant evidence that at least 19 species of bats are dependent on caves for roosting sites, especially as maternity roosts (table 7.2; chapter 11). A few caves exist in areas with volcanic cliffs and crevices, but the great majority are in limestone. We have found that areas with extensive limestone and many caves often have a remarkably high diversity of bats: for example, we documented 17 species in the Peñablanca Protected Landscape, Cagayan Province; 18 species in Caramoan National Park, Camarines Sur Province; and 20 species in the Tayabas Caves area of Quezon Province (Balete, Heaney, et al., unpublished data). Not all of these species are dependent on caves, but most of them are. Caves clearly are critical habitat for Luzon bats; much further research is needed to document impacts in these locales and develop strategies for effective conservation.

TABLE 7.2. Luzon bats that current data indicate are heavily dependent on caves to maintain their populations. Data from Heaney et al., 2010; chapter 11.

Eonycteris robusta	*Rhinolophus philippinensis*
Eonycteris spelaea	*Rhinolophus rufus*
Rousettus amplexicaudatus	*Rhinolophus subrufus*
Emballonura alecto	*Rhinolophus virgo*
Hipposideros antricola	*Miniopterus australis*
Hipposideros bicolor	*Miniopterus schreibersii*
Hipposideros lekaguli	*Miniopterus tristis*
Hipposideros pygmaeus	*Myotis macrotarsus*
Rhinolophus arcuatus	*Chaerophon plicatus*
Rhinolophus macrotis	

In addition, there is one currently known species of small mammal that we suspect is restricted to forest over limestone: the undescribed species of *Apomys* from the vicinity of Callao Caves in Cagayan Province. There is very little information about this species, but we found it only in an extensive area of limestone and not in adjacent lowland forest (chapter 10).

Thus it is clear that all of the habitats on which the native mammals of Luzon depend are currently being reduced. For the fauna to be effectively protected and preserved, adequate areas of each type of habitat must be retained in a natural or nearly natural state. Loss of lowland forest habitat has been greatest overall, but such a small amount is left that the present rate of loss has declined; nonetheless, protection of what remains is crucial. Current activities pose substantial threats to highland (montane and mossy) forest, forest over ophiolites, and forest over limestone, but the paucity of information available at this time means that we can do little more than point to the immediate need for focused conservation efforts and more research.

Areas of Endemism

In most continental areas, including the United States and Europe, mammal species usually are widely distributed (or were, prior to human disturbance) and localized areas of endemism are few. In continental mountainous areas in the tropics, local endemism is much more common and is an important factor in determining the extent of biodiversity (e.g., Graham et al., 2014; Stanley, 2011). Strikingly, even in comparison with the continental tropical mountain areas, we know of no other region that has as many endemic species of mammals in an area equal to or smaller than Luzon (chapter 3), and we know of no place that has as many distinct centers of endemism (fig. 7.3).

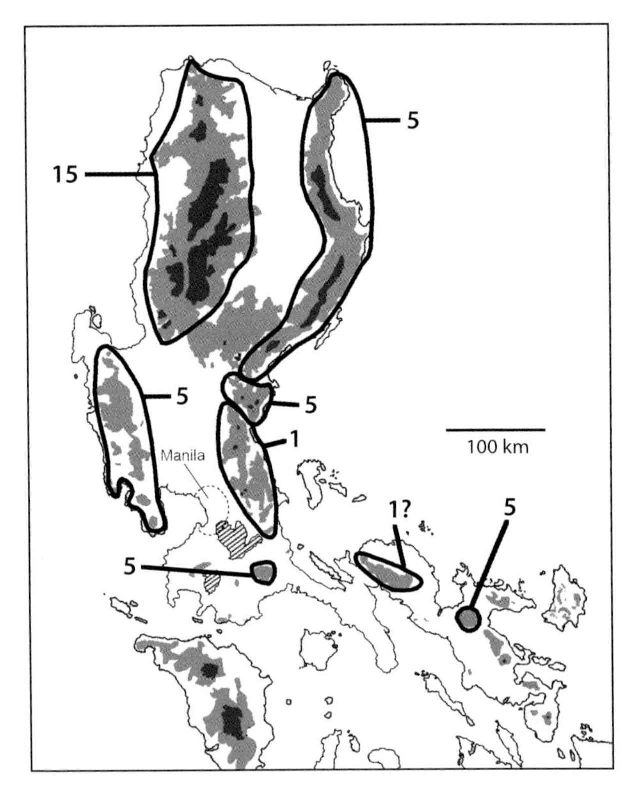

FIG. 7.3. The areas of mammal endemism on Luzon and the number of species endemic to each, including species that have been formally described and those that are currently under study. Data from Heaney et al. (submitted); see also table 3.2.

Given that the presence of these areas of endemism is such a highly distinctive aspect of the Luzon mammal fauna, the successful conservation of this fauna requires well-managed protected areas in each of the eight centers of endemism. Fortunately, designation of one or more such areas in each of these centers has either been done or is under consideration as part of an expansion of the protected-area system (e.g., Ambal et al., 2012; Conservation International et al., 2006; Department of Environment and Natural Resources, 2012). Two challenges remain at this point. First, some parks have not been successfully managed, and destruction of natural habitat has been extensive. The worst example is Mt. Data National Park (fig. 1.4), which has been reduced to about 80 ha of forest out of the original 5512 ha, but logging, clearing for agriculture, and mining are also expanding in many of the other protected areas (FAO, 2008; Lapitan et al., 2010; van der Ploeg et al., 2011). An enhanced protection program is clearly needed; too many of the parks exist primarily on paper.

Hunting

Hunting has been part of traditional Philippine cultures since prehistoric times. When we conducted our studies in Balbalasang-Balbalan National Park from 2000 to 2003, we learned that the Banao people manage their hunting carefully, with a clear understanding that overhunting will reduce populations of their deer and wild pigs in the future (Heaney et al., 2000; Rickart et al., 2011a). Hunting of large mammals by the

Aeta, the traditional occupants of the Sierra Madre, was probably also done on a sustainable basis, in part because the human population density was low (Griffin and Griffin, 2000). In many areas, a variety of techniques were used in the past to snare small mammals and capture bats, although there is scant information on the extent of these practices. Recognizing that the data are quite limited, we suspect that traditional hunting in the Philippines, conducted by residents using long-established methods and intended for local consumption, most likely was sustainable and had little negative impact on wildlife populations.

With a few exceptions, perhaps including parts of the northern Central Cordillera and the Sierra Madre, this is no longer the case. Reports we have received from local hunters (chapter 9) indicate that deer are now locally extinct in about a quarter of the areas where we conducted detailed surveys, declining and scarce in about half, and moderately common in about a quarter. Warty pigs were described to us as relatively common and stable in about a quarter of our survey areas (generally the same places where deer were common) and uncommon and declining in three-quarters of the others. Most of our survey areas are officially designated as protected, and they often contain much of the forest cover that exists in their vicinity, yet deer and warty pigs have been overhunted within them. One such place is Mt. Banahaw, where deer may be approaching local extinction and warty pigs are scarce due to current overhunting (Heaney et al., 2013b; Scheffers et al., 2012). Similar methods are used to hunt monkeys and civets, but because these probably are non-native species on Luzon (chapter 4), we have less concern about them. A principal worry is a shift from hunting for strictly local use to market hunting for bushmeat to be sold elsewhere. We have no information on the overall extent of market hunting on Luzon, but we do know that it does occur (e.g., Scheffers et al., 2012). In other tropical regions, market hunting has decimated wildlife populations. Furthermore, the consumption of bushmeat is thought to be responsible for some emergent diseases that have recently shifted from wildlife to human hosts in parts of the world. Although no such host transfers have been documented in the Philippines, the potential is sufficient reason for concern.

Not only large mammals are subject to heavy hunting pressure. The giant flying foxes on Luzon (genera *Acerodon*, *Desmalopex*, and *Pteropus*) are intensively hunted using firearms, nets, and fishhooks dangled from ropes, and these taxa have declined precipitously as a result (chapter 11; Heaney et al., 1997; Mildenstein et al., 2005; Scheffers et al., 2012; Stier and Mildenstein, 2005). *Acerodon jubatus* appear to have declined most dramatically. Although a few aggregations of 50,000–100,000 individuals remain in isolated portions of the Northern Sierra Madre (van Weerd et al., 2003), similar reports from many parts of Luzon in the early 1900s contrast sharply with the most current groupings of not more than about 5000 and usually only a few hundred individuals (Mildenstein, pers. comm.). Hunting of *Acerodon* and *Pteropus* can be especially intensive, because they roost colonially in the tops of large trees. This habit of aggregating in open, visible places makes them easy to find, and entire roosts may be eliminated over a brief span of time by persistent hunters armed with modern weapons.

Bats that roost in caves are also especially vulnerable, and they are heavily hunted over most of Luzon. Because certain caves serve as maternity colonies at specific periods during the year, hunting at those times can be especially damaging. Hunting sometimes consists of netting bats with fishing nets or sweepnets, but often they are slaughtered wholesale by building fires in the caves. Smoke inhalation and anoxia kill the bats, which roost along the ceiling and upper walls. Little research has been conducted to document the impact of hunting, but our observations and those of others imply a huge decline in bat populations in many areas because of it (Mould, 2012; Sedlock et al., 2014a). All of the cave-dependent bats listed in table 7.2 are subject to hunting.

Status and Prospects for Protected Areas and Wildlife

The Philippines has one of the oldest systems of protected areas in Southeast Asia, dating from the early 1900s, and it has continued to expand to this day, with broad funding from international agencies (Catibog-Sinha and Heaney, 2006; Leones and Navarro, 2012; Mallari et al., 2001; extensive details are available in Ong et al., 2002). Fortunately, discoveries of previously unknown species of mammals and other organisms are often highlighted in decisions to establish new protected areas (e.g., Heaney and Balete, 2012), so the protected-area system is increasingly responsive in providing official places of refuge for much of the biodiversity on Luzon (Ambal et al., 2012; Department of Environment and Natural Resources, 2012).

The potential benefits of the system are huge. The existence of these protected areas is a crucial requirement for an effective long-term strategy for biodiversity conservation, as well as for other affiliated benefits, especially protection of the watersheds that are the source of clean, consistent water for agriculture, industry, and home use. It is widely understood, however, that this is only one of the necessary steps. We will list some additional components briefly, focusing on mammals, with the recognition that these are complex topics that require much more substantial development than can be covered in a volume such as this.

While the official designation of protected areas is essential, this does not guarantee that actual protection on the ground will follow. We have found that some level of commercial logging (i.e., for sale, not for personal use) and conversion to commercial agriculture takes place in many—and perhaps nearly all—protected areas, and at times the scale is quite large. Mt. Data, Mt. Pulag, Mt. Banahaw, and the Northern Sierra Madre are the most prominent examples (Heaney et al., 2006b; van der Ploeg et al., 2011), but certainly they are not the only ones. While we accept the potential for sustainable harvesting of both lumber and wildlife within protected areas when this harvest is adequately managed, we have found that there is insufficient active management in most protected areas, and at times it seems entirely absent. Increased efforts toward both the education of local residents and the enforcement of laws regarding protected areas would benefit the system greatly. Protected areas are most effective where local people (especially members of indigenous groups) are actively involved in conservation management; prime examples are Mt. Isarog and Balbalasang-Balbalan (Heaney and Regalado, 1998; Heaney et al., 2000).

A closely related issue involves the construction of roads. Roads often provide a means of boosting local rural economies by increasing access to markets for their products; this is especially likely to be the case when the roads are built or improved in existing rural communities where deforestation took place long ago. Unfortunately, illegal logging, hunting, mining, and agriculture are often made possible by the construction of roads into previously inaccessible areas—including protected areas—thus creating serious negative impacts on wildlife. The roads themselves, and the forest clearing that may accompany them, may also cause serious degradation of watersheds, resulting in increased erosion, floods, and landslides. Careful planning and management of road expansion should conducted to avoid the degradation of existing forest, so that both watersheds and biodiversity can be protected (Laurence et al., 2014).

The Philippines' current system of protected areas was often focused on preserving existing areas of old-growth forest. Because lowland forest generally was already badly degraded and therefore excluded from consideration, it is poorly represented in the present protected-area system. Perhaps 30% of the original highland forest still exists, and many parks preserve portions of this habitat, but probably only a small percentage of the original lowland forest still remains. Expansion of the protected-area system to include lowland forest should be a major goal. This may be difficult, however, because lowland timber is valuable, and agriculture is often possible in this habitat type (though sometimes only on a marginal scale). Nonetheless, given that a majority of the mammals that have been severely impacted by loss of their habitat on Luzon are bats that require lowland forest, which is also necessary to sustain a great many other organisms, this is an essential element of the protected-area system that is sorely in need of expansion. Second-growth forest that is protected from degradation may often be the best option for increasing the area of lowland forest, since old growth is often absent. Our experience suggests that several decades of natural regeneration often produces forest that is good habitat for a large and gradually increasing number of species that live in lowland forest.

There are currently some protected areas that include forest over limestone: Caramoan National Park and the Peñablanca Protected Landscape are two prominent examples. Given the number of species of bats that depend on caves, and recognizing that many other organisms do as well, expanding protection of sites within this habitat appears to be another high priority.

Perhaps the least-known of the habitats discussed in this volume is forest over ophiolites, where ultrabasic and ultramafic soils predominate. We know so little about the biota of these areas that not much can be said with confidence, but mining for minerals found in ophiolites is growing rapidly. For example, some is taking place in the Zambales Mountains, where five species of mammals are endemic, including two (*Apomys brownorum* and *Rhynchomys tapulao*) that occur only near the peaks of the mountains, where mining is currently intensive and expanding. It seems prudent to quickly begin designating protected areas in such places, in order to reduce the likelihood of extinction for species that have especially small areas of distribution, as well as to make conducting additional research a high priority.`

There are 10 species of native mammals on Luzon that are officially listed as threatened by the Philippine government (table 7.3; Wildlife Conservation Society of the Philippines, 1997), including three that are Endangered, six that are Vulnerable, and one that is considered to be threatened because of trade. The data presented in this volume, much of which is new (especially in chapters 9, 10, and 11), lead us to agree with most of these assessments, but also prompt us to suggest changes for several of them, increasing the level of concern in some instances and decreasing it in a few others. Specifically, *Cervus mariannus* should be considered Endangered; they have been eliminated throughout much of their former range on Luzon and now persist only in increasingly smaller areas, due to intensive hunting. Strict protection is needed immediately. Given the evidence available, we suggest that *Acerodon jubatus* require careful monitoring. The current listing of Endangered is appropriate, but if present trends continue, within less than a decade they could become Critically Endangered. We also suggest that bats that are dependent on both lowland forest and caves (tables 7.1 and 7.2) require intensive study, both to determine the number of species present (i.e., taxonomic studies) and their distribution, and to assess the size of their populations and trends in population size. We suspect that some of these are likely to be Threatened currently, but we lack sufficient data to determine which ones.

On the other hand, *Pteropus hypomelanus* and *P. pumilus* are subject to hunting, but both species are wide-

TABLE 7.3. Native Luzon mammals that are listed by the Philippine Department of Environment and Natural Resources as threatened, as of 2014.

Species	Status
Acerodon jubatus	Endangered
Pteropus hypomelanus	Endangered
Pteropus pumilus	Endangered
Archboldomys luzonensis	Vulnerable
Cervus mariannus	Vulnerable
Crateromys schadenbergi	Vulnerable
Desmalopex leucopterus	Vulnerable
Phloeomys cumingi	Vulnerable
Sus philippensis	Vulnerable
Pteropus vampyrus	Other Threatened Species

spread over most of the Philippines, tolerate disturbed forest well, and are fairly common in many places; we think they should be considered Vulnerable, rather than Endangered. *Archboldomys luzonensis* are confined to a small area on Mt. Isarog, but they are fairly common on the mountain and occur mostly in habitat that is too cold, wet, and steep to be attractive for logging or agriculture. In addition, the park is currently stable and well managed. Therefore we think that this species should not be listed as Threatened.

The Philippines also had some of the first laws regulating the taking of wildlife, especially large mammals and birds. The current legal framework, the Wildlife Act (Republic Act 9147) of 2004 is a detailed document that establishes strict regulation and penalties for taking wildlife without proper permits. This document has been important in allowing active protection of wildlife—especially confiscation of large shipments of endangered species—and the legal capacity for enforcement is an essential element in this control. As with protected areas, however, we have seen limited evidence of enforcement over most of Luzon, including within officially protected areas and other forested areas administered by the national government. A stepped-up imposition of fines has been cited as an effective deterrent by some studies (e.g., Scheffers et al., 2012), and this should be used as one of the tools to reduce overhunting. The establishment and enforcement of hunting seasons (so that local residents could benefit from wildlife) might be considered, along with the current (but largely unenforced) prohibition of commercial market hunting.

Prospects for the Future

When we began our first studies of mammals on Luzon in the late 1980s, illegal logging on a massive scale was rampant. The boundaries of national parks were often unmarked and ignored. Conflicts arose between local farmers who wished to protect their watersheds and powerful interests who wanted to maximize the extraction of lumber. Subsistence hunting by local people who had been impoverished by governmental corruption was conspicuous and seemed to be increasing steadily. Human population growth was exceptionally high, placing an ever greater strain on the environment. The economy was in a tailspin; people were desperate, and the future looked grim (Heaney and Regalado, 1998). Research on biodiversity was tolerated but largely ignored by governmental agencies, many of which appeared to focus solely on resource extraction.

The problems we see today are still serious. Too many species face a continuing loss of habitat and overhunting. Respect for the value of protected areas for wildlife conservation, as well as for watershed protection, is low. Too few people on Luzon are aware that their home supports one of the greatest concentrations of unique biodiversity of any place on Earth, and therefore they are unable to put a value on protecting it.

But the Luzon we see today is also a place with a far more hopeful future than what we encountered in the 1980s (Ambal et al., 2012; Posa et al., 2008). The economy has improved greatly, and there is less of a sense of desperation. Much of this is fueled by one of the most dramatic social changes in the history of the Philippines: recent reports from national and international agencies state that fully 25% of the workforce in the Philippines—which we take to mean about 20% of the adults—are employed outside of the country and send home remittances. Many of these overseas workers are from middle- and lower-income families; the money they send home often allows the family to open a small business and put a couple of kids through college, so they will be able to get an even better job overseas. For rural families, this money from overseas often means that they are able to move down by the highway or to a nearby town, leaving behind a hard life of subsistence farming. We have seen this process of "highwayization" all over Luzon and measure its prevalence by the ever-increasing number of Western Union stores that we see in small towns throughout Luzon. Simultaneously, the average family size in the Philippines has dropped to about half of what it was in the 1980s; at ca. 2.6 children per woman, it is still far above the stable point, but its reduction has allowed families to give more resources to each child and place fewer demands on the environment. We suspect that the shift toward so many people working overseas is a component of this reduction in family size, as an emphasis on education and on women being part of the workforce has increased. Whether these changes are beneficial to society and to the country's culture is beyond our scope, but we believe that the process—and its impact on the environment and on the unique wildlife of Luzon—needs to be recognized.

Because of the improving economy, we have also begun to notice an unexpected phenomenon. In some rural areas where we have worked, we have seen places that had been marginal for farming now left to regenerate into second-growth forest. There is still far too little of this, but the trend, if it expands, may be one of the most important prospects for improving environmental conditions and wildlife populations in the future. Research on this topic is greatly needed.

We also see an upwelling of interest in the environment and "the outdoors." The Wildlife Conservation Society of the Philippines (WCSP; recently renamed the Biodiversity Conservation Society of the Philippines) is an organization of researchers, students, workers in non-governmental organizations, and governmental employees who share an interest in biodiversity research, education, and management. The Society was founded in 1992, with 26 members; annual meetings now typically host about 250 participants, and there are about 950 members overall (Posa et al., 2008). The Philippine Department of Environment and Natural Resources (DENR) actively supports research on all aspects of biodiversity and solicits information from the WCSP in setting its policies and management goals. The number of ecotourists visiting national parks has increased dramatically, and this has the potential to become an industry that will promote conservation for its own benefit. Although too little information reaches Philippine students about their country's biodiversity, they are increasingly likely to learn at least a bit about their own fauna and flora, not just about organisms from North America, Africa, and mainland Asia.

While we recognize that these topics are presented here only briefly and are painted with a broad brush, we believe that they are critically important in determining the future of biodiversity on Luzon, and in the Philippines more broadly. Most of the issues are far beyond the ability of biologists to influence; the rise and fall of immigration and the receipt of remittances from those working overseas may determine much of the future of the country, but the situation is not something that biologists can do more about than carefully note the importance of these developments and try to document their impact. We do believe that biologists can nevertheless make a significant difference by increasing society's knowledge about the natural world. This may help guide official efforts, such as influencing the future development of the protected-area system; promote ecotourism, by increasing awareness of the wonders of the fauna in the forest through which mountaineers hike; and—especially—help students and other citizens of the Philippines understand that they live in an extraordinary place, with a biodiversity and habitats that provide immense benefits to all and may serve as the basis for great national pride. It is to that end that this book is primarily intended.

8

Synthesis: Island Biogeography Theory and the Mammals of Luzon

As discussed in the previous chapters, and documented in further detail in chapters 9–11, the mammal fauna of Luzon is remarkable in many respects. These aspects, which we summarize in this chapter, help us understand the processes by which these mammals have evolved and the ecological processes that have maintained their diversity. But the Luzon mammal fauna can also serve as a model system more broadly, allowing us to advance a general understanding about the dynamics of biological diversity of organisms on islands worldwide. That, too, is part of the story we discuss in this chapter.

We need to emphasize that, in many respects, this volume is a beginning to the study of Luzon mammals, not an end. One of the most important things we attempt here is to identify questions that will help lead scientific research forward in the future. Island biogeography has played a crucial role in the development of a data-based understanding of evolution, ecology, and conservation since the earliest days of the science of biology, and it remains at the forefront today.

The Island of Luzon

By any definition, Luzon is a tropical oceanic island (chapter 2). It lies well below the Tropic of Cancer, which passes through the southernmost part of Taiwan at 22.5° N. Temperatures on Luzon at sea level are warm throughout the year, with mean monthly high temperatures averaging 24°C–28°C. As in other tropical areas, the difference between daily high and low temperatures is greater than the difference between the monthly mean daily highs in the coldest and warmest month; in other words, the daily fluctuations exceed the annual fluctuations, taken on a monthly basis (fig. 2.3). In this respect, Luzon has a highly stable environment.

There is substantial variation in other respects, however. In some parts of the island, especially along the eastern edge, all months of the year are wet, with at least 20 cm of rain each month, even during the 3- to 4-month-long dry season that is typical in the lowlands. But in other areas (especially the Cagayan Valley and the western edge of the island), there is almost no rain during the dry season, creating a seasonal tropical environment (fig. 2.3). More dramatically, Luzon is topographically rugged; there is a conspicuous decrease in temperature and increase in rainfall as elevation increases (figs. 2.4 and 2.5), and habitats and plant communities change concordantly. Thus most of the environmental variation on Luzon is associated with the variation along elevational gradients on its many mountain ranges and isolated mountains.

It would be a mistake to think of Luzon as a fixed entity, unchanging through time. Rather, it has a remarkably dynamic history, having originated 26–30 million years ago as a series of small islands that gradually grew and combined, forming the area of the modern Central Cordillera about 15 million years ago, as a result of volcanic activity associated with the subduction zones that bracket Luzon (chapter 4). Other islands emerged and grew over time (though some eroded and disappeared), with a burst of activity that began about 7 million years ago and progressively produced more islands. These merged to form the island that we think of as "modern Luzon" less than 1 million years ago (figs. 4.6 and 5.3). Luzon remains one of the most geologically active places on earth, with major eruptions adding both new land and environmental heterogeneity.

The Large Mammals of Luzon

The current native mammal fauna of Luzon is typical of mammals on oceanic islands in Southeast Asia: lots of bats and non-flying small mammals, and only a few medium-sized and large species. Current evidence indicates that three of these species—long-tailed macaques, palm civets, and Malay civets—were all introduced from elsewhere in Asia within the past 4000 years. Even in forested areas, where the monkeys and civets are fully naturalized, only two out of the five large species—Philippine brown deer and Philippine warty pigs—are native; there are no native primates or mammalian carnivores on Luzon, and none are known from the (scanty) fossil record. That, however, tells only part of the story. At some time during the Pleistocene (i.e., the period when glaciers periodically covered much of the Northern Hemisphere and sea levels globally dropped to about 120 m below the present levels; fig. 4.7), two species of elephants, a rhinoceros, and a close relative of the tamaraw (the dwarf water buffalo of Mindoro Island) colonized the Philippines from the Asian mainland, along with brown deer and warty pigs. We do not know the habitat in which the extinct species lived, but it most likely was drier and more open than the seasonal forest that covered the Cagayan Valley (where most of the fossils have been found) prior to a few thousand years ago. Nor do we know why these large mammals became extinct, a topic that deserves serious investigation. But we have learned from this limited but direct fossil record that the history of the Luzon mammal fauna has been dynamic, with colonizations and extinctions involving large mammals at some time within the past few million years.

The Bats of Luzon

The Luzon mammal fauna is highly diverse, with at least 57 species of bats and 50 species of native non-flying mammals. The bats are most diverse in lowland tropical rainforest, with few species living above 1000 m elevation (chapter 3). Seven different families of bats are represented (chapter 11); this is only one less than Borneo, which is six times larger than Luzon and has often been connected as part of mainland Southeast Asia (Payne et al., 1985). Only two (4%) of the 57 bat species are restricted to Luzon alone, and 38 (67%) are (or appear to be) members of species that are widespread in Indo-Australia; the rest are widespread in the Philippines (table 5.1). For a few groups, there has been significant speciation within the Philippines, most notably in the *Haplonycteris-Otopteropus-Alionycteris* clade (fig. 11.2.1.2) that contains three species (or possibly four; Heaney et al., 1998). But most species of bats on Luzon are closely related to bats that occur not only elsewhere in the Philippines, but far beyond the archipelago (Heaney, 1991; Heaney and Rickart, 1990). This directly implies—indeed, it can be the case only if—these bats are able to maintain gene flow over large distances and among many islands separated by wide and deep sea channels. The limited studies of bat genetics support this conclusion: although the species vary in their rate of gene flow (i.e., the rate of movement by individual bats among islands, followed by successful reproduction), most of the species have high rates of gene flow among islands (Heaney and Roberts, 2009; Heaney et al., 2005; Roberts, 2006a). The small number of genera of bats endemic to the Philippines (5 out of 32, or 16%; chapter 11) implies that diversification within the archipelago has been limited, especially since the largest of the endemic clades (the *Haplonycteris* clade referred to above) contains only three or four species. The bat fauna of Luzon, and of the Philippines more broadly, has been formed largely (though not entirely) through direct colonization by many species that represent genera and families from outside the Philippines. The current diversity of bats on Luzon is largely (though not completely) a direct consequence of colonization, not speciation within the island or archipelago.

Having said this, we hasten to make two points. First, we repeat a comment made often in chapter 11: many of the insectivorous bats are poorly studied, and it is likely that many of the taxa currently recognized as widespread single species (e.g., *Rhinolophus arcuatus*) actually represent several closely related species, sometimes distributed on different islands, but sometimes occurring together on a single island (Ingle and Heaney, 1992; Patrick et al., 2013; Sedlock and Weyandt, 2009; see also Esselstyn et al., 2012b, regarding *Hipposideros*). When documented, the number of these cryptic species and the patterns of phylogenetic relationships among them will add a much-needed level of precision to the general statements given above, as well as help us understand the timing and geographic circumstances under which bats are able to diversify within the Philippines. Second, there is evidence that gene flow within species of bats *between* islands is less than gene flow *within* islands. For example, gene flow between Luzon and the islands that made up Greater Mindanao (fig. 4.7) is less than gene flow within Luzon, even though the San Bernardino Strait is only 15 km wide (Heaney and Roberts, 2009; Heaney et al., 2005; Roberts, 2006a). Thus permanent sea channels do matter—but they usually do not matter enough to entirely eliminate gene flow within most of the species of bats that have been studied to date.

We also can say that the geological history and present-day topographic diversity of Luzon seem to have played little role in shaping its bat fauna. The lowland forest once covering most of the island most likely formed a continuous habitat for most bat species. It is noteworthy that just two species of bats appear to show geographic divergence within Luzon, and these are the only two that occur primarily in upland forest: *Haplonycteris fischeri* and *Otopteropus cartilagonodus* (Heaney and Roberts, 2009; Roberts 2006a; chapter 11). Though genetic studies have been limited to few taxa thus far, species that live in lowland forest on Luzon seemingly do not have disjunct genetic clades within the island. We look forward to future studies that will test this apparent pattern.

The Native Small Mammals of Luzon

We think of the native small mammals on Luzon as falling into two groups, in terms of their histories and patterns of diversification. The first are the New Endemics, including one species of shrew (*Crocidura grayi*; pp. 64–65), plus *Crunomys fallax*, *Rattus everetti*, *Bullimus luzonicus*, and *Abditomys*/*Tryphomys*. The ancestor of each of these lineages arrived on Luzon within the past 2 million years (or perhaps as many as 4 million years ago, in the case of *Abditomys*/*Tryphomys*), and each has undergone either little speciation (in *Abditomys*/*Tryphomys*) or none (apparently all of the others; fig. 5.4). These lineages are similar to those of the bats in showing little or no diversification within Luzon. The New Endemics are notable for having arrived on Luzon recently (on a geological timescale); all of them have their closest relatives on Greater Mindanao and apparently came from that source. Thus they appear to be recent colonists that have added to species richness and taxonomic diversity on Luzon but have undergone little or no speciation. They are mostly habitat generalists, occurring from low to high elevation (with the exception of the very poorly known *Crunomys fallax*), and although they do penetrate into mature forest, they typically are most abundant in habitat that has been disturbed to some degree by either natural events (e.g., typhoons or volcanic eruptions) or human causes (chapter 6).

The second group is made up of the Old Endemics, which are the members of the two highly diverse endemic lineages of Luzon mammals: cloud rats and earthworm mice. The ancestor of the cloud rats arrived on proto-Luzon about 14 million years ago (fig.

5.4), just as the Central Cordillera was forming a large island with a substantial upland area. Diversification began about 10 million years ago and continued up to the present time, producing 12 species on Luzon and 6 others that subsequently colonized other Philippine islands from Luzon (fig. 5.1). The ancestor of the earthworm clade arrived about 7 million years ago and began diversifying on Luzon by about 5.5 million years ago, giving rise to at least 36 species on Luzon, some of which we have yet to formally describe (fig. 5.2; chapters 5 and 10), and at least 16 species that colonized other islands from Luzon and often diversified within those islands (e.g., Steppan et al., 2003). Thus these two lineages arrived on Luzon when it was made up largely of the Central Cordillera and have undergone vast speciation subsequently, radiating to form communities that include up to 6 cloud rats and 7 earthworm mice living in the same area (i.e., syntopic). Our studies are by no means complete, but those conducted thus far indicate that the rate of speciation has not slowed; rather, it seems that this rate in the most diverse genus, *Apomys*, has held steady since their clade appeared (Justiniano et al., 2015). The rate of diversification in the cloud rat lineage on Luzon is high, at about one per million years, and very high in the earthworm mouse lineage, at about five per million years (chapter 5). We note that there is some evidence (for *Archboldomys* and *Batomys*) that colonization from proto-Luzon to central Bicol (where Mt. Isarog and Mt. Malinao are located today) occurred when that area was a separate island, and they have persisted subsequent to the merger of modern Luzon into a single island.

Members of the two Old Endemic lineages are most abundant and diverse at medium to high elevation in mature or old-growth montane and mossy forest. The restriction of these species to high elevations is intimately associated with their diversification, and nearly all species occur in areas that are topographically and climatically isolated from their closest relative. In other words, the geological history of Luzon has shaped the geographic circumstances that create climatic and habitat variation, which have been key to the promotion of speciation (chapter 5). Most taxa in these endemic clades are tolerant of at least moderately disturbed secondary forest, a few penetrate fully into very heavily disturbed anthropogenic habitats, and several occur down to sea level. They show no evidence of being poor competitors with the New Endemics, and indeed appear to be superior in all but heavily disturbed habitat (chapter 6).

The Non-Native Small Mammals of Luzon

Asian house shrews (*Suncus murinus*), house mice (*Mus musculus*), and five species of rats (*Rattus argentiventer*, *R. exulans*, *R. norvegicus*, *R. nitidus*, and *R. tanezumi*) on Luzon are closely associated with humans, our habitations, and our disturbances. None of these species are native to Luzon, and all them appear to have arrived in association with the influx of people from Taiwan that began about 4000 years ago (chapter 4). Three of these taxa (*S. murinus*, *M. musculus*, and *R. norvegicus*) are tightly associated with high-density human habitations (mostly in or immediately adjacent to buildings) and are found virtually nowhere else. One (*R. argentiventer*) is associated elsewhere in the Philippines with open grassy agricultural areas, but it seems to be barely present on Luzon. The final species (*R. nitidus*) formerly inhabited high-elevation ricefields in the Central Cordillera but is now either quite uncommon or extinct (chapters 4 and 10). Only two of the non-natives (*R. exulans* and *R. tanezumi*) occur widely in heavily and moderately disturbed habitat, and they fare poorly as disturbed habitat regenerates into forest and the native small mammals move back in. In undisturbed habitat, or in habitat that has regenerated into high-quality forest, they are nearly or entirely absent. Thus we consider these non-native small mammals, as a group, not to have been very successful in becoming naturalized on Luzon: two are so rare as to be functionally extinct, three are highly restricted to the most intensively anthropogenic habitats, and two occur widely but do poorly in the face of competition from native species. It is only in the places where humans have thoroughly altered the natural environment that the non-native species predominate. That two of them (*R. exulans* and *R. tanezumi*) are, in fact, abundant and widespread on Luzon is a testament to the abundance of humans and our anthropogenic habitats.

The Dynamics of Diversity on Luzon

The following is a summary of the patterns we believe are major components of the long-term dynamics of mammalian diversity on Luzon (chapter 5).

The early islands that eventually became part of Luzon were probably rather small, may often have eroded away as others were produced nearby, and probably rarely developed the type of high-elevation montane forest that forms the habitat for most of the cloud rats and earthworm mice today. By about 15 million years ago, volcanic islands had begun to merge into a large island that is now represented by the Central Cordillera, and substantial highland habitat began to be available. Other islands continued to develop—most through volcanic activity, but some as uplifted ophiolites and/or limestone regions—and neighboring islands often merged. Luzon's rate of growth has undoubtedly fluctuated since its origin, but the overall pattern has been one of progressively increasing area, elevation, and topographic complexity, with some islands appearing and disappearing as subduction zones were formed, shifted in location, disappeared, and reappeared. Overall, the island's history has been one of progressive growth as a result of continued subduction of ocean crust.

Among the living species of mammals on Luzon, the oldest lineages are the cloud rats and earthworm mice. The ancestor of the cloud rats arrived on Luzon about 14 million years ago, which was 10–15 million years after the first islands were produced by volcanic eruptions. Speciation in this clade began by about 10 million years ago and has continued at a fairly rapid pace since then. The ancestor of the earthworm mice arrived about 7 million years ago; their speciation began by about 5.5 million years ago and has continued at a very high rate since then, with no evidence of tapering off. Subsequent colonization of Luzon by a few small mammals in roughly the past 4 million years has added to total diversity, but current evidence indicates only a little speciation by these taxa (fig. 5.4). Thus, of the 50 species of native small mammals estimated to be present on Luzon, 44 (88%) are members of two lineages that arrived long ago and have undergone extensive diversification within the area of modern Luzon, while the remaining 5 native small mammal lineages, which arrived more recently, have not speciated much. Colonization has not been common, by any definition, but following colonization, the native small mammal fauna shows evidence of both persistence and speciation as prominent features, with speciation accounting for about seven times as much species richness as direct colonization.

We see clear evidence that the members of the Old Endemic lineages are competitively superior to the New Endemics in natural habitat, with the more newly arrived species faring best in somewhat disturbed habitats, though they do fairly well across all but the most highly anthropogenic habitats. Similarly, the non-native small mammals do poorly in secondary or mature forest and are competitively inferior to both the Old Endemic and New Endemic species in old-growth and mature secondary forest. Indeed, of the seven species of non-native small mammals on Luzon, two are functionally extinct, three are very highly restricted to the most completely anthropogenic habitats (i.e., dwellings, large buildings, barns, etc.), and only two occur in newly regenerating secondary forest. The abundance and diversity of non-native small mammal species on Luzon today thus appear to be largely a function of the extent of human disturbance, with native species heavily predominating in anything that approximates natural habitat.

Much of the speciation that has taken place among the small mammals is clearly linked to the complex geological history of Luzon. Luzon's continued growth in area and in topographic complexity is very roughly paralleled by the increasing richness of the small mammal fauna. We note this as a correlation; how much of the speciation would have taken place had Luzon remained much smaller is an open question, but it does appear that the progressively increasing size and topographic complexity of the island has played a positive role in the progressively increasing species richness of the non-volant small mammal fauna.

The contrast between the non-volant mammals and the bats is striking. The bat fauna is composed largely of species that have colonized sufficiently recently that most are members of genera that occur elsewhere and often are closely related to species outside the Philippines. Unfortunately, we currently lack estimates of when any of the bats arrived in the Philippines, but it seems as though the ancestors of most of the extant species arrived much more recently than the ancestors of most of the non-volant mammals. Because the preponderance of bat species are not endemic to Luzon or the Philippines, because most appear to be members of species widespread in Southeast Asia, because there are few endemic genera, and because we see evidence of only a few endemic clades of bats with the Philippines (and none on Luzon), we suspect that colonization greatly exceeds speciation as the source of the current species of bats, perhaps by a ratio of as much as five or ten to one. We note that this is a crude estimate and point to this topic as being another that is worthy of much additional research.

The contrast between non-volant mammals and bats—in which the non-volants have 7 times as many species present through speciation as by direct colonization, and the bats have perhaps 5 or 10 times as many by colonization as by speciation—is possible only because bats are able to fly between islands. The rate of colonization by bats between islands in the Philippines and from places outside the archipelago must be high to maintain so many widespread species; the colonization rate by small mammals must be quite low to produce the high levels of endemism on Luzon. Efforts to quantify this qualitative observation would surely produce many insights into the impact of varying rates of colonization and speciation and may show the manner in which these two processes interact over evolutionary time.

We believe the data on bats may imply an important aspect of the dynamics of long-term species richness that deserves emphasis. We certainly see evidence that colonization rates among some bats are high. Unlike the murid rodents, we find less evidence of long-term persistence of lineages among bats. *Desmalopex* and the members of the small *Alionycteris-Haplonycteris-Otopteropus* clade probably represent lineages that have persisted in the Philippines for a long time, since they are endemic at the level of genus or above. But there are no other endemic genera, which may indicate the lack of a long-term persistence of lineages among the other bat species. If this is the case (which surely requires further investigation), it implies that many bat species that colonize the Philippines, and Luzon in particular, must become extinct on an evolutionary timescale (i.e., hundreds of thousands to millions of years). In other words, the level of species richness for bats appears to be the result of colonization rates that are high (on an evolutionary timescale), speciation that is substantially lower, and persistence rates that are also substantially lower than those among the non-volant native mammals (i.e., extinction rates for bats are higher). All three processes (colonization, speciation, and extinction) are present among bats, but we think that species richness has been driven much more by colonization and extinction than by speciation.

Mammals and Models: Insights and Extrapolations

From the early 1970s until recently, most studies of island biogeography were conducted in the context of a model that strongly emphasized the roles of colonization and extinction. This model, called the equilibrium model (MacArthur and Wilson, 1963, 1967), hypothesized that on any given island, the rates of colonization and extinction are equal over time, such that the number of species is usually nearly constant (i.e., in equilibrium). If this were not the case, the number of species would either decline to zero (if extinction predominated) or increase continuously (if colonization predominated). This model also assumes that the rates of ongoing colonization and extinction are typically sufficiently high to have nearly constant turnover in the species that are present. This turnover will occur rapidly enough that speciation is unlikely and can be largely ignored as a process. The model further assumes that the species arriving from outside the island (most often from a larger island or a continental source) replace species on the island that are less capable of surviving. In other words, newly arriving species are typically competitively superior to those that are resident on the island, and it is often competition that drives the rate of extinction. Finally, because the model deals with processes that operate over relatively short periods of time, it treats islands as fixed entities, without dynamic histories of their own.

This model does a poor job of describing the processes that we see as influencing the mammals of Luzon. Even with bats, there is clear evidence that speciation has taken place, and there are indications that at least some lineages (*Desmalopex* and the *Alionycteris-Haplonycteris-Otopteropus* clade) have been in the Philippines for a lengthy period of time (i.e., millions of years, not thousands of years). Among the non-volant small mammals, colonization has been rare, speciation has predominated, lineages are ancient, persistence is more conspicuous than extinction, and endemic species seem to be consistently competitively superior to non-native species in any but heavily disturbed habitat. Luzon itself has been highly dynamic in its growth, changing dramatically during the past 5 million years, which is the period of time during which much of its mammalian diversity has developed (fig. 5.4).

Nonetheless, in several often-overlooked paragraphs, MacArthur and Wilson (1967:173–175) noted that on large islands and archipelagoes that are at the edge of the dispersal ability of a given group of organisms, and where colonization therefore is rare, speciation and diversification may predominate; they referred to this as a radiation zone. Although the authors did not develop the concept further, they did list eight examples—and among them were "the murid rodents of Luzon." MacArthur and Wilson understood that these animals fell outside of the framework of their equilibrium model and left it to others to develop another general model for such circumstances.

Our own efforts to develop a model, based in large part on our gradually improving understanding of mammalian biogeography in the Philippines, led us to postulate (Heaney, 2000) that species richness on oceanic islands can result from colonization alone (fig. 8.1, zone A), but probably most often a given level of species richness results from an intermediate situation in which some species on the given island are recent colonists (and are native but not endemic), others are the descendants of colonists that have persisted and become distinct endemic species but have not diversified, and still others are the descendants of lineages in which each of several ancestral species has given rise to many species (zone C). There might also be a situation in which most species are non-endemic colonists and all of the rest are endemic species closely related to those in the source area (zone B). We also postulated that island archipelagoes are likely to be subject to major geological change on the same timescale as colonization and speciation, so their carrying capacity varies a great deal. Thus any given island is unlikely to maintain a constant (i.e., equilibrial) number of species. In other words, we hypothesized that most island biotas are likely to be in a nearly constant state of change (i.e., non-equilibrial) in species richness over the long periods of time that are relevant to the islands and their biotas.

This simple model seems to describe what we now see in the Luzon mammal fauna. The bats—with many species shared with external source areas (i.e., non-endemic native species), a substantial percentage of endemic species closely related to those outside Luzon and/or the rest of the Philippines, and just a few small endemic clades—appear to fall in zone C but are very close to the edge of zone B. The murid rodents, however—with a few recent colonists that are not endemic species, a few endemic species that have arrived recently (on a geological timescale), and a great many species that are the descendants of just a few clades that

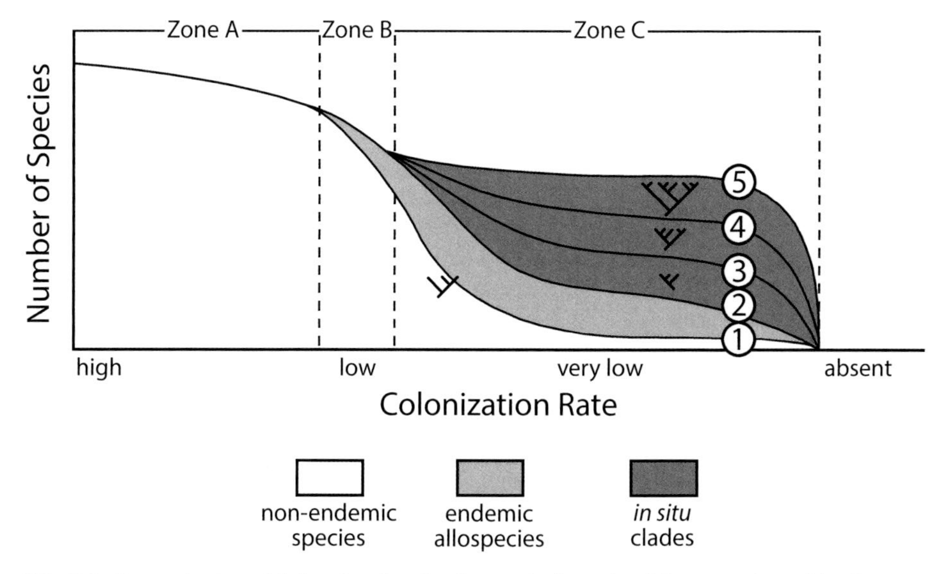

FIG. 8.1. Conceptual model showing the development of species richness on large islands or archipelagoes that experience different rates of colonization due to varying degrees of isolation. In zone A, the island is near enough to a source that all species on the island are also present in the source area(s), often a continental area but possibly a larger and/or older island, due to high rates of gene flow (i.e., frequent movement of individuals between the relevant areas). In zone B, gene flow to some species is low enough (or perhaps absent following a single colonization event) that they differentiate from their close relative in the source area, but no diversification has taken place within the island, due to insufficient time or to inadequate space or topographic diversity to allow speciation (i.e., gene flow is maintained within the island). In zone C, some colonizing species have diversified, some among them greatly and others less so; the amount of diversification may be influenced by how long the lineage has been on the island. For speciation to take place, gene flow among populations on the island must be interrupted, often by isolation in patches of habitat. Modified from Heaney (2000).

have diversified greatly—clearly fall into the middle of zone C, as presciently implied by MacArthur and Wilson (1967) in their description of a zone of radiation.

Although our model is helpful in visualizing the impact of any given colonization rate on the likelihood of speciation and the combined effect of these rates on species richness, it is still quite simple and therefore limited. We have been aware that a more comprehensive model is needed (Heaney, 2007; Heaney et al., 2013a), a situation that has been largely (but not entirely) met by Whittaker et al. (2008, 2010). Their General Dynamic Model (fig. 8.2) is limited in that it was explicitly developed in the context of oceanic islands that are formed over geological hotspots. The best-known example of such an archipelago is Hawaii. The big island of Hawaii is the youngest and largest, since it sits over the single plume of magma that produces volcanic materials, while the other islands are progressively older and more heavily eroded, correlated with their distance from the hotspot. These kinds of islands are generated as the Pacific Plate slowly moves to the northwest over the magma plume, which remains stationary; as the plate moves, the magma plume forms a new island that eventually is carried away by the moving plate, and a newer island forms. These hotspot islands thus have a definite life cycle: each builds up over a period of a million years or so, then moves away from the hotspot and gradually erodes down and disappears below the waves within another 4–6 million years or so.

Whittaker et al. (2008, 2010) postulate that immi-

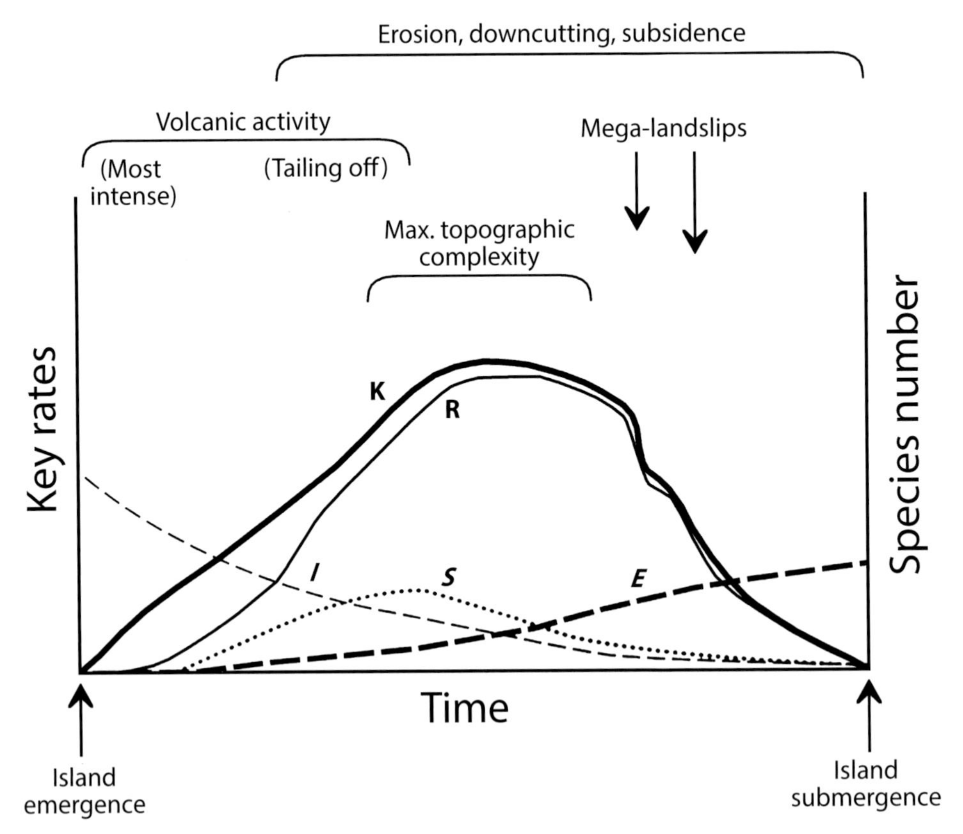

FIG. 8.2. General Dynamic Model of island biogeography developed by Whittaker et al. (2008). This is a conceptually based, graphical representation of key rates and properties. The time axis should be considered as some form of log functions, because island building is typically much more rapid than island decline. This graph shows the postulated relationships between biological characteristics and the ontogeny of a single island, where *I* is immigration rate, *S* is speciation rate, and *E* is extinction rate (with each rate referring to the number of species per unit of time). K is the potential carrying capacity (defined as the number of species), and R is realized species richness. Redrawn from Whittaker et al. (2010).

gration will be high during the early history of the island, when the island grows and becomes capable of supporting increasing numbers of species, but speciation quickly comes to predominate as the number of endemic species rises and the island becomes large and topographically diverse. The rates of both colonization and speciation then decline as the island ceases to grow and open niches become much less available. Extinction becomes increasingly common as the island shrinks and then disappears from erosion.

We find this model to be thought provoking and generally successful as a framework for understanding the mammal fauna of Luzon. The notions of long time spans as crucial to the development of the fauna, largely independent rates of colonization and speciation, the role of island size and topographic diversity in influencing all of the biological processes, and especially the importance of speciation, all represent crucial improvements over the equilibrium model of MacArthur and Wilson (1967). Much current research is being done within this framework involving oceanic islands and a variety of taxa (e.g., Borges and Hortal,

2009; Cameron et al., 2013; Cardoso et al., 2010), and the long-term dynamics of biotas on oceanic islands is becoming much better known as a result.

We note, however, that this model was developed within the framework of the geological dynamics of archipelagoes that form from geological hotspots. As Whittaker and Palacios (2007) have explicitly recognized, archipelagoes such as the Philippines (as well as the Lesser Antilles, Marianna Islands, and Solomon Islands; Nunn, 2009) that form along plate margins as a result of subduction have a different type of geological history. The details discussed in chapter 4 (and shown in figs. 4.5, 4.6, and 5.3) represent a single, specific example: the development of Luzon, a large plate-margin island. Here we use Luzon as the basis for a more generalized model of the development of plate-margin islands (figure 8.3).

In this simple conceptual model, the early history of a hotspot island and a plate-margin island may be quite similar: volcanic eruptions build an island that becomes progressively higher and larger. The difference between the two becomes apparent at the time that the hotspot island moves away from the magma plume and volcanic activity on the island permanently ceases; in many such archipelagoes, the growth phase lasts for only a million years, and sometimes less. On the other hand, subduction zones often persist over tens of millions of years, so that a given island may continue to grow over a very long period of time, as is the case with Luzon, which had its origin 26–30 million years ago. The geological fate of a hotspot island is to erode and disappear; the geological fate of a plate-margin island often is to persist and grow as the subduction zone continues to produce volcanic materials.

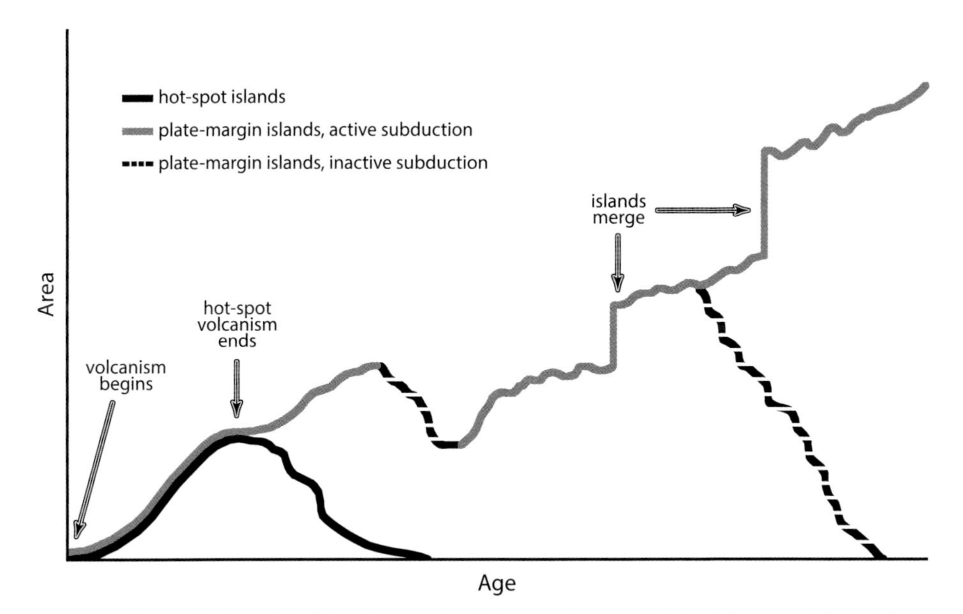

FIG. 8.3. Conceptual model of the history of a plate-margin oceanic island (gray and dotted lines) compared with that of a hotspot island. Hotspot islands typically build up in size and elevation over a relatively short period of time, then erode down and eventually disappear (Whittaker et al., 2008). Plate-margin islands grow irregularly due to volcanism produced by subduction zones, which may become inactive for a time (or permanently so), but which usually remain active for many millions of years. The many small islands produced along the subduction zone, which frequently occur in a roughly linear array along this zone, often gradually merge into progressively larger and more topographically diverse islands of internally varied geological age.

As noted in chapter 4, however, there are several common exceptions to this pattern of growth. Sometimes subduction zones become inactive and either disappear entirely or resume activity only after a long period of time. For example, the Northern Sierra Madre existed as an island (or set of nearby islands) from about 25 to 30 million years ago, but the volcanic activity ceased (due to a change in subduction), the island(s) disappeared entirely, and they did not reappear until about 5 million years ago. Because subduction zones sometimes disappear entirely (Hall, 2012), we include that as one possible fate of a plate-margin oceanic island in figure 8.3.

Another exception to the pattern of simple continuous island growth results from the linear nature of subduction zones (fig. 4.5). Rather than having all of the volcanic material emerge at a single vent (or from vents that are very nearby), as in the case of hotspot islands, subduction zones produce a string of islands all along the subduction zone. These islands apparently are often not evenly spaced; in the Philippines they tend to be clumped (fig. 4.5). Also, some islands emerge as uplifted limestone or as ophiolites (chapter 4). Regardless of these details, it is apparent that typically *many* islands form along subduction zones. They each originate at different times, grow over time, and ultimately merge over time (figs. 4.5, 4.6, and 5.4). We show this as occurring at several places on the graph in figure 8.3, but presumably mergers could occur at virtually any time and between islands of varied sizes.

The hypothetical, generalized history of a large plate-margin island has profound implications for the dynamics of species richness for any group of organisms that is able to colonize the island. Such an island will persist much longer than a typical hotspot island, will become much larger and more topographically diverse, and will often grow by joining islands that may have unique endemic biotas at the time of the merger. The internal dynamics will often differ, with individual volcanoes erupting, becoming larger and higher for a time, and then eroding down as other volcanoes erupt at varying distances on the same island (fig. 4.1), leading to the development of topographically complex sets of highland habitats, such as montane and mossy forest (chapter 2), that may function as sky islands that increase the likelihood of speciation (chapter 5). The long period of existence of large plate-margin islands provides the timeframe that seems necessary for the extensive diversification of early colonizers. On Luzon, the development of the highly morphologically and ecologically diverse communities of cloud rats and earthworm mice (figs. 5.1 and 5.2), with large numbers of species in each clade currently occurring fully sympatrically, seems possible only with the passage of many millions of years. These long periods of evolutionary history on a single, large, plate-margin island such as Luzon may help to explain why the native species are superior competitors to the recently introduced, non-native rats and mice. The native species have evolved in their native habitat and, to put it into the simplest possible terms, they have gotten to be good at using the available resources and dealing with the local climatic (e.g., typhoons) and geological (e.g., volcanic eruptions) vicissitudes.

While figure 8.3 and this discussion provide a rather brief and superficial treatment of what are undoubtedly a complex set of issues surrounding the dynamics of biological diversity on oceanic islands, we offer it as one step in the direction of developing a more comprehensive model of island biogeography. We believe that the General Dynamic Model (fig. 8.2) presents the best current model for considering the types of variation in major biological processes that influence species richness on oceanic islands. If the geological processes that it incorporates can be generalized to include plate-margin islands and their more complex histories, we think such a model would come close to realizing the potential for a genuinely comprehensive conceptual model that will allow a generation or more of research scientists to explore the complexities of island life. It is worth noting that the challenges of understanding these evolutionary and biogeographical processes have long perplexed our predecessors, as is apparent in a letter Charles Darwin sent to Alfred Russel Wallace (who was then in the Aru Islands near New Guinea) in May 1857, over a year before they jointly first described the process of evolution by natural selection: "One of the subjects . . . which [has] cost me much trouble, is the means of distribution of all organic beings found on oceanic islands; & any facts on this subject would be most gratefully received" (quoted in van Whye and Rookmaaker, 2013:131).

The Philippine Islands support what may be the greatest concentration of unique biodiversity on earth. These islands surely are likely to remain one of the best places to understand the processes that influence life on islands around the globe. The mammal fauna of the archipelago is extraordinary in many ways; the mammals and the habitats they occupy (fig. 8.4) deserve to be celebrated, enjoyed, and protected by the citizens of the Philippines as one of the finest parts of their natural heritage.

FIG. 8.4. The valley of the Tanudan River (below the municipal center of Barlig, Mountain Province) displays the steep terrain, lush rice terraces, pine forest, and dense montane forest that characterizes the Central Cordillera, the geologically oldest part of Luzon.

Part II

NATURAL HISTORY OF LUZON MAMMALS

9

Large Mammals

Like other oceanic islands, Luzon has few large mammals that are native in modern times and only a limited number of species that are known to have existed in the past (chapter 4). Elephants, rhinoceros, and dwarf water buffalo—known from fossils found on Luzon—have long since disappeared without a trace, leaving behind wild pigs and deer as the only current native large mammals. Other large mammals that are present on Luzon are long-tailed macaques, palm civets, and Malay civets, but a combination of genetic and archeological data now indicate that these three species most likely were imported by humans within about the past 4 thousand years, with domestic mammals (water buffalo, domestic pigs, goats, cattle, horses, etc.) imported more recently.

Although deer, wild pigs, monkeys, and the two civet species formerly were common throughout most of the island and are still widespread on Luzon, all have been heavily reduced by the combination of loss of habitat (mostly through logging followed by conversion to agriculture) and current overhunting (chapter 7). The native warty pigs (*Sus philippensis*), and especially deer (*Cervus mariannus*), have been heavily impacted, with local extinctions and huge reductions reported over nearly all of Luzon, including in many officially protected areas. If current trends were to continue, either of these species, and particularly the deer, could become exceedingly rare within a few decades; they are greatly in need of improved protection. We note that the places where we have seen the most stable populations are those where traditional societies have remained strong and in control of their land. Improved management may involve giving more control to such indigenous ethnic groups of people where traditional societies exist, rather than the application of strict, externally imposed limits that may be needed in areas where such societies no longer are extant.

We find it intriguing that three of the widespread large mammals on Luzon—monkeys and the two species of civets—are exotic taxa that have become fully naturalized (chapter 6). Although all of them will live in close proximity to humans and their disturbances, they also have successfully penetrated fully into natural forest. Perhaps they have been able to do so because they have no natural counterparts in the native Luzon fauna. Domestic pigs, non-native rats and mice, and non-native shrews have had far, far less success, even though all these latter species have been present on Luzon for a similar length of time. The ecological dynamics of these three successful large, invasive species is a topic that deserves much research in the future.

A: *Macaca fascicularis*. B: *Paradoxurus philippinensis*. C: *Viverra tangalunga*. D: *Cervus mariannus*. E: *Sus philippensis*.

Macaca fascicularis long-tailed macaque

(Raffles, 1821). Trans. Linn. Soc. Lond., 13:246. Type locality: Bengkulen, Sumatra, [Indonesia].

DESCRIPTION: Total length 890–1200 mm; tail 440–600 mm; hind foot 120–150 mm; weight 3.5–6.5 kg. A large monkey, with a long tail. Adult males are larger than females. No other non-human primate occurs on Luzon.

DISTRIBUTION: Burma [= Myanmar] to Timor; throughout the Philippines.

EVOLUTION AND ECOLOGY: Morphological and genetic data show that Philippine populations differ little from those on Borneo. Archeological data indicate that monkeys first appeared on Luzon about 4000 years ago and most likely were introduced by humans from

Borneo, or possibly Palawan (chapter 4). They occur widely in agricultural areas near forest, as well as in second growth, secondary, and primary forest from sea level to at least 2250 m elevation. They are found in lowland, montane, and occasionally mossy forest (fig. 3.8), often moving long distances up and down mountains in a single day. Long-tailed macaques feed on a wide range of fruits, leafy materials, seeds, insects, other invertebrates, and (rarely) small vertebrates.

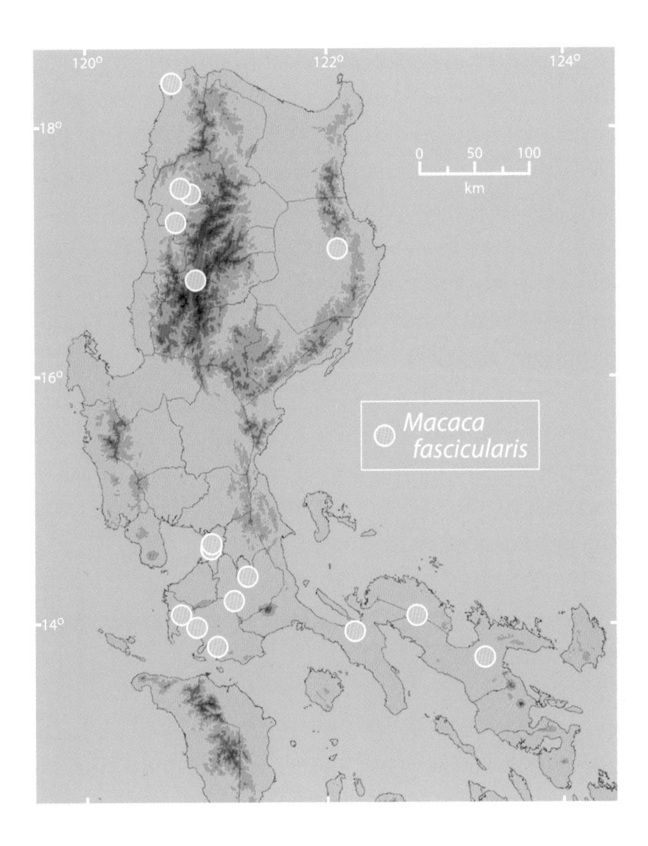

They forage primarily during daylight, in trees and often on the ground. They give birth to a single young each year. They are locally common to uncommon and are hunted heavily for food, for use as pets, and for medical research. Philippine eagles sometimes prey on young monkeys.

STATUS: Widespread in Asia; probably non-native on Luzon but occur throughout much of the island. During our surveys, local residents and hunters reported that they were relatively common and stable in Balbalasang and on Mt. Banahaw, Mt. Irid, and Mt. Palalli; uncommon and declining on Mt. Amuyao, Mt. Pulag, Saddle Peak, Mt. Tapulao, and in Caramoan National Park and the Tayabas Caves area; and locally extinct on Mt. Data. Overhunting is the primary cause for their decline. For several decades, they have been captured in the wild and bred in captivity for export, mostly for medical research; the impact of this on Luzon's macaque populations has not been documented.

REFERENCES: Danielsen et al., 1994; Duya et al., 2007; Fooden, 2006; Heaney et al., 1991, 2004, 2006a, 2006c; Rabor, 1986; Rickart et al., 1993; Scheffers et al., 2012; Thomas, 1898.

Paradoxurus philippinensis common palm civet

Jourdan, 1837. Comptes Rendus Hebdomadaires Séanc. Acad. Sci. Paris 5:521–524. Type locality: Philippines.

DESCRIPTION: Total length 770–840 mm; tail 340–410 mm; hind foot 62–70 mm; ear 34–39 mm; weight 1.5–3 kg. A fairly large, cat-like animal, with a snout that is longer and more pointed than that of a true cat. The fur is usually dark, sometimes nearly black but most often dark brown, with black spots running down the back in several rows. The face sometimes is black on the anterior portion, with paler fur posterior to the eyes and mouth, giving it a masked appearance. The tail is longer than the head and body and is usually black. The legs are proportionately short for the size of the torso. *Viverra tangalunga* are larger overall (weight 2.5–4 kg); have long legs and a longer hind foot (89–105 mm); and have alternating bands of black and pale brown fur on the tail (except for a black stripe that runs from the base to the tip along the top).

DISTRIBUTION: Borneo, the Mentawi Islands, and the Philippines; found throughout the Philippines, including all parts of Luzon.

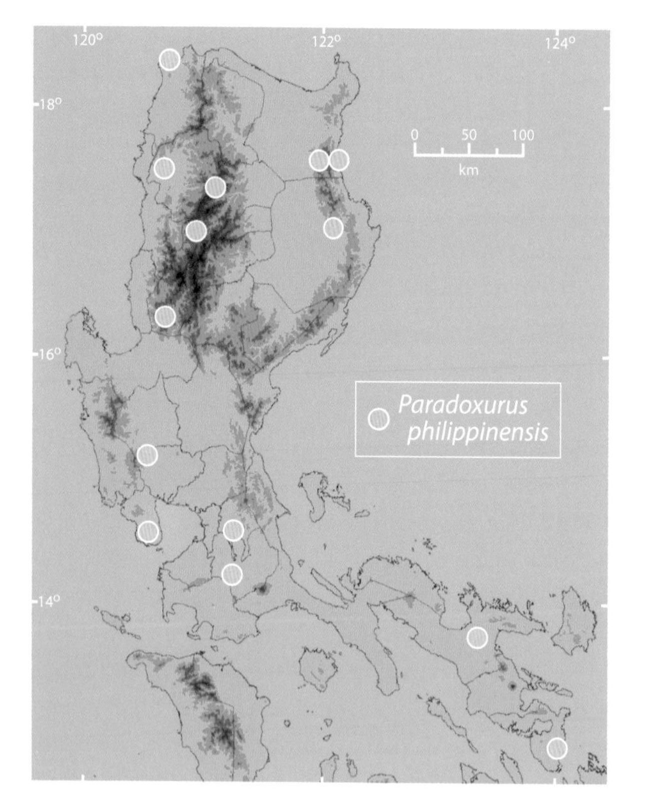

EVOLUTION AND ECOLOGY: Philippine populations were formerly included in *Paradoxurus hermaphroditus*, but genetically based phylogenetic studies show that these two species are distinct. The genus *Paradoxurus* is inferred to have originated about 4.5 million years ago, followed by diversification into three main clades, beginning about 4 million years ago. The ancestor of Bornean and Philippine populations diverged from their closest relative (in the Mentawi Islands of Indonesia) by 2.5 million years ago. Palawan populations are weakly genetically distinct, but those from Luzon appear to be indistinguishable from those on Borneo. Lack of archeological evidence for this species on Luzon prior to about 4000 years ago probably indicates that they were introduced from Borneo at that time. If so, they have become naturalized and occur in agricultural and forested areas from sea level up to at least 2400 m elevation, in lowland, montane, and mossy forest. They are active mostly at night, both on the ground and frequently in trees, feeding on a wide range of fruits, invertebrates, and vertebrates. Scats found on Mt. Isarog that were believed to have come from this species contained the remains of fruits (including *Elaeocarpus* and *Freycinetia*) and mammals (*Apomys*, *Phloeomys*, and *Rattus*), and one individual was seen feeding on ripe coffee fruits. Rural residents often capture them in snares for personal consumption and for sale as bushmeat, and they occasionally keep them as pets. Due to the popularity of "civet coffee" (from beans consumed by palm civets and "harvested" from their scats), there is increasing demand for them to be kept in captivity for the production of "processed" coffee beans. In many areas on Luzon, they are reported by local people to be the most common carnivore.

STATUS: Common and geographically widespread; probably non-native on Luzon.

REFERENCES: Alcala and Brown, 1969; Balete and Heaney, 1997; Bartels, 1964; Esselstyn et al., 2004; Heaney et al., 1991, 1999, 2004, 2006c; Heideman et al., 1987; Hoogstraal, 1951; Patou et al., 2008, 2010; Rabor, 1986; Rickart, 2003; Rickart et al., 1993; Scheffers et al., 2012; Thomas, 1898; Timm and Birney, 1980; Veron et al., 2014.

Viverra tangalunga
Malay civet, *tangalung*

Gray, 1832. Proc. Zool. Soc. Lond., p. 63. Type locality: Sumatra, [Indonesia].

DESCRIPTION: Total length 760–905 mm; tail 209–350 mm; hind foot 89–105 mm; ear 40–42 mm; weight

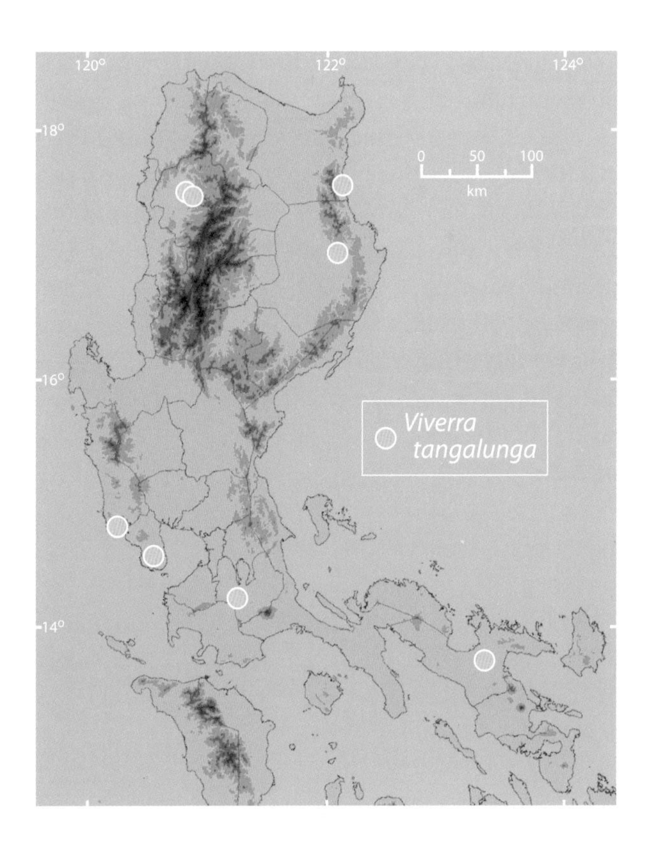

2.5–4 kg. The snout is long and rather pointed; the legs and tail are long. The torso is covered by gray-brown fur, with black spots that form rows over the back; the tail has alternating bands of black and nearly white fur. The legs are usually black. The throat has fairly wide bands of alternating cream and black fur. *Paradoxurus philippinensis* are smaller (weight 1.5–3 kg); have shorter legs and hind feet; and lack distinct alternating bands of pale and dark fur on the throat.

DISTRIBUTION: Malay Peninsula to Sulawesi and Amboina; found throughout the Philippines.

EVOLUTION AND ECOLOGY: The genus *Viverra* is inferred to have originated during the Pliocene, ca. 4.6 million years ago. Phylogenetic analyses involving Philippine *V. tangalunga* suggest possible natural dispersal from Borneo to Palawan during the Pleistocene, and this is supported by the presence of late Pleistocene fossils from Palawan. *Viverra tangalunga* on Luzon and most other Philippine islands, however, are not distinguishable morphologically or genetically from those on Borneo, and it is likely that they were introduced by humans to the Philippines (except Palawan). They are now thoroughly naturalized; they live in mixed forest/agricultural mosaic, and primary and secondary lowland, montane, and mossy forest from sea level to at least 1500 m elevation (fig. 3.8). They are active primarily at night, foraging for small vertebrates and arthropods, and occasionally for fruits. They are probably more carnivorous than *Paradoxurus philippinensis*, and we suspect that they are the primary mammalian predators of small mammals in lowland and montane forest. Unlike *P. philippinensis*, they rarely climb trees. They are often reported by local people to be less common than *P. philippinensis*. They are said to feed on chickens, so they are hunted (usually with snares) and eaten by rural people. This species has a karyotype (on Leyte) of $2n = 36$, FN = 64.

STATUS: Widespread in Asia; probably non-native on Luzon, though widespread and moderately common.

REFERENCES: Esselstyn et al., 2004; Gaubert and Cordeiro-Estrela, 2006; Heaney et al., 1991, 1998, 1999, 2004, 2006b, 2006c; Rabor, 1955; Rickart, 2003; Rickart et al., 1993; Timm and Birney, 1980; Veron et al., 2014.

Sus philippensis
Philippine warty pig

Nehring, 1886. Sitzb. Ges. Naturf. Fr., Berlin, p. 83.
Type locality: Luzon Island.

DESCRIPTION: The only wild pig on Luzon Island. They have sparse, bristly, black or dark brown hair that is longest over the shoulders and the back of the neck and head. Adult males on Luzon often have tufts of white hair on the side of the jaw and scattered in the dark hair over the rest of the body. Tusks and "warts" (i.e., projections of skin-covered bone on the snout) are usually conspicuous.

DISTRIBUTION: Occur only in the Greater Luzon and Greater Mindanao Faunal Regions. Formerly widespread on Luzon, currently declining rapidly.

EVOLUTION AND ECOLOGY: Member of a clade of three species restricted to the Philippines, thus showing extensive speciation within the archipelago. Late Pleistocene fossils of *Sus philippensis*, ca. 50,000–68,000 years ago, were recovered from an archeological site in the Cagayan Valley, along with those of an extinct dwarf *Bubalus* and *Cervus mariannus*, as well as some native small mammals. *Sus philippensis* remains were also found at another archeological site in Cagayan Valley, dated ca. 4000 BP, indicating its importance to Neolithic people at that time; these warty pig remains were found along with the remains of the first documented domestic pigs (*S. scrofa*) in the Philippines. They formerly were abundant from sea level to at least 2800 m elevation, in virtually all habitats. They live in small groups that include adults of both sexes and their young. They probably feed on a wide range of plant and animal material, but details are lacking. They are common in montane and mossy forest from 925 m to 2150 m elevation in Balbalasang, Kalinga Province; in contrast, they occur mainly among tall *Saccharum* stands and regenerating vegetation on the slopes of Mt. Pinatubo. Local people have told us that the litter size is typically two, much smaller than in domestic pigs.

STATUS: Now common on Luzon only in remote forests. They are heavily hunted and declining rapidly in many areas, and they may be threatened by hybridization with domestic pigs in some areas. They are extinct on Marinduque. During our surveys, local residents and hunters described them as relatively common and stable on Mt. Irid, Mt. Palalli, Mt. Pinatubo, and Mt. Tapulao, as well as in Balbalasang; uncommon and declining on Mt. Amuyao, Mt. Banahaw, Mt. Cetaceo, Mt. Labo, Mt. Natib, Mt. Palay-palay, Mt. Pulag, and Saddle Peak, as well as in Caramoan and Peñablanca National Parks and in the Tayabas Caves area. They are often consumed locally, and sometimes are sold in markets.

REFERENCES: Danielsen et al., 1994; Garcia and Deocampo, 1995; Heaney et al., 1991, 1999, 2004, 2006b; Lucchini et al., 2005; Mijares et al., 2010; Mona et al., 2007; Oliver, 1992; Oliver and Heaney, 2008; Oliver et al., 1993b; Piper et al., 2009; Rabor, 1986; Scheffers et al., 2012.

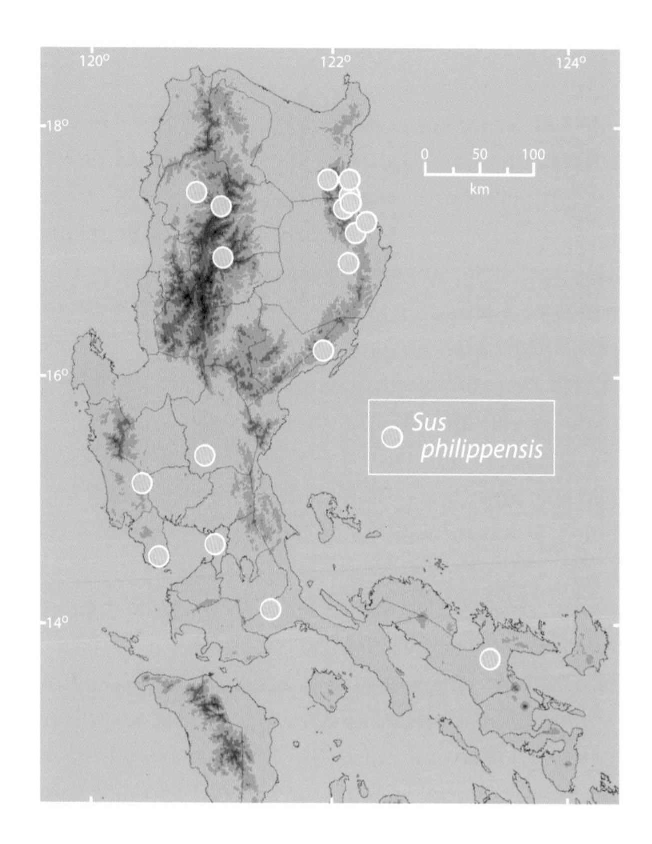

Cervus mariannus
Philippine brown deer, Philippine deer

Desmarest, 1822. Mammalogie. In Encycl. Meth., 2:436. Type locality: Mariana Islands.

DESCRIPTION: The only deer native to Luzon. The pelage of adults is medium to dark brown, with a coarse texture; fawns have pale spots on the back and sides, but adults rarely show such spots. Antlers are present only on adult males; they are shed and regrown each year.

DISTRIBUTION: Originally restricted to the Philippines, but introduced into the Mariana (Guam, Rota, and Saipan) and Caroline (Pohnpei) Islands. They occurred throughout Greater Luzon, Greater Mindanao, and Mindoro until recently, but are now highly restricted in their distribution.

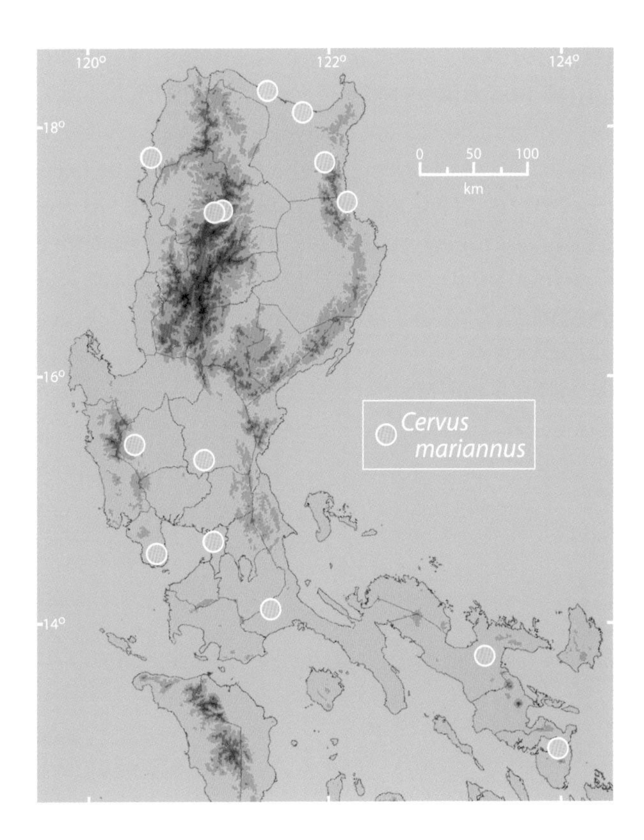

EVOLUTION AND ECOLOGY: Member of the subgenus *Rusa*, which sometimes is recognized as a distinct genus. *Cervus alfredi*, from Greater Negros-Panay, is closely related. Fossil remains of *C. mariannus*, dated to the late Pleistocene (ca. 50,000–68,000 years ago), were discovered at an archeological site in the Cagayan Valley, along with the remains of *Sus philippensis*, *Bubalus* (an extinct species of dwarf water buffalo), and some native small mammals. This species has also been found in a Neolithic midden (ca. 4500 BP), along with remains of both *Sus philippensis* and the first recorded remains of domestic pigs (*S. scrofa*) in the Philippines. Substantial variation in size and color exists, even within limited areas; further taxonomic and genetic study is needed. The Philippine brown deer formerly occurred from sea level to at least 2900 m elevation in primary and secondary forest. They are currently common in montane and mossy forest at 925–2750 m elevation in remote parts of the Central Cordillera (e.g., Balbalasang, Kalinga Province) and the Sierra Madre (e.g., Mt. Cetaceo and Mt. Anacuao) on Luzon, and in the high mountains of Mindanao. They are frequently associated with grassy clearings in forest, and local people often set fires on mountainsides to encourage the growth of the grassy and herbaceous vegetation that is eaten by the deer. They are active mostly at night; their loud barking calls are often heard near dawn and dusk.

STATUS: Locally common in remote, isolated areas, but heavily hunted and declining over most of their range. Captive individuals are sometimes kept in backyard farms or in deer farms. Local extinctions have been reported on Catanduanes and Biliran Islands. At the times of our surveys, local hunters reported that they were locally extinct on Mt. Data, Mt. Palay-palay, and Mt. Palalli, and in Caramoan National Park; declining and scarce on Mt. Amuyao, Mt. Banahaw, Mt. Cetaceo, Mt. Labo, Mt. Natib, and Saddle Peak; and were moderately common on Mt. Bali-it, Mt. Irid, Mt. Pinatubo, Mt. Pulag, and Mt. Tapulao. They are a common, non-native pest species on the Marianna and Caroline Islands, where they were introduced.

REFERENCES: Balete et al., 2013a, 2013b; Danielsen et al., 1994; Grubb and Groves, 1983; Heaney et al., 1991, 1998, 1999, 2004, 2006b, 2013a, 2013b; Oliver et al., 2008; Piper et al., 2009; Pitra et al., 2004; Rabor, 1986; Rickart et al., 1993; Sanborn, 1952; Scheffers et al., 2012; Taylor, 1934; Wiles et al., 1999.

10

A Guide to the Small Mammals

Of the 122 species of mammals now known to live on Luzon, 57 are bats, making them the most diverse group overall. The small mammals that do not fly, however, amount to almost as many. There are 48 native species by our latest count (and the number is rising) and 7 species that are not native, for a total of at least 55 non-volant small mammals on Luzon. The origin of this diversity is mostly a story of rare colonization events followed by hugely successful diversification (chapter 5). To best reflect this fundamental aspect of Philippine small mammal diversity, this chapter presents the following species accounts, grouped by their history.

Section 1 concerns the shrews (Family Soricidae). They arrived recently (in geological terms), and speciation for them has resulted primarily from their having slowly spread out over the archipelago, as recounted by Jacob Esselstyn (pp. 64–65). Only two species are present on Luzon: one native and one non-native. They play an interesting but minor role in the story of small mammal diversity on Luzon.

A far greater and more complex story is evident among the other small mammals: members of the Family Muridae (referred to here as murids), widely known as "mice" and "rats," though few of the native species look or behave like anyone's idea of a rat or mouse. Only the non-native species conform to those traditional notions of rats and mice, and all of them live with and near people. The vast majority of the native murids live in natural habitats and rarely interact with humans—yet they are part of one of the most interesting stories that can be told about mammalian diversity in the Philippines.

Sections 2 and 3 provide essential information on the two groups of Old Endemic murid species—the descendents of just two species that arrived in the Philippines roughly 14 and 8 million years ago and have radiated into the highly diverse habitats on Luzon (chapter 5). The first of these, the giant cloud rats and their smaller relatives, are familiar to many rural people on Luzon as well as to some suburban people, because the two large species that live in the lowlands and can readily coexist with humans (and are often hunted for their meat). The second, the earthworm mice, are small and less conspicuous, but they are even more diverse, despite their dependence on one primary and distinctive source of food that gives them their name—earthworms.

Section 4 presents a summary of the New Endemic species of murids. They are all native species, and most are members of endemic genera, but these particular five genera most likely represent four separate colonizations from outside the Philippines (chapter 5). Their ancestors arrived in the Philippines far more recently than the Old Endemics—perhaps 2–4 million years ago—and they have undergone far less diversification. Unlike the Old Endemics, which have their primary diversity high in the mountains, the New Endemics tend to occur in the lowlands.

Section 5 includes the non-native murid species: the animals accidentally brought by humans from Asia. They are the pests of homes and crops that are all too familiar to most people. These six species each have distinctive ecologies, living in rather different ways with humans and the habitats we create. Their interactions with the native small mammals are described in chapter 6, and it is worth emphasizing that on Luzon, when it comes to competition between the native and non-native species, it is the native species that win when the habitat looks even marginally like natural forest.

In the descriptions of species that follow, we emphasize the means to distinguish between species that occur in a given part of Luzon. For example, when discussing a mouse from the Mingan Mountains, we provide key features that allow the reader to distinguish that species from other mice from the same area. For comparisons of species that are closely related but live in different parts of Luzon, we refer the reader to the technical descriptions of the species that have been published elsewhere, which are listed in each species account.

SECTION 1

Order Soricomorpha: Shrews and Their Relatives
Family Soricidae: Shrews

The Order Soricomorpha is one of the largest and most diverse among the Mammalia, with 4 families and well over 400 species. The most diverse and widespread among these is the Family Soricidae, with 26 genera and more than 375 species that occur on every continent except Australia and Antarctica. Only two species live on Luzon, however. One is a non-native species that lives principally in and around human structures (*Suncus murinus*), and the other lives in forested and brushy regions throughout Luzon (*Crocidura grayi*).

Crocidura grayi.

Nine additional species of *Crocidura* live elsewhere in the Philippines. Most of them are the descendants of species that entered the Philippines from the south about 3–4 million years ago and diversified as they spread through the islands. One other species lives in the Batanes Islands and has not changed substantially since it arrived in the much more recent past, either from Taiwan or southern China. All of these shrews are small animals, many of them inconspicuous and rarely seen but often abundant in both mature and regenerating forest, feeding on insects and other invertebrates.

REFERENCES: Esselstyn and Goodman, 2010; Esselstyn et al., 2009, 2011; Hutterer, 2005.

Crocidura grayi
Luzon shrew

Dobson, 1890. Ann. Mag. Nat. Hist., ser. 6, 6:494. Type locality: Luzon Island.

DESCRIPTION: Total length 125–149 mm; tail 47–64 mm; hind foot 13–17 mm; ear 8–10 mm; weight 8.5–12 g. The smallest native non-flying mammal on Luzon (although several bats weigh less). Like all *Crocidura*, they have tiny eyes; a long, slender, flexible snout; and ears that are partly hidden within the fur. The fur is dark gray or brown, slightly paler ventrally than dorsally. The basal half of the tail has scattered long hairs. *Suncus murinus* are similar but larger (total length more than 170 mm; weight 26 g or more). *Mus musculus* are similar in size (total length 151–172 mm; weight 12–21 g) but the tail is longer (79–91 mm) and lacks long, scattered hairs; the eyes are larger; the ears are longer (11–14 mm) and project far beyond the fur; the snout is not flexible; and there is a single pair of upper and lower incisors. Members of the genera *Archboldomys* and *Soricomys* are superficially shrew-like rodents, with a total length of more than 167 mm and a tail more than 60 mm long.

DISTRIBUTION: Occur only on Greater Luzon and Mindoro, where they are widespread in hilly and mountainous regions, and on Calayan Island in the Babuyan Islands (north of Luzon). The Mindoro population is sometimes recognized as a separate subspecies, *C. grayi halconus*.

EVOLUTION AND ECOLOGY: Current genetic evidence suggests that *Crocidura grayi* reached Luzon about 2 million years ago. They are most closely related to *C. beatus*, which occur on Mindanao, Leyte, Samar, and associated islands. *Crocidura grayi* are present in lowland, montane, and mossy forest (including karst and ultramafic soil areas), from near sea level to at least 2700 m elevation. Abundance varies greatly, though on any given mountain they are usually most abundant at medium to high elevations in montane and mossy forest. They are present in primary and secondary forest, but they are often most common beside trails or in disturbed patches in the forest. This species often occurs together with the endemic shrew-like rodents of the genera *Archboldomys* and *Soricomys*, but in such places, *C. grayi* tend to be more abundant in more-disturbed habitat, and the shrew-mice in the less-disturbed habitat. Luzon shrews are sometimes present in areas with mixed agriculture and regenerating shrubby vegetation at middle or high elevations, but they are absent

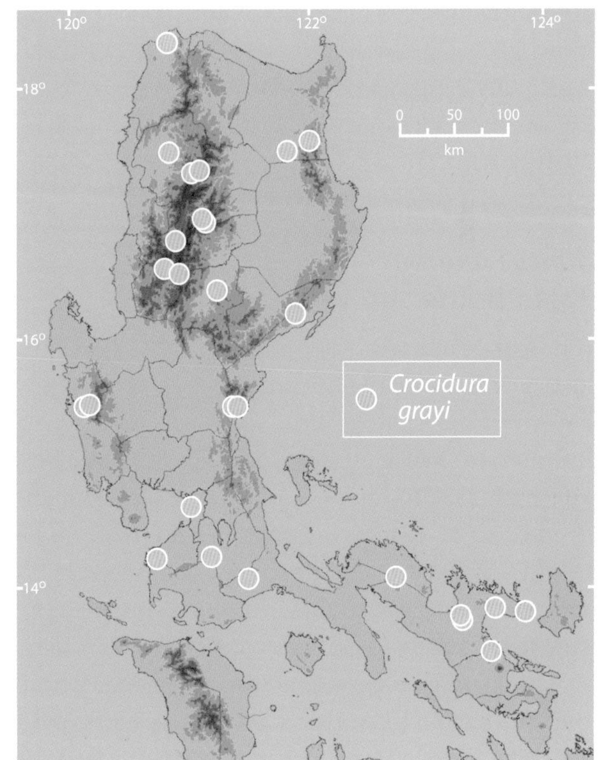

from such areas at low elevation. They rarely co-occur with *Suncus murinus*, which tend to be associated with heavy disturbance, but *S. murinus* sometimes are present in mature forest when *C. grayi* are absent. Of five pregnant females, four had one embryo and one had two; average litter size (from embryo and placental-scar counts) is about 1.2. They are active both day and night, and they feed primarily on small invertebrates, including earthworms, insects, and spiders. When population levels are high, they can be seen foraging along the surface of the ground, and their high-pitched squeaks can sometimes be heard. They produce a distinctive, musky odor that can be detected by some humans. We have sometimes found their bodies on trails, when they have been killed but then discarded by predators. The karyotype of specimens from Mt. Isarog is $2n = 38$, FN = 58.

STATUS: Stable; a widespread and common species.

REFERENCES: Alviola et al., 2011; Balete et al., 2011, 2013a, 2013b; Duya et al., 2007, 2011; Esselstyn and Brown, 2009; Esselstyn et al., 2009, 2011; Heaney and Ruedi, 1994; Heaney et al., 1991, 1999, 2004, 2010, 2013a, 2013b; Rickart, 2003; Rickart et al., 1991, 2011a, 2011b, 2013.

Suncus murinus
Asian house shrew

(Linnaeus, 1766). Syst. Nat., 12th ed., 1:174. Type locality: Java, [Indonesia].

DESCRIPTION: Males are larger than females. Males: total length 177–195 mm; tail 58–69 mm; hind foot 20–21 mm; ear 14–15 mm; weight 30–49 g. Females: total length 174–182 mm; tail 54–58 mm; hind foot 18–19 mm; ear 12–14 mm; weight 26–33 g. Like other Philippine shrews, Asian house shrews have tiny eyes; a long, slender, flexible snout; and ears partially hidden within the fur. The fur is dark gray dorsally and slightly paler ventrally. Scattered long hairs occur over much of the length of the tail, which tapers visibly from the base to the tip. *Crocidura grayi* are similar but much smaller (total length less than 150 mm; weight 12 g or less). House shrews are sometimes mistaken for house mice (*Mus musculus*), which are smaller (total length 151–172 mm; weight 12–21 g); have larger eyes; and lack visible hair on the tail.

DISTRIBUTION: Throughout most of Asia and Indo-Australia; throughout the Philippines, especially in anthropogenic habitats.

EVOLUTION AND ECOLOGY: Not native to the Philippines, probably having arrived along with people at an unknown time within the past several thousand years. They are abundant in urban and agricultural areas throughout the Philippines, often inside houses, where they may be seen at night as they pursue insect pests. They are usually absent from forest, but on islands and isolated mountains with few native mammals (including Mt. Irid, Mt. Labo, and Mt. Natib on Luzon), they may occur in disturbed and primary forest, especially in montane and mossy forest. They are found from sea level to at least 1690 m elevation on Luzon. When both *Crocidura grayi* and *Suncus murinus* are present in an area, *S. murinus* tend to be in heavily disturbed areas and/or buildings, and *C. grayi* in forest.

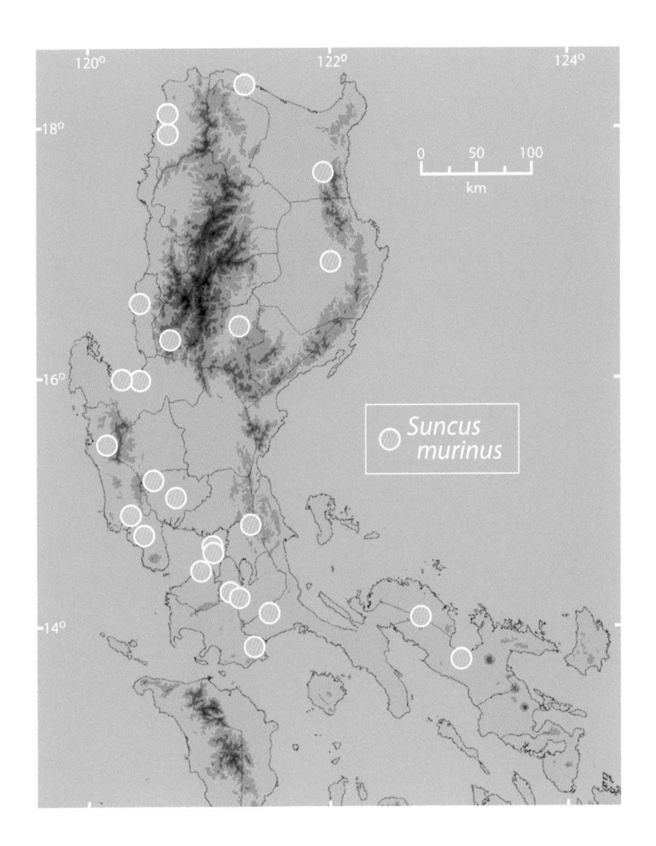

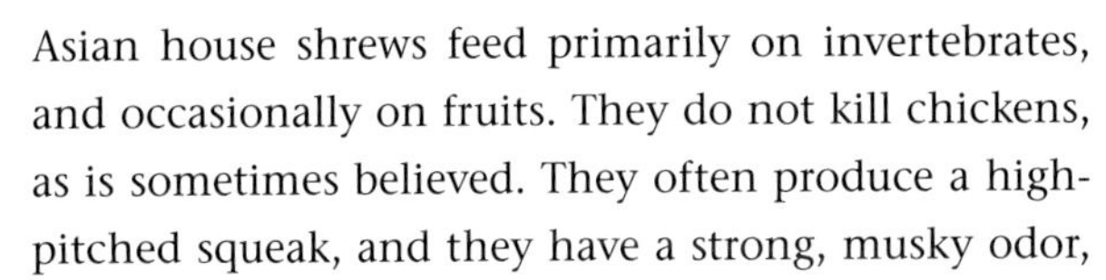

Asian house shrews feed primarily on invertebrates, and occasionally on fruits. They do not kill chickens, as is sometimes believed. They often produce a high-pitched squeak, and they have a strong, musky odor,

Suncus murinus, photographed on Negros Island. Courtesy of Paul Heideman.

making them conspicuous when they are present in a house. They usually are active only at night. Females typically become pregnant after they have reached a weight of about 28 g; typical litter size is about three.

STATUS: Abundant non-native species.

REFERENCES: Alviola et al., 2011; Balete et al., 2009, 2013a; Esselstyn et al., 2004; Heaney et al., 1989, 1998, 1999, 2006c, 2013b; Ong et al., 1999; Rabor, 1977, 1986; Reginaldo and de Guia, 2014; Reginaldo et al., 2013; Rickart et al., 1993, 2013.

SECTION 2

Old Endemics: Cloud Rats

One of the most remarkable instances of evolutionary diversification among Philippine mammals is found in the cloud rats. They range in size from 2.7 kg (six pounds) to 15 g (about half an ounce), representing both the largest and smallest species of native rodents on Luzon. They feed on a range of foods, from bamboo shoots and tender young leaves to fruits and seeds; all of them appear to be entirely herbivorous. Most are strikingly colored in some fashion: some are rusty reddish-orange, others are pure black, and still others have contrasting areas of black and white.

The ancestor of this group arrived in the Philippines roughly 14 million years ago (chapter 5) and since then has given rise to 5 genera and 18 species (12 on Luzon and at least 6 elsewhere in the Philippines). Only a few of the species live in the lowlands, but from four to six species in this group can be found living together (sympatrically) within a single area of cloud forest in the Central Cordillera. As such, they are major contributors to mammalian diversity on Luzon, both in terms of total species and in the number of species found in a single place.

The most recently discovered genus in this group is the smallest: the species of *Musseromys* weigh only 15–

A: *Carpomys melanurus*. B: *Batomys granti*. C: *Phloeomys pallidus*. D: *C. phaeurus*. E: *Crateromys schadenbergi*. F: *Musseromys gulantang*.

22 g. The first species was described in 2009, and the three others in 2014. Their heads are broad, and their jaws heavily muscled; we think all of them feed on nuts with thick, hard coats that cover seeds high in energy and nutrients. These tiny mice live in the forest canopy, climbing on vines and through dense vegetation at night. To aid them in this activity, they have vibrissae (i.e., whiskers) on the sides of the snout that are extraordinarily long; those of *M. gulantang* reach nearly to their ankles. Each species also has a second set of vibrissae that grows out of a patch of bare skin behind each eye, helping to create a sensory sphere that surrounds each mouse and probably allowing it to move through the canopy on the darkest of nights.

Their closest relatives are the dwarf cloud rats (*Carpomys*), which are beautiful animals that live in the mossy forest's canopy in the high mountains of the Central Cordillera. Though the two known species of *Carpomys* were discovered and described long ago, we "rediscovered" both in recent years, one after a period of 57 years without being seen by a scientist, and the other after 112 years.

Species of *Crateromys* and *Phloeomys* that have been maintained in zoos often live for a decade or more. This substantial longevity is matched by a low rate of reproduction: our limited data indicate that all of the cloud rats species, from largest to smallest, have only one or two babies per litter and have only one litter per year. These surely must be among the murids that are least similar to the rats and mice that live with humans, in terms of their appearance, their food, and their reproductive patterns. Five of the species on Luzon have been discovered only recently, and all remain rather poorly known in many aspects of their behavior and ecology—an excellent opportunity for further research on a remarkable set of animals.

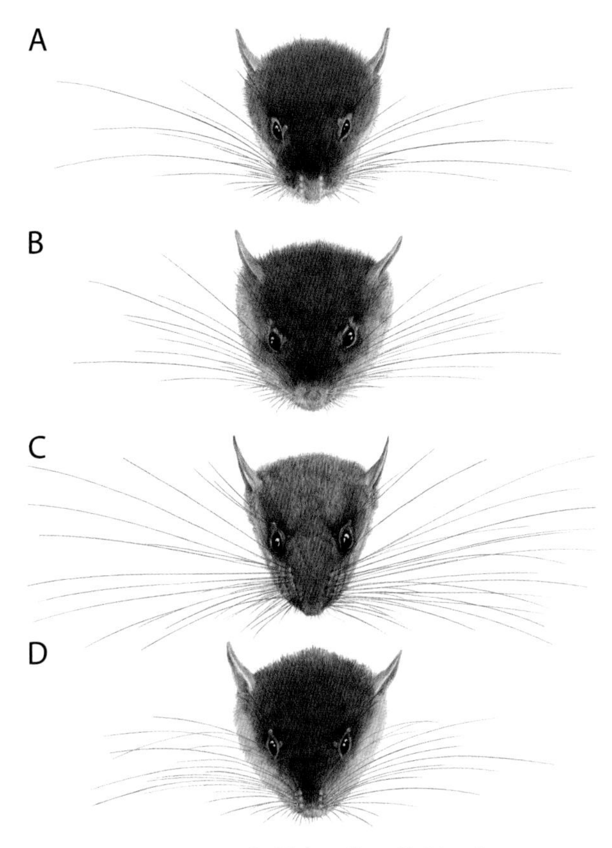

A: *Musseromys anacuao*. B: *M. beneficus*. C: *M. gulantang*. D: *M. inopinatus*.

Batomys dentatus
large-toothed batomys

Miller, 1910. Proc. U.S. Natl. Mus., 38:400. Type locality: Haights-in-the-Oaks, 7000 feet elevation, Mountain Province [= Atok Municipality, Benguet Province], Luzon.

DESCRIPTION: Total length 380 mm; tail 185 mm; hind foot 36 mm; weight unknown. A moderately large rat, with long, dense, slightly woolly fur that is dark brown dorsally and somewhat paler (buffy) on the underparts. The tail is slightly shorter than the length of the head and body and is hairy along its entire length. The anterior (basal) two-thirds of the tail is dark brown, and the posterior third of the tail (on the one known specimen) has unpigmented (white) skin and hair. The vibrissae are unusually long, extending beyond the ears. There is a narrow ring of bare skin around the eye. *Batomys granti* are similar but smaller in all dimensions; have somewhat shorter and less woolly fur; rarely have any white on the tail (when it is present, it is only at the very tip); and have vibrissae that do not extend beyond the ears. Both species of *Carpomys* have a short (34 mm or less) and broad hind foot; a short, broad head; and yellowish-brown fur dorsally. *Bullimus luzonicus* and *Rattus everetti* are substantially larger; both have scaly tails with unpigmented tips and with virtually no hair visible.

DISTRIBUTION: Known only from the type locality in Atok Municipality, Benguet Province, Luzon.

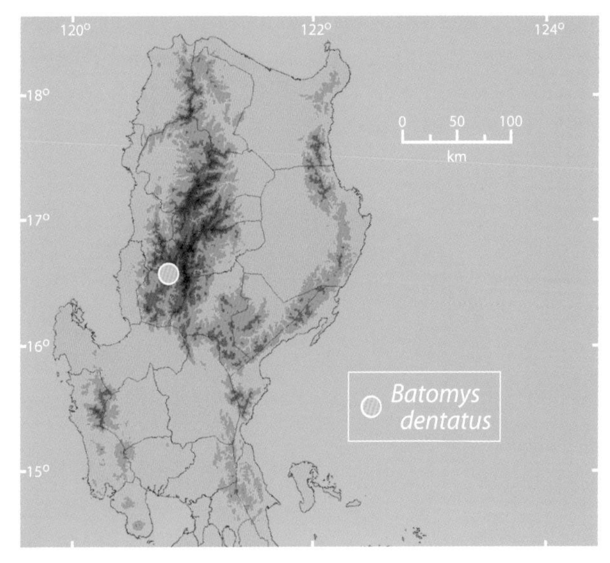

EVOLUTION AND ECOLOGY: Very poorly known in all respects. The single known specimen was taken at ca. 2100 m elevation, probably in mossy oak forest, on the rolling plateau where the type locality is situated. We did not encounter this species in our studies elsewhere in the Cordillera, where we documented *Batomys granti*.

STATUS: Unknown. Targeted surveys in the Central Cordillera are needed.

REFERENCES: G. Miller, 1910; Musser et al., 1998.

Batomys granti
Cordillera hairy-tailed rat, Cordillera batomys

Thomas, 1895. Ann. Mag. Nat. Hist., ser. 6, 16:162. Type locality: "Highlands of Northern Luzon," subsequently defined by Thomas (1898) as the plateau of Mt. Data, 7000–8000 feet elevation, Lepanto [now in Mountain Province], Luzon Island.

DESCRIPTION: Total length 342–355 mm; tail 148–163 mm; hind foot 35–39 mm; ear 22–23 mm; weight 182–226 g. A medium-sized rat, with dense, soft fur that is dark brown on the upperparts, often with a reddish tint, and paler brown on the underparts. The tail is slightly shorter than the length of the head and body and is covered entirely with short brown hairs. There is a narrow ring of bare skin surrounding the eye. The vibrissae do not extend beyond the ears. *Batomys dentatus* are larger (total length 380 mm); have slightly longer and woolly fur; have vibrissae that extend beyond the ears; and have white skin and hair on the posterior one-third of the tail. Both species of *Carpomys* have a short (less than 35 mm) and broad hind foot; a broad head, with a short snout; and weigh less (less than ca. 165 g). Other rodents of similar size in the Cordillera have scaly tails, with virtually no visible hair on the tail.

DISTRIBUTION: Fairly widespread in the Central Cordillera of northern Luzon. Previous reports from Mt. Isarog refer to a separate species, *Batomys uragon*, now recognized as distinct from *B. granti*.

EVOLUTION AND ECOLOGY: The genus *Batomys* is most closely related to *Crateromys*. It is clear that *B. uragon* from Mt. Isarog is the closest relative of *B. granti*, but relationships within the genus are much in need of

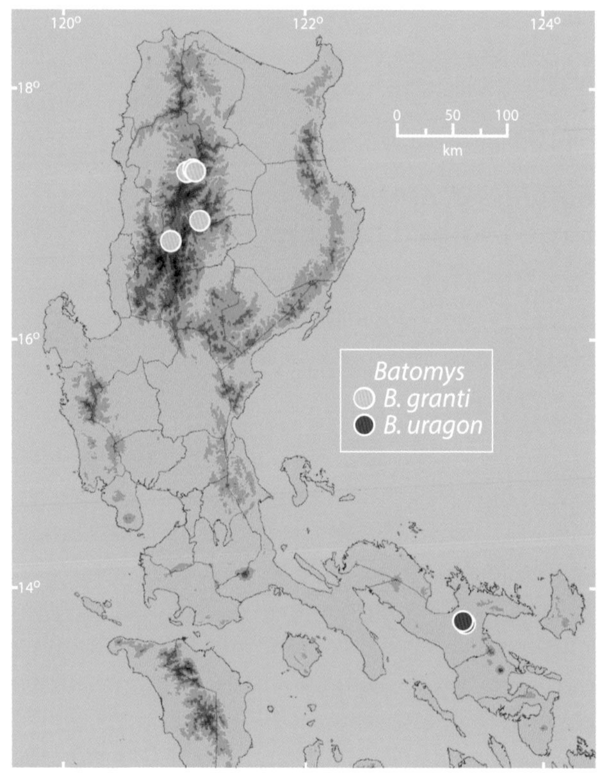

study, especially the species from Greater Mindanao. *Batomys granti* occur from 1600 m to 2530 m elevation, in primary and secondary montane and mossy forest, and are usually more common in mossy forest than in montane forest. They appear to be absent from the highest peaks (e.g., above 2600 m elevation, as on Mt. Pulag). They feed on plant material (including seeds, fruits, and perhaps leaves), foraging at night both on the ground and in trees (up to at least 5 m above the ground). We captured one individual in a leaf-nest built into the base of a deep pit. Females have two pairs of inguinal mammae. The standard karyotype is $2n$ = 52, FN = 52, and is indistinguishable from that of *B. uragon* from Mt. Isarog.

STATUS: Widespread and moderately common at higher elevations in the Central Cordillera.

REFERENCES: Heaney et al., 2004, 2011b; Jansa et al., 2006; Musser et al., 1998; Rabor, 1955; Rickart and Heaney, 2002; Rickart et al., 2011a, 2011b; Sanborn, 1952; Thomas, 1898.

Batomys uragon
Mt. Isarog hairy-tailed rat, Mt. Isarog batomys

Balete et al., 2015. Proc. Biol. Soc. Washington, 128. Type locality: Mt. Isarog, 4 km N, 22 km E Naga City, 1750 m elevation, Camarines Sur Province, Luzon, 13°40′ N, 123°22′ E.

DESCRIPTION: Total length 315–356 mm; tail 128–171 mm; hind foot 35–39 mm; ear 20–22 mm; weight 135–200 g. A medium-sized rat, with dense, soft fur that is dark brown on the upperparts, often with a reddish tint, and paler brown on the underparts. The tail is slightly shorter than the length of the head and body and is entirely covered with short brown hairs. There is a narrow ring of bare skin surrounding the eye. The vibrissae do not extend beyond the ears. *Chrotomys gonzalesi* have a mid-dorsal pale (nearly white) stripe flanked by two broad dark (nearly black) stripes. *Rattus everetti* have a long, scaly tail that has virtually no visible hair. Other rodents on Mt. Isarog differ greatly in size.

DISTRIBUTION: Known only from Mt. Isarog, Camarines Sur Province. (For map, see *Batomys granti* species account.)

EVOLUTION AND ECOLOGY: Specimens of *Batomys* from Mt. Isarog were formerly included in *B. granti*, but our recent studies have shown them to be distinct. They occur from 1350 m to 1800 m elevation in primary montane and mossy forest, and are more common in mossy forest than in montane forest. They feed on plant material (including seeds and perhaps leaves); stomachs contained finely masticated plant matter, and a captive individual accepted seeds and rejected earthworms and insects. They forage at night, often in well-defined runways on the ground, usually on steep terrain. A density of 3.0 individuals/ha was documented on Mt. Isarog at ca. 1700 m elevation. *Apomys musculus* and *Archboldomys luzonensis* were both more numerous there, but the estimated biomass for *B. uragon* (591 g/ha) was the highest for small mammals at the study site. Home-range size for one female, based on a small number of captures, was at least 0.11 ha. Females have two pairs of inguinal mammae. This species has a standard karyotype of $2n$ = 52, FN = 52 (reported as *B. granti*).

STATUS: Moderately common on Mt. Isarog, in habitat that is within a national park and largely undisturbed.

REFERENCES: Balete and Heaney, 1997; Balete et al., 2015; Heaney et al., 1999; Rickart et al., 1991; Rickart and Musser, 1993.

Carpomys melanurus
greater dwarf cloud rat, black-tailed dwarf cloud rat

Thomas, 1895. Ann. Mag. Nat. Hist., ser. 6, 16:162. Type locality: "Highlands of Northern Luzon," subsequently defined by Thomas (1898) as the plateau of Mt. Data, 7000–8000 feet elevation, Lepanto [now in Mountain Province], Luzon Island.

DESCRIPTION: Total length 360–367 mm; tail 180–183 mm; hind foot 32–34 mm; ear 20 mm; weight 165 g. A beautiful dwarf cloud rat, with a broad, blunt head; dark fur around the eyes gives the appearance of a mask. The tail is long and covered by short, nearly black hair; the hind feet are rather short and broad. The fur is soft, dark brown on the back and nearly white on the abdomen. *Carpomys phaeurus* are smaller in all dimensions (total length less than 340 mm); the hair on the tail is dark reddish-brown. *Batomys dentatus* have a longer (36 mm), narrower hind foot; the last third of the tail is white. *Batomys granti* have a longer (over 35 mm) and narrower hind foot; a prominent band of pale skin around the eye; and are larger (weight 182–226 g).

DISTRIBUTION: Known only from Mt. Pulag (Benguet Province) and Mt. Data (Mountain Province) in the Central Cordillera of northern Luzon. Trapping in apparently suitable habitat elsewhere on Luzon has not yielded any captures.

EVOLUTION AND ECOLOGY: *Carpomys* is most closely related to *Musseromys*, with which it last shared a common ancestor about 4 million years ago. These two genera are most closely related to *Batomys* and *Crateromys*, and last shared a common ancestor with them about 4 million years ago. Their ecology is poorly known. The two known specimens are from mossy forest, from about 2200 m to 2300 m elevation. The specimen from Mt. Pulag was captured in a trap placed on top of a moss-covered branch, ca. 10 cm in diameter and 4 m above the ground, on which a well-worn trail was visible. They are probably nocturnal and primarily arboreal, probably feeding on seeds. Females have two pairs of inguinal mammae and an elongate urinary papilla that looks like a penis sheath. Our recent surveys on Mt. Data failed to document this species, suggesting that it may be locally extinct, due to habitat destruction.

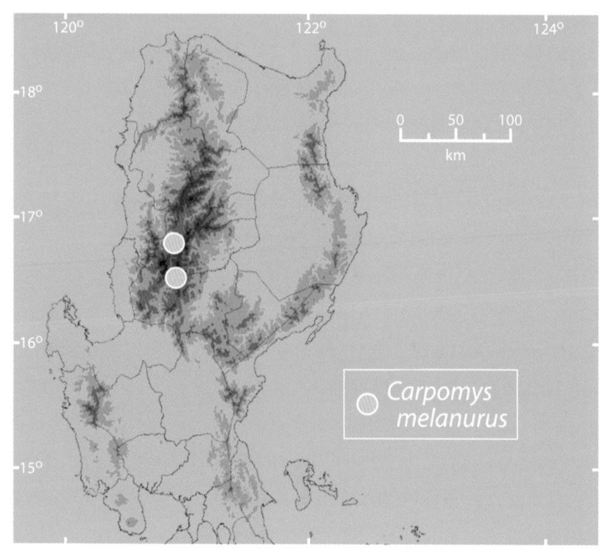

STATUS: Unknown; may be locally common, but also may be limited in distribution. Focused studies are needed.

REFERENCES: Heaney et al., 2006b; Largen, 1985; Rickart et al., 2011b; Thomas, 1898.

Carpomys phaeurus
lesser dwarf cloud rat, brown-tailed dwarf cloud rat

Thomas, 1895. Ann. Mag. Nat. Hist., ser. 6, 16:162.
Type locality: "Highlands of Northern Luzon," subsequently defined by Thomas (1898) as the plateau of Mt. Data, 7000–8000 feet elevation, Lepanto [now in Mountain Province], Luzon Island.

IDENTIFICATION: Total length 322–336 mm; tail 156–161 mm; hind foot 31–32 mm; ear 19 mm; weight 123 g. A beautiful dwarf cloud rat, with a broad, blunt head; a trace of dark fur around the eyes gives the appearance of a mask. The tail is long and covered by short, dark reddish-brown hairs; the hind feet are rather short and broad. The fur is soft, dark brown on the back and nearly white on the abdomen. *Carpomys melanurus* are larger in all dimensions (total length over 360 mm) and have a longer tail (over 180 mm) that is covered by nearly black hair. *Batomys dentatus* are larger (total length ca. 380 mm) and have a longer tail (ca. 180 mm) that is dark at the base and white on the last third. *Batomys granti* have a longer (35 mm or more) and narrower hind foot; adults weigh more (over ca. 180 g).

DISTRIBUTION: Widespread in the Central Cordillera of northern Luzon.

EVOLUTION AND ECOLOGY: The genus *Carpomys* includes two species that are similar in appearance, differing most conspicuously in size. Both were captured on Mt. Data in the 1890s, but there are so few specimens that the extent of their geographical overlap and the differences in ecology between them are unknown. The few known specimens were captured in mossy forest at ca. 2150 m to 2500 m elevation; we captured one on the trunk of a large (DBH = ca. 1 m; ca. 9 m tall) moss-covered oak tree, ca. 5 m above the ground, in mossy forest. They are probably nocturnal and primarily arboreal, probably feeding on seeds. Whitehead (in Thomas, 1898) described local people digging them out from nests among tree roots. Females have two pairs of inguinal mammae.

STATUS: Poorly known, but seemingly widespread in mossy forest, a habitat that is extensive but declining slowly from urbanization and agriculture (chapter 7).

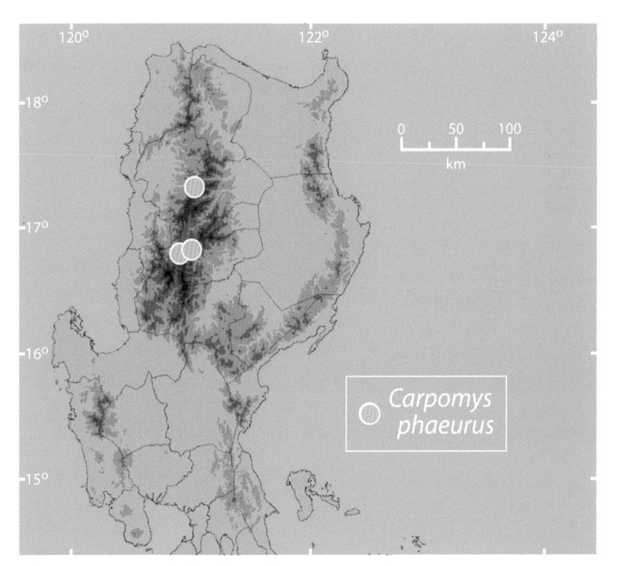

The population on Mt. Data may be extinct. Focused studies are needed, using relevant capture techniques.

REFERENCES: Heaney et al., 2004, 2006b; Largen, 1985; Rabor, 1955; Rickart et al., 2011a; Sanborn, 1952; Thomas, 1898.

Crateromys schadenbergi
Luzon bushy-tailed cloud rat

(Meyer, 1895). Abh. Mus. Dresden, 6:1. Type locality: Mt. Data, NW Luzon, Philippines.

DESCRIPTION: Total length 603–760 mm; tail 300–390 mm; hind foot 72–73 mm; ear 30–32 mm; weight 1.35–1.55 kg. A spectacular animal, with long, dense, soft, flowing fur covering the entire body. The pelage is usually entirely black, but some individuals have bands or patches of white fur on the head and forelimbs, some have white underfur, and some are mostly gray. They have a short, broad head, with powerful jaws for chewing. The hind feet are broad, relative to their length, and appear to be capable of grasping. The length of the tail varies from slightly to much longer than the length of the head and body. Like other members of the cloud rat–group, females have a long urinary papilla that looks like a penis sheath. *Phloeomys pallidus* are larger and stouter overall (weight up to 2.7 kg); have shorter, coarser fur; have shorter, somewhat bristly hair on the tail; and have a longer (80 mm or more) hind foot.

DISTRIBUTION: Known only from the central and southern portions of the Central Cordillera of northern Luzon; there are no reports from the northern portion.

EVOLUTION AND ECOLOGY: *Crateromys* is most closely related to *Batomys*; we currently estimate that they last shared a common ancestor about 4 million years ago. Four species of *Crateromys* are known, each from a different part of the Philippines. The Luzon species is morphologically most similar to *C. heaneyi* (from Panay), but details about relationships are currently

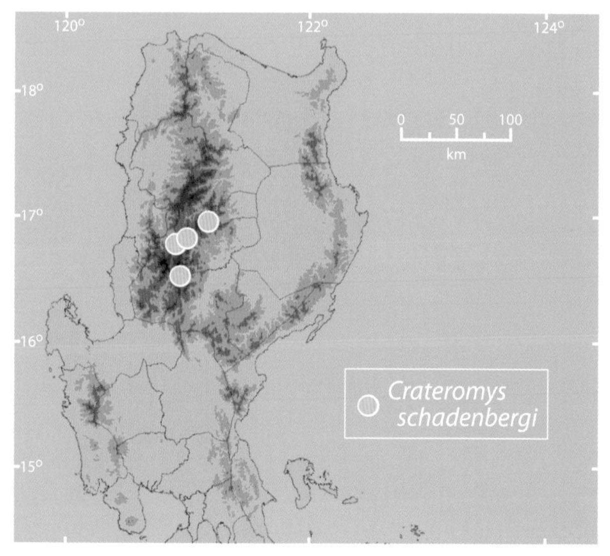

lacking. The ecology of the species is poorly known. They occur only from about 2000 m to 2740 m elevation in mossy forest (occasionally pine forest) in the central and southern Central Cordillera, but they have not been found in similar habitat in the northern Central Cordillera. They are nocturnal and primarily arboreal, and local hunters report that they feed on tender vegetation and buds. Females have two pairs of inguinal mammae and are reported by hunters to give birth once each year to one or two young. They are reported by hunters to sleep in cavities in trees or in holes among tree roots. They are hunted for their meat and fur and occasionally are kept as pets.

STATUS: Locally common in high-elevation oak forest, and occasionally found in adjacent pine forest. Hunting for this species is widespread, including in some places where they are common, such as Mt. Amuyao and Mt. Pulag. They are locally extinct on Mt. Data, having apparently disappeared by 1946. Detailed field studies are needed.

REFERENCES: Feiler, 1999; Heaney et al., 2006b; Oliver et al., 1993a; Rabor, 1955, 1986; Sanborn, 1952.

Musseromys anacuao
Sierra Madre tree-mouse

Heaney et al., 2014. Amer. Mus. Novitates, 3802:20. Type locality: 0.2 km E Mt. Anacuao peak, 1725 m elevation, Dinalungan Municipality, Aurora Province, Luzon Island, 16.25527° N, 121.88896° E.

DESCRIPTION: Total length 160–165 mm; tail 82–86 mm; hind foot 18 mm; ear 15–16 mm; weight 17–21 g. A small mouse, with dense, soft, rusty-brown fur dorsally and buffy reddish-brown fur ventrally. The head is broad and large, relative to the body; the rostrum is short and blunt. The vibrissae are long; a second, smaller set of vibrissae arise from a small area of bare skin posterior to the eye. The tail is equal to or slightly longer than the length of the head and body; there are long hairs for about 15 mm at the tip of the tail. The hind foot is broad, with large plantar pads. *Apomys musculus* are larger (total length 182–213 mm); have longer hind feet (20–25 mm); lack visible hair on the tail; and do not have a bare patch of skin with long vibrissae posterior to the eye.

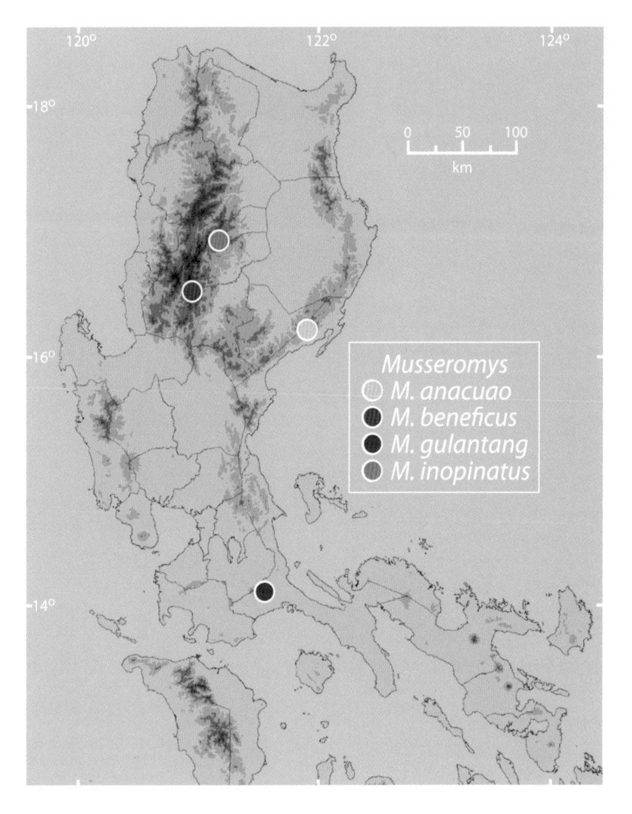

DISTRIBUTION: Currently known only from Mt. Anacuao, but should be sought at similar elevations elsewhere in the Northern Sierra Madre.

EVOLUTION AND ECOLOGY: Most closely related to *Musseromys beneficus*, which occur in the southern Central Cordillera. We captured them on steep slopes at 1760 m elevation on Mt. Anacuao, about 300 m from the peak, in mature mossy forest that was dominated by elaeocarps, oaks, laurels, myrtles, and coniferous podocarps that reached 7–12 m in height. The trees supported heavy loads of epiphytic moss, vines, orchids, ferns, and other small plants. All individuals were captured at night, from 1 to 3 m above the ground, on the mossy, leaning trunks of large trees, small branches, or hanging vines, using fried coconut with peanut butter as bait. We did not capture them in montane forest at 1500 m, 1300 m, or 1125 m elevation, or in lowland forest at 940 m elevation. *Apomys musculus* and *Rattus everetti* were captured in the canopy in the same area. Females have two pairs of inguinal mammae. One female captured in May had a single embryo.

STATUS: Uncertain because of insufficient data. Focused studies using appropriate arboreal trapping techniques are needed throughout the Northern Sierra Madre.

REFERENCES: Heaney et al., 2013a, 2014b.

Musseromys beneficus
Mt. Pulag tree-mouse

Heaney et al., 2014. Amer. Mus. Novitates, 3802:17. Type locality: Mt. Pulag National Park, 1.15 km S, 1.35 km E Mt. Pulag peak, 2695 m elevation, Benguet Province, Luzon Island, 16.58816° N, 121.90960° E.

DESCRIPTION: Total length 157–163 mm; tail 82 mm; hind foot 18 mm; ear 16 mm; weight 18–22 g. A small mouse, with long, dense, soft, dark rusty-brown fur dorsally and buffy reddish-brown fur ventrally. The head is broad and large, relative to the body; the rostrum is short and blunt. The vibrissae are long; a second, inconspicuous set of vibrissae arise from a small area of bare skin posterior to the eye. The tail is slightly longer than the length of the head and body; there are long hairs for about 25 mm at the tip of the tail. The hind foot is broad, with large plantar pads. *Musseromys inopinatus* are larger (weight 17–19.5 g), with a longer tail (85–88 mm) that is noticeably longer than the length of the head and body. *Apomys musculus* are similar in head and body length but have a longer tail (97–121 mm); have dark brown fur dorsally and nearly white fur ventrally; lack visible hair on the tail; and do not have a bare patch of skin with vibrissae posterior to the eye.

DISTRIBUTION: Currently known only from Mt. Pulag, but should be sought at similar elevations elsewhere in the southern Central Cordillera. (For map, see *Musseromys anacuao* species account.)

EVOLUTION AND ECOLOGY: Most closely related to the species from Mt. Anacuao in the Northern Sierra Madre, not to a species in the Cordillera. We captured the three known specimens at 2695 m elevation on Mt. Pulag, in mature mossy forest dominated by oaks, laurels, myrtles, and coniferous podocarps with heavy loads of epiphytic moss, orchids, ferns, and other small plants; canopy vines (*Smilax* sp.) were common. All individu-

als were captured in a single trap over a period of four days, set ca. 3 m above the ground on a branch thickly covered with moss. All were captured at night, using fried coconut with peanut butter as bait. *Apomys musculus* and *Crateromys schadenbergi* were captured in the canopy in the same area. An adult female had two pairs of inguinal mammae and a single uterine scar. An adult male (weight 22 g) had large (11 × 8 mm) scrotal testes.

STATUS: Uncertain because of insufficient data. Focused studies using appropriate arboreal trapping techniques are needed throughout the Central Cordillera.

REFERENCES: Heaney et al., 2014b; Rickart et al., 2011b, submitted.

Musseromys gulantang Banahao tree-mouse

Heaney et al., 2009. Bull. American Mus. Nat. Hist., 331:216. Type locality: S side Mt. Banahaw, 620 m elevation, Barangay Lalo, Tayabas Municipality, Quezon Province, Luzon Island, 14°3′6.5″ N, 121°32′22.5″ E.

DESCRIPTION: Total length 178 mm; tail 101 mm; hind foot 20 mm; ear 16 mm; weight 15.5 g. A small mouse, with rusty-orange fur dorsally and bright orange fur ventrally. The head is broad; has a short, broad muzzle; and is large relative to the size of the body. The vibrissae are unusually long, reaching nearly to the ankles; a second set of vibrissae arise from a patch of bare skin posterior to the eye. The ears are elongate and come to blunt tips; the tail is long, with long hairs near and at the tip; the hind foot is broad. *Apomys musculus* are similar in size but have dark brown fur dorsally and nearly white fur ventrally; lack visible hair on the tail; have a muzzle with the typical shape; and have vibrissae of the usual position and length.

DISTRIBUTION: Known only from Mt. Banahaw, Quezon Province, Luzon. (For map, see *Musseromys anacuao* species account.)

EVOLUTION AND ECOLOGY: The genus *Musseromys* is most closely related to *Carpomys*, with which it last shared a common ancestor about 4 million years ago. *Musseromys gulantang* is known from a single specimen captured in fairly mature second-growth lowland forest, at 620 m elevation. This species is one of the few cloud rats that live in lowland forest; all of this species' close relatives (*Carpomys*, *Batomys*, and *Crateromys*) on Luzon, including the three recently discovered species of *Musseromys*, occur only in montane and mossy forest. The single known specimen was caught at the point of intersection of two large vines that hung down from the canopy, about 1 m above the ground. We suspect that the species is nocturnal and largely arboreal. Its proportionately heavy jaw musculature implies that it chews on hard materials, and we found chewed

seeds of *Elaeocarpus* sp. (which have a thick, hard coat surrounding a rich, nutritious seed) beneath the area where the specimen was captured. This leads us to suspect that the species feeds primarily on seeds, perhaps especially on those with hard coats.

STATUS: Uncertain. The type locality is within Mts. Banahaw–San Cristobal National Park. Similar second-growth forest is common in nearby parts of Luzon. Detailed, focused field studies are needed.

REFERENCES: Heaney et al., 2009, 2013b, 2014b.

Musseromys inopinatus
Amuyao tree-mouse

Heaney et al., 2014. Amer. Mus. Novitates, 3802:14. Type locality: 1.0 km N, 1.0 km W Mt. Amuyao peak, 2150 m elevation, Barlig Municipality, Mountain Province, Luzon Island, 17.02213° N, 121.11791° E.

DESCRIPTION: Total length 163–166 mm; tail 85–88 mm; hind foot 18–19 mm; ear 17 mm; weight 17–19.5 g. A small mouse, with dense, soft, rusty-brown fur dorsally and buffy reddish-brown fur ventrally. The head is broad and large, relative to the body; the rostrum is short and blunt. The vibrissae are long; a second, rather inconspicuous set of vibrissae arise from a small area of bare skin posterior to the eye. The tail is longer than the length of the head and body; there are long hairs for about 10 mm at the tip of the tail. The hind foot is broad, with large plantar pads. *Musseromys beneficus* are larger (weight 18–22 g), with a shorter tail (82 mm) that is only slightly longer than the length of the head and body. *Apomys musculus* are similar in length of the head and body but have a longer tail (97–121 mm); dark brown fur dorsally and nearly white fur ventrally; lack visible hair on the tail; and do not have a small, bare patch of skin with vibrissae posterior to the eye.

DISTRIBUTION: Currently known only from Mt. Amuyao, but should be sought at similar elevations elsewhere in the southern Central Cordillera. (For map, see *Musseromys anacuao* species account.)

EVOLUTION AND ECOLOGY: One of four species of *Musseromys*, which speciated during the past 4 million years. We captured *M. inopinatus* from 1650 m to 2300 m elevation on Mt. Amuyao, in mature and lightly disturbed montane and mossy forest dominated by oaks, laurels, and myrtles, with coniferous podocarps included at 2300 m elevation. The trees have heavy loads of epiphytic moss, orchids, ferns, and other small plants. All individuals were captured from 0.5 to 5 m above the ground, on the horizontal trunks of trees, small branches, or hanging vines, at night or in the late afternoon. *Apomys musculus*, *Batomys granti*, and *Rattus everetti* were captured in the canopy in the same area. Females have two pairs of inguinal mammae.

STATUS: Uncertain because of insufficient data. Focused studies using appropriate arboreal trapping techniques are needed throughout the Central Cordillera.

REFERENCES: Heaney et al., 2014b; Rickart et al., 2011a, submitted.

Phloeomys cumingi
bugkun, southern Luzon giant cloud rat

(Waterhouse, 1839). Proc. Zool. Soc. Lond., p. 108.
Type locality: Luzon Island.

DESCRIPTION: Total length 671–752 mm; tail 274–314 mm; hind foot 74–85 mm; ear 34–37 mm; weight 1.85–2.05 kg. A very large rodent, with rather rough, dark brown fur (sometimes fading to rusty-red) over the entire body, including the tail. The eyes and ears are proportionately small. *Phloeomys pallidus* are larger (weight up to 2.7 kg) and have white or silvery fur over much or all of the body.

DISTRIBUTION: Luzon Faunal Region only. Recorded on Catanduanes and Marinduque Islands, the Bicol Peninsula, and in the Southern Sierra Madre and Mingan Mountains.

EVOLUTION AND ECOLOGY: The genus *Phloeomys* is the most basal member of the cloud rat clade; it diverged from the ancestor of the other cloud rat genera about 10 million years ago. There are two species in the genus, both occurring only in the islands of Greater Luzon; the two overlap in the southern Sierra Madre and Mingan Mountains, but no information is available on the details of their distribution or their ecology in these areas. They occur in mature and disturbed lowland forest, from sea level to nearly 1100 m elevation, and appear to be fairly common in many areas. Females have two pairs of inguinal mammae. Lactating females are usually associated with a single young; litters of two are reported to be uncommon. They nest in hollow trees (including strangler figs) and hollow logs. They are active only at night, foraging mostly in trees (but also sometimes on the ground) for herbaceous material, including tender young leaves, bamboo shoots, and flowers. They are reported to eat some agricultural crops, including sweet potatoes, corn [= maize], and sayote. They reportedly wander into suburban areas at times, but are often not seen because they are active only at night. They reportedly produce a guttural call that consists of a single note. In zoos, captives of either this species or *P. pallidus* have lived for nearly 14 years. *Phloeomys cumingi* have a standard karyotype of $2n = 44$, FN = 66.

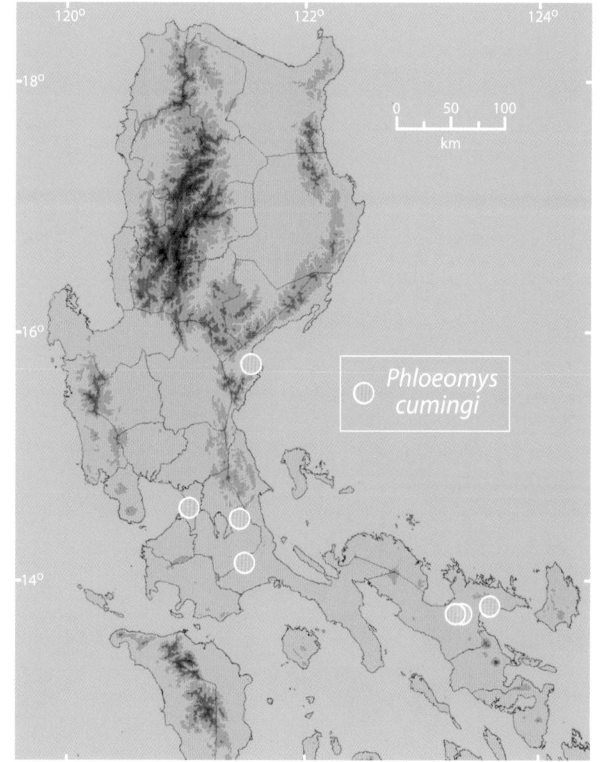

STATUS: Moderately widespread and common, but heavily hunted and subject to habitat destruction (Oliver et al., 1993a).

REFERENCES: Balete et al., 2013b; Heaney et al., 1991, 1999, 2013b; Nowak, 1999; Oliver et al., 1993a; Rickart and Musser, 1993; Rickart et al., 1991; Scheffers et al., 2012.

Phloeomys pallidus
bu-ot, northern Luzon giant cloud rat

Nehring, 1890. Sitzb. Ges. Naturf. Fr., Berlin, p. 106.
Type locality: Luzon Island.

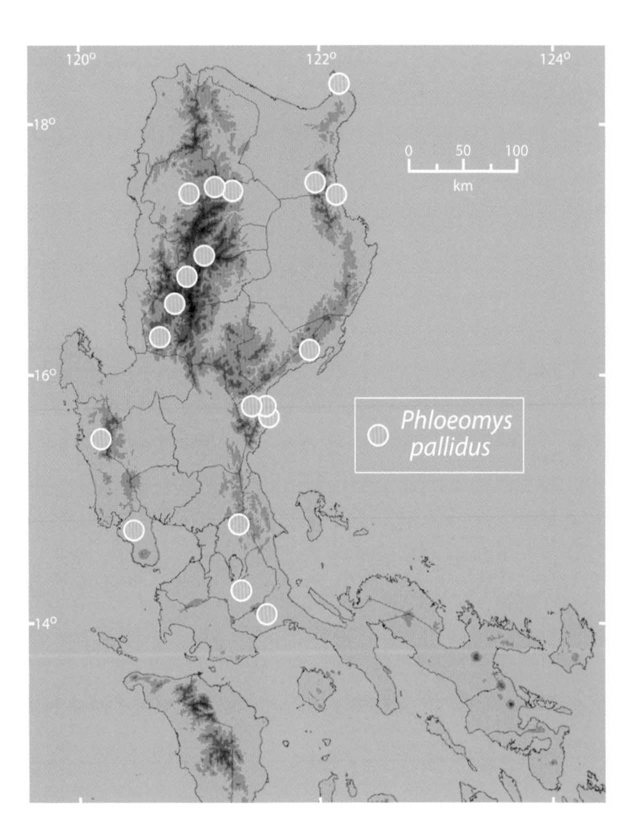

DESCRIPTION: Total length 707–770 mm; tail 320–349 mm; hind foot 80–90 mm; ear 34–39 mm; weight 2.2–2.7 kg. A very large, strikingly colored rodent, with a long, hairy tail. The fur color varies greatly, sometimes nearly entirely white but usually mostly white or very pale gray-brown, with some black or dark brown fur around the eyes, on the ears and tail, and sometimes on the neck and forelimbs. The hind feet are broad, with stout claws. *Phloeomys cumingi* are slightly smaller (weight 2.1 kg or less) and have dark brown and/or rusty-red fur over the entire body. *Crateromys schadenbergi* have long, soft, very dense fur (mostly black) covering the entire body, including the tail; are smaller (up to ca. 1.5 kg); and have a shorter (ca. 72 mm), proportionately broader hind foot.

DISTRIBUTION: Widespread in northern and central Luzon.

EVOLUTION AND ECOLOGY: The largest member of the subfamily Murinae. Their distinctive molar teeth, which are high-crowned and possess a series of "plates" surrounded by tough enamel, led early researchers to suggest that this species is distantly related to all other rodents, but recent DNA-based studies confirm that it is related to *Crateromys* and the other cloud rats. The species occurs from sea level to about 2300 m elevation, in primary and secondary lowland, montane, and mossy forest; in heavily disturbed scrub; and in some suburban areas with extensive vegetation. They are active at night, foraging in trees and on the ground, feeding on tender leaves, bamboo shoots, and other vegetation, probably including some agricultural crops. They nest in hollow trees and logs. Females have two pairs of inguinal mammae and usually give birth to a single young, though sometimes they have twins. A photograph from the Bronx Zoo from June 2014 showed a juvenile reported to be two weeks old that was fully furred, with eyes wide open, sitting upright. Where they occur with *C. schadenbergi*, *Phloeomys pallidus* are found in lowland and montane forest, whereas *C. schadenbergi* occur at higher elevations in mossy forest, but the two coexist in some places (e.g., Mt. Amuyao and Mt. Pulag). The two species of *Phloeomys* overlap in the Southern Sierra Madre and Mingan Mountains, but no data are available on differences in their habitat use. Local hunters report that they produce a loud, guttural call that consists of a single note, repeated after intervals of seconds to minutes. Captive *Phloeomys* (of uncertain species) are reported to live for nearly 14 years. This species has a karyotype of $2n = 40$, $FN = 60$ (reported as *P. cumingi*).

STATUS: Widespread and apparently common in primary and secondary forest over most of Luzon, but hunted in most areas, sometimes to local extinction.

REFERENCES: Balete et al., 2013a; Heaney et al., 2006b; Jotterand-Bellomo and Schauenberg, 1988; Oliver et al., 1993a; Pasicolan, 1993; Rabor, 1955; Rickart et al., 2011a, 2011b, 2013; Sanborn, 1952; Thomas, 1898.

SECTION 3

Old Endemics: Earthworm Mice

Throughout Luzon, above 500 m elevation in forest of almost any type and in any condition, the most abundant and diverse small mammals are usually the members of the earthworm mouse clade. As the name clearly implies, all of these species feed enthusiastically, and often primarily, on earthworms. Though this food source seems odd and surprising, worms are typically abundant and diverse on Luzon, and they generally become increasingly common with rising elevation. These mice follow the same pattern, increasing in diversity and abundance with higher elevation. In some places, we have found as many as seven species living in an area of a hectare or less, each using the habitat in a unique fashion (chapter 3).

This great diversity has been achieved quite rapidly. Our molecular data suggest that the original ancestor of this clade arrived in the Philippines (probably from somewhere in continental Asia) about 8 million years ago, and most of the speciation has occurred within the past 2 million years (chapter 5). We formally recognize 30 species in 5 genera on Luzon, but we suspect that about another half dozen remain that are "hidden" within the arboreal branch of the forest mice (*Apomys* subgenus *Apomys*); those studies are ongoing. We have discovered and formally named 15 of the 30 species from this clade on Luzon during the past 25 years, and the great majority in the past 5 years, which is evidence of how much has been learned recently—and an indication of how much is probably waiting to be learned in the future.

The earthworm mice show less variation in body size than the cloud rats, but they make up for it in the shapes of their bodies and the microhabitats they exploit. *Apomys* is the most diverse genus on Luzon, with 14 species that we currently recognize (one of which has yet to be formally named). These look the most like typical mice, with large eyes and ears and a long tail. Some of these mice live mostly in the forest canopy, and others almost entirely on the surface of the ground. They feed on a range of invertebrates, as well as on seeds and probably fruits; they are the only members of this clade with diets that include both animal and plant matter. *Archboldomys* and *Soricomys* are small, shrew-like animals with dark fur and small eyes; they forage in the leaf-litter for worms and other invertebrates. *Chrotomys* are more robust animals that

A: *Apomys musculus*. B: *Rhynchomys soricoides*. C: *Soricomys montanus*. D: *Chrotomys whiteheadi*. E: *Archboldomys maximus*. F: *C. silaceus*.

dig into the top layers of the soil; most species in this genus have boldly contrasting black and white stripes that run down their backs, with golden fur on their sides. *Rhynchomys* hop about on trails that they create on the surface of the ground and probe for earthworms with their elongate snouts. The smallest are *A. musculus*, some of which weigh only 16 g as adults, and the largest are *R. soricoides*, which reach 225 g.

Although most small mice in continental areas have high rates of reproduction, all of the earthworm mice have litters of one or two (rarely three) young, and our limited evidence indicates that they have only one litter per year. This implies that they must experience low rates of mortality from predators and be relatively long lived. A study of the ecology of this set of species would surely be of great value: their evolution in a unique feeding niche and their exceptionally low rate of reproduction present a wonderful case of diversification and specialization in a unique environment.

Most small mammals show substantial changes in their coat color as well as their body size as they become mature. Juveniles typically have grayer and rather duller fur than adults; as they grow, they often take on richer, brighter, and more lustrous tones. *Chrotomys whiteheadi* are a good example, although they differ from other genera in having contrasting dark and pale stripes even from an early age. These changes in size and color can cause much confusion, since the illustrations in this book and nearly all others show only adult coloration, and often from a tenth to half of the individuals in a population are young. Keeping these changes of color and size in mind will make the reader's use of this book as a field guide much more successful than otherwise would be the case.

Young *Chrotomys whiteheadi*, with the changes in coat color that take place as individuals mature from (*top to bottom*) juveniles to subadults to young adults. Compare these drawings with the one in the species account for *C. whiteheadi*.

Genus *Apomys*

Diversity among small mammals on Luzon reaches its peak in the genus *Apomys*, usually referred to as Philippine forest mice. Our recent, intensive studies have led to formal recognition of 13 species on Luzon, compared with the 4 that were known only a few years ago. Six are currently recognized from elsewhere in the Philippines, but it is apparent that many more are present on Greater Mindanao, Greater Negros-Panay, Mindoro, and some small, isolated oceanic islands. Many field surveys and additional detailed taxonomic studies are needed to reveal the actual extent of species richness and the full evolutionary history of this remarkably diverse genus.

The genus *Apomys* diverged from the other members of the earthworm mouse clade about 4.5 million years ago (chapter 5), during a time of rapid geological change. Two distinctive subgenera within *Apomys* are recognized; these shared a common ancestor as recently as about 3.2 million years ago (fig. 5.4). The nominate subgenus *Apomys* is currently represented on Luzon by only two named species, *A. microdon* and *A. musculus*. These are small, long-tailed mice that are active primarily in the forest canopy. Our preliminary, unpublished studies suggest that each of these may represent a species-group rather than a single species, with geographically isolated populations in geologically distinct portions of Luzon. In this volume, we continue to deal with each of them as a single species, simply because our studies are incomplete. In addition, we include here our initial information on a third species of this subgenus, known currently only from a single specimen from a limestone area in the Cagayan River Valley. Further studies of this subgenus most likely will add substantially to our understanding of the processes that have produced mammalian diversity on Luzon.

The second subgenus of *Apomys* is *Megapomys*. Compared with the small *Apomys*, these mice are larger, have a shorter tail (relative to their head plus body

A: *Apomys* (subgenus *Apomys*) *musculus*. B: *Apomys* (subgenus *Megapomys*) *sacobianus*.

Representative members of subgenus *Megapomys*. A: *Apomys abrae*. B: *A. brownorum*. C: *A. zambalensis*.

length), and are active on the surface of the ground. *Megapomys* has been the subject of much recent study and includes eight species that we have formally named and described since 2011; we suspect that few species in this subgenus still remain to be discovered on Luzon. Their history of diversification (chapter 5) tells us much about the processes that have produced mammalian diversity within Luzon and also provides some clues about the role of colonization between islands in the speciation process.

Because of the exceptional diversity within *Apomys*, we present the species accounts not simply in alphabetical order, but rather grouped by subgenus.

REFERENCES: Heaney et al., 1998, 2011a, 2014a; Justiniano et al., 2015; Musser, 1982b; Steppan et al., 2003.

Subgenus *Apomys*: Arboreal Forest Mice

Apomys microdon small Luzon forest mouse

Hollister, 1913. Proc. U.S. Natl. Mus., 46:327. Type locality: Biga [= Viga], Catanduanes Island, Philippines.

DESCRIPTION: Total length 227–256 mm; tail 124–145 mm; hind foot 26–28 mm; ear 17–19 mm; weight 30–42 g. A small mouse, with large eyes and ears. The tail is longer than the length of the head and body. The pelage is soft, brown with a slight orange tint dorsally and bright orange-brown with some areas of white hair ventrally. *Apomys musculus* are smaller (total length 182–213 mm; weight 16–24 g); have more yellowish dorsal and ventral fur; and typically occur at higher elevations. Other species of *Apomys* are substantially larger. *Musseromys* species are smaller (total length 178 mm or less; weight 22 g or less) and have a short, broad snout; short and broad hind feet; and a tail with long hairs near and at its tip. *Archboldomys* and *Soricomys* have tails that are shorter than or equal to the length of the head and body, and dark brown fur dorsally and ventrally.

DISTRIBUTION: Occur only within the Luzon Faunal Region, where they are widespread.

EVOLUTION AND ECOLOGY: *Apomys microdon* and its closest relative, *A. musculus*, last shared a common ancestor about 1.7 million years ago. These two species (or perhaps species-groups) are broadly sympatric over large parts of Luzon, with *A. microdon* occurring from sea level to 2025 m elevation and *A. musculus* from about 1460 m to 2730 m, but the actual zone of syntopic over-

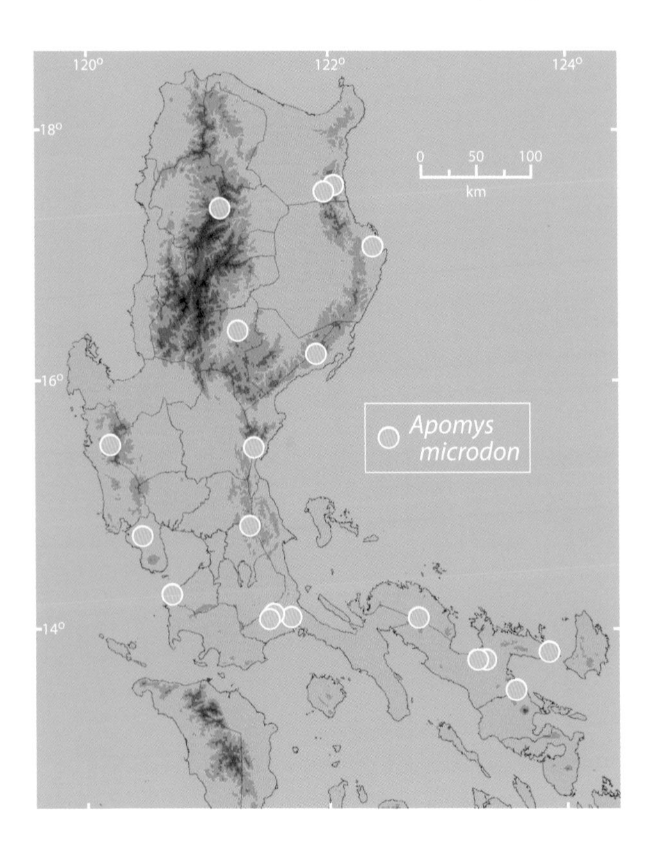

lap on any given elevational gradient is usually narrow or absent. *Apomys microdon* occur in secondary and primary lowland and montane forest, and occasionally in mossy forest. They are active at night, most often in trees but sometimes on the ground, feeding on seeds and occasionally on insects and earthworms. We have found them nesting in the axils of *Pandanus* and in clumps of climbing bamboos; the nests were made largely with dry bamboo leaves, with a few leaves from broad-leafed trees. This species is frequently sympatric with one or more species of the subgenus *Megapomys*, as well as with other small mammals that feed on earthworms and insects (e.g., *Crocidura*, *Archboldomys*, *Chrotomys*, *Rhynchomys*, and *Soricomys*), but it and *A. musculus* are the only members of the earthworm mouse clade that are arboreal. Females have two pairs of inguinal mammae. Average litter size appears to be 2.0: four pregnant females had three, two, two, and one embryos, and another had two placental scars. Adult males have small testes, up to about 4 × 8 mm.

STATUS: Widespread, stable, and persistent in secondary forest, but deforestation has removed some of its habitat.

REFERENCES: Alviola et al., 2011; Balete et al., 2009, 2011, 2013a, 2013b; Danielsen et al., 1994; Duya et al., 2011; Heaney et al., 1999, 2004, 2013a, 2013b; Hollister, 1913; Justiniano et al., 2015; Reginaldo and de Guia, 2014; Rickart et al., 1991, 2011a, 2011b, 2013; Steppan et al., 2003.

Apomys musculus least Philippine forest mouse

Miller, 1911. Proc. U.S. Natl. Mus., 38:403. Type locality: Camp John Hay, 5000 feet elevation, Baguio, Benguet Province, Luzon.

DESCRIPTION: Total length 182–213 mm; tail 97–121 mm; hind foot 20–25 mm; ear 14–17 mm; weight 16–24 g. A small mouse, with large eyes and ears. The tail is longer than the length of the head and body. The fur is soft, brown with a slight yellow-orange tint dorsally and pale yellow-orange with some areas of white fur ventrally. *Apomys microdon* are larger (total length 227–256 mm, weight 30–42 g) and have ventral fur that is brighter and more orangish; other species of *Apomys* are substantially larger. *Musseromys* species are smaller (total length 157–178 mm), with a short, broad snout; short (18–20 mm) and broad hind feet; and a tail with long hairs near and at the tip. *Archboldomys* and *Soricomys* have dark brown fur dorsally and ventrally and tails that are equal to or shorter than the length of the head and body.

DISTRIBUTION: Occur only on Luzon, where they are widespread.

EVOLUTION AND ECOLOGY: Most closely related to *Apomys microdon*; they last shared a common ancestor about 1.7 million years ago. The two species are widespread in the mountains of Luzon, but typically *A. musculus* occur at higher elevations than *A. microdon*. They are common to abundant in primary and secondary montane and mossy forest, from 1460 m to 2730 m elevation, and they are sometimes found in mixed shrubby and grassy habitats. They are normally active at night, though occasionally during the day in the rainy season, usually in the forest canopy but sometimes on or near the ground. We have found them in nests made from dry bamboo leaves, which are hidden in clumps of climbing bamboo and in tree hollows. They are omnivorous, feeding on seeds, insects and

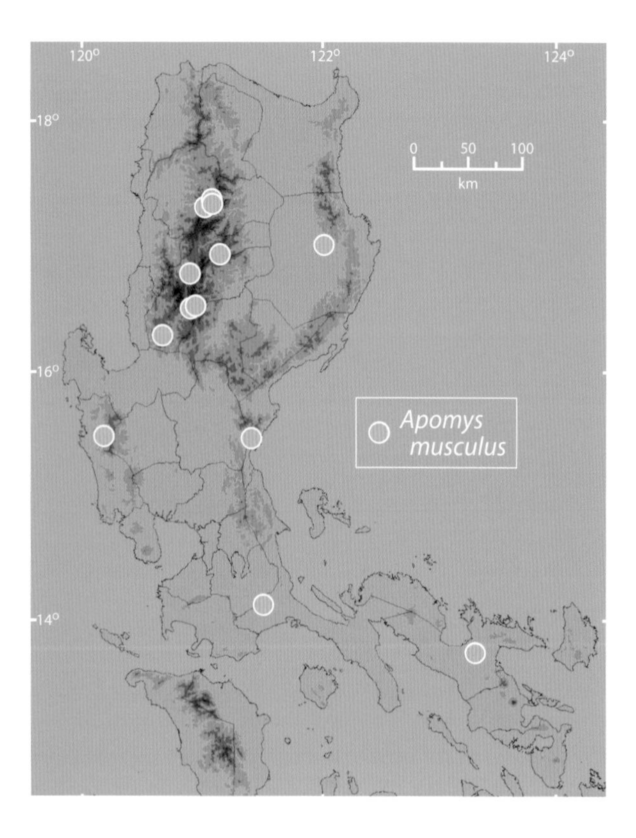

other invertebrates, and earthworms. In mossy forest on Mt. Isarog, they occurred at a density of 8.8/ha, the highest for any of the five species of small mammals in that study area , but their biomass (163.2 g/ha) was the second lowest. The mean distance moved between captures was 42 m; maximum distance ranged from 150 m to 226 m; and mean home-range size was estimated at 0.22 ha (range = 0.01–0.56 ha). Based on numbers of embryos and uterine scars, 10 females had a mean litter size of 2.0 (range = 1 to 3) and two pairs of inguinal mammae. Males have small testes, up to 5 × 11 mm. We have occasionally found scats of civets (probably *Viverra tangalunga*) that contained bones and teeth of this species. *Apomys musculus* have a karyotype of $2n$ = 42, FN = 52.

STATUS: Widespread and moderately common at medium to high elevations in both primary and secondary forest. Studies of taxonomy and phylogeny are needed to resolve questions regarding the extent of diversification.

REFERENCES: Balete & Heaney, 1997; Balete et al., 2008, 2009, 2011, 2013a, 2013b; Duya et al., 2007, 2011; Heaney et al., 1999, 2004, 2013a, 2013b; Justiniano et al., 2015; Reginaldo et al., 2013; Rickart and Musser, 1993; Rickart et al., 1991, 2011a, 2011b.

Apomys sp.
Cagayan Valley karst mouse

DESCRIPTION: Total length 181 mm; tail 97 mm; hind foot 21 mm; ear 13 mm; weight 17 g. A small mouse. The fur is soft, bright orange-brown dorsally and pale orange ventrally. The tail is longer (115%) than the length of the head and body and is entirely dark gray. *Apomys musculus* are similar in size but have dorsal fur that is dark brown and a tail that is white ventrally. *Mus musculus* have gray or grayish-brown fur and are smaller (up to 175 mm total length), with a shorter tail (up to ca. 91 mm) that also is entirely dark gray. *Crunomys fallax* are similar in size (total length 185 mm) but have a shorter tail (79 mm) and shorter ear (10

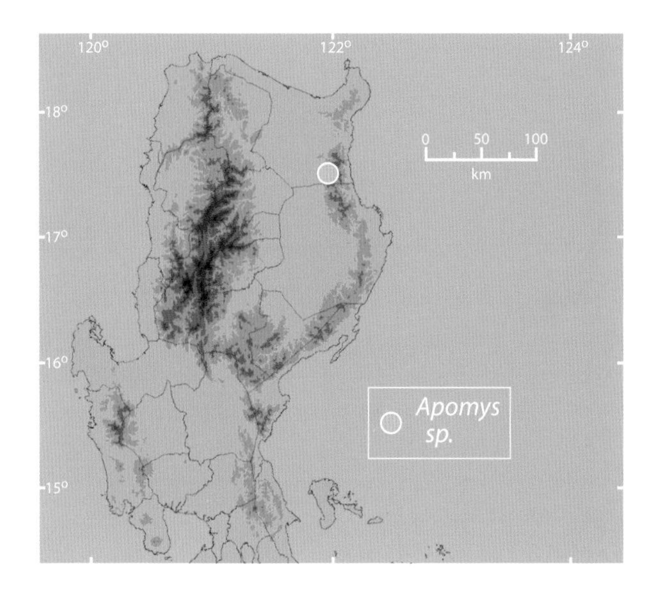

mm); the dorsal fur contains some bristly hairs. Other rodents are substantially larger.

DISTRIBUTION: Known from a single specimen captured near Callao Cave, Peñablanca Municipality, Cagayan Province.

EVOLUTION AND ECOLOGY: The first and only known specimen, a young adult male, was captured in June 2011 in a region of karstic limestone not far from Callao Cave. The trap was on the ground, in regenerating molave (*Vitex*) forest with mixed grass and shrubs, at ca. 200 m elevation. Additional surveys in nearby karst failed to obtain more specimens. Our preliminary studies indicated that it probably represents a new species that is a member of the subgenus *Apomys*. At this time, it is the only known species of mammal in the Philippines that may be entirely associated with karst, a habitat that elsewhere in Asia often supports local endemics.

STATUS: Unknown; both ecological studies in the field and phylogenetic studies in the museum are needed.

REFERENCES: Unpublished data and specimen in FMNH.

Subgenus *Megapomys*: Large Forest Mice

Apomys abrae
Cordillera pine forest mouse

(Sanborn, 1952). Fieldiana: Zool., 33(2):133. Type locality: Massisiat, 3500 feet elevation, Abra Province, Luzon Island.

DESCRIPTION: Total length 258–291 mm; tail 121–148 mm; hind foot 35–38 mm; ear 21–23 mm; weight 59–79 g. An attractive mouse, with large eyes and ears. The fur is soft, brown dorsally and white ventrally, with white fur on the upper sides of the forefeet and hind feet. The tail is about equal (usually 97%–101%) to the length of the head and body. *Apomys datae* are similar in size but heavier (weight 67–105 g); usually have a tail a bit shorter (usually 89%–95%) than the length of the head and body; and have dark hairs extending over much of the tops of the forefeet and hind feet. Other species of *Apomys* in the Cordillera are smaller (weight less than 25 g) and have tails substantially longer than the head and body.

DISTRIBUTION: Occur only within the Central Cordillera of northern Luzon, but are widespread within that region.

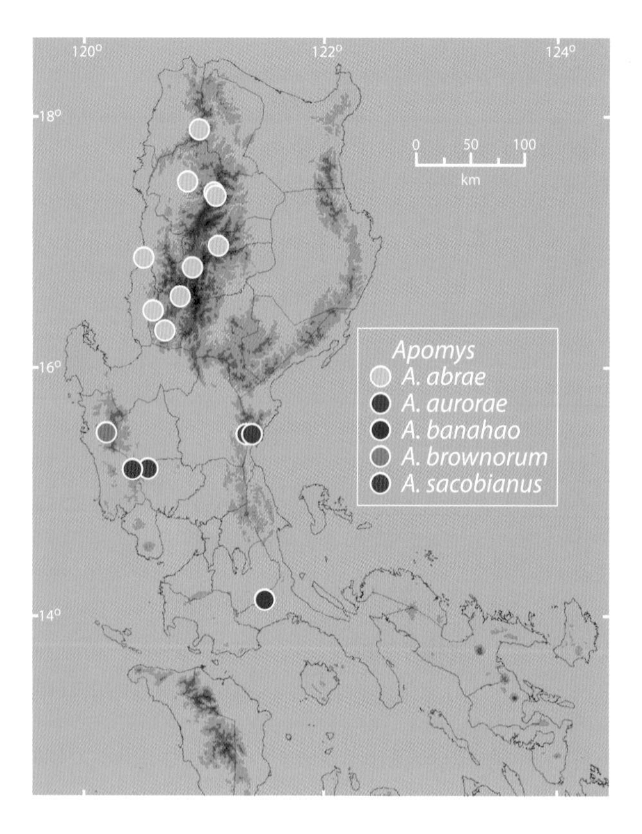

EVOLUTION AND ECOLOGY: A member of the diverse subgenus *Megapomys*. It is most closely related to *Apomys datae*, which occurs sympatrically with it in the Central Cordillera; the two species last shared a common ancestor about 400,000 years ago (fig. 5.4). Current data imply some hybridization, resulting in mitochondrial introgression between the two. *Apomys abrae* are common in pine forest with broad-leafed undergrowth, at 900 m to about 2200 m elevation, and are moderately common in primary and secondary broad-leafed montane forest from ca. 1000 m to 1600 m elevation. In regions of overlap with *A. datae*, *A. abrae* occur in drier and more open habitat than *A. datae*, and they often occur at elevations below the range of *A. datae*. *Apomys abrae* are absent from the highest peaks in the Cordillera (such as Mt. Pulag), even when pine forest is present. They are nocturnal and forage on the surface of the ground. They are omnivorous, consuming seeds, insects, and earthworms. Females have two pairs of inguinal mammae. Four pregnant females had an average of two embryos (range = 1–3). This species has a karyotype of $2n = 44$, FN = 54, similar to that of *A. datae*.

STATUS: Widespread and abundant in a common habitat; stable.

REFERENCES: Heaney et al., 2004, 2011; Justiniano et al., 2015; Musser, 1982b; Musser and Heaney, 1992; Rabor, 1955; Reginaldo and de Guia, 2014; Rickart et al., 2011a, 2011b; Sanborn, 1952; Steppan et al., 2003.

Apomys aurorae
Aurora forest mouse

Heaney et al., 2011. Fieldiana: Life and Earth Sci. 2:45. Type locality: 2 km S, 2 km W Mt. Mingan peak, 1305 m elevation, Dingalan Municipality, Aurora Province, Luzon Island, 15.46456° N, 121.38421° E.

DESCRIPTION: Total length 262–295 mm; tail 129–153 mm; hind foot 33–37 mm; ear 18–21 mm; weight 58–92 g. A large, attractive mouse, with large eyes and ears. The fur is soft, rusty-brown dorsally and white or nearly white with a pale orange wash ventrally. The tail is sharply bicolored (dark dorsally and nearly white

ventrally) and long, roughly equal to the length of the head and body. The tops of the hind feet are nearly white, with only scattered dark hairs. *Apomys minganensis* usually are somewhat smaller (total length 246–279 mm), have a tail shorter than the length of the head and body, with scattered dark hairs and pigmented scales ventrally; darker fur dorsally; and dark hair on the dorsal surface of the hind feet. Other *Apomys* in the Mingan Mountains are much smaller (total length 256 mm or less; weight 42 g or less). *Rhynchomys* sp. are much larger (total length 312–328 mm; weight 158–176 g).

DISTRIBUTION: Recorded only from Mt. Mingan, but probably widespread in the Mingan Mountains. (For map, see *Apomys abrae* species account.)

EVOLUTION AND ECOLOGY: Diverged about 800,000 years ago from the common ancestor of the clade that includes *Apomys iridensis* and *A. zambalensis* (fig. 5.4). *Apomys aurorae* are distantly related to *A. minganensis*, which also occur in the Mingan Mountains, last sharing a common ancestor about 2.0 million years ago. Aurora forest mice have been documented from 733 m to 1677 m elevation, in lightly disturbed and primary lowland and montane forest, and in the lowermost portions of mossy forest. They were the most abundant small mammal species up to 1476 m elevation in lowland and montane forest and were quite uncommon in mossy forest at higher elevations. They showed a strong preference for live earthworm bait rather than fried coconut, but probably they are at least somewhat omnivorous. Nearly all were captured at night, and all were captured on the ground surface (not in trees). They appear to be replaced by *A. minganensis* in mossy forest; a study of interactions and possible competitive exclusion would be of great interest. No information on reproduction is available.

STATUS: Recently discovered and generally poorly known, but apparently abundant and probably widespread in the Mingan Mountains and able to tolerate at least a small amount of habitat disturbance.

REFERENCES: Balete et al., 2011; Heaney et al., 2011; Justiniano et al., 2015.

Apomys banahao
Banahao forest mouse

Heaney et al., 2011. Fieldiana: Life and Earth Sci. 2:28. Type locality: Mt. Banahaw, 1465 m elevation, Barangay Lalo, Quezon Province, Luzon Island, 14.06635° N, 121.50855° E.

DESCRIPTION: Total length 250–287 mm; tail 111–133 mm; hind foot 33–37 mm; ear 22–24 mm; weight 71–92 g. An attractive mouse, with large eyes and ears. The fur is soft and dense, dark brown with slight rusty tints dorsally and white with a slight gray wash ventrally. The tail is sharply bicolored (dark brown dorsally and nearly white ventrally) and relatively short (86%–103% of the length of the head and body). The hind feet have dark hair in the center of the dorsal surface, surrounded by white hair. *Apomys magnus*, which overlap with this species at middle elevations, are much larger (total length 272–305 mm; weight 92–128 g); have a tail equal to or longer than the length of the head and

body; have fur that is rusty-brown dorsally, with prominent black guard hairs; and have hind feet covered almost entirely by white hair dorsally. Other *Apomys* at high elevation on Mt. Banahaw are much smaller (total length 256 mm or less; weight 42 g or less). *Rhynchomys banahao* are much larger (total length 305–320 mm; weight 135–155 g).

DISTRIBUTION: Recorded only from Mt. Banahaw and the adjacent Mt. Banahaw de Lucban; they should also be sought on Mt. San Cristobal. (For map, see *Apomys abrae* species account.)

EVOLUTION AND ECOLOGY: Diverged about 1.0 million years ago from its common ancestor with *Apomys brownorum*. *Apomys banahao* are more distantly related to *A. sacobianus* and *A. lubangensis*, last sharing a common ancestor about 1.3 million years ago (fig. 5.4). Banahao forest mice have been documented in old-growth montane and mossy forest, from 1465 m to 2030 m elevation on Mt. Banahaw and at 1500 m on Mt. Banahaw de Lucban. They were the most abundant small mammal species at those elevations, occurring at some of the highest densities measured for any small mammal on Luzon. Other species that were found with them included *Crocidura grayi*, *A. magnus*, *A. microdon*, *A. musculus* (the latter two being arboreal mice), *Bullimus luzonicus*, *Rattus everetti*, and *Rhynchomys banahao*. *Apomys banahao* showed a slight preference for live earthworm bait rather than fried coconut and probably are omnivorous. Nearly all were captured at night, and nearly all were on the ground surface (not in trees). *Apomys magnus* overlap with *A. banahao* in montane forest at ca. 1465 m elevation, and also at adjacent lower elevations; a study of ecological interactions and possible competitive exclusion would be of great interest. Nine pregnant females carried from one to two embryos each (mean = 1.9); each had two pairs of inguinal mammae. This species has a karyotype of $2n = 48$, FN = 54, which represents a unique arrangement among *Apomys*.

STATUS: Apparently abundant and stable within their habitat, all of which lies within Mts. Banahaw–San Cristobal National Park.

REFERENCES: Heaney et al., 2011, 2013b; Justiniano et al., 2015.

Apomys brownorum
Tapulao forest mouse

Heaney et al., 2011. Fieldiana: Life and Earth Sci. 2:45. Type locality: Mt. Tapulao peak, 2024 m elevation, Barangay Salasa, Palauig Municipality, Zambales Province, Luzon Island, 15°28′54.8″ N, 120°07′10.4″ E.

DESCRIPTION: Total length 230–255 mm; tail 107–116 mm; hind foot 31–36 mm; ear 21–22 mm; weight 60–84 g. An attractive mouse, with large eyes and ears. The fur is soft, with a rich tone of brown dorsally and pale grayish with a slight orange wash ventrally. The tail is sharply bicolored (dark brown dorsally and nearly white ventrally) and relatively short (82%–90% of the length of the head and body). The hind feet are covered dorsally with dark hairs, nearly to the toes. *Apomys zambalensis*, which occur at slightly lower elevations, are much larger (total length 294–311 mm; weight 74–125 g); have a tail equal to or longer than the length of the head and body; and have fur that is a rich orange-

brown dorsally and nearly white with a slight orange wash ventrally. Other *Apomys* at high elevation in the Zambales Mountains are much smaller (total length 256 mm or less; weight 42 g or less). *Rhynchomys tapulao* are much larger (total length 290–308 mm; weight 129–156 g).

DISTRIBUTION: Recorded only from Mt. Tapulao, but possibly widespread on the high peaks of the Zambales Mountains. (For map, see *Apomys abrae* species account.)

EVOLUTION AND ECOLOGY: Diverged about 1.0 million years ago from its common ancestor with *Apomys banahao* (fig. 5.4). *Apomys brownorum* are more distantly related to *A. sacobianus*, which also occur in the Zambales Mountains, last sharing a common ancestor about 1.3 million years ago; the latter are found only at much lower elevations (365–1080 m) in heavily disturbed volcanic habitat on Mt. Pinatubo. Tapulao forest mice have been documented only near the peak of Mt. Tapulao at ca. 2024 m elevation, in old-growth and regenerating mossy forest (the latter at the site of a mining operation). They were the most abundant small mammal species at that elevation; other species at the site included *Crocidura grayi*, *A. microdon*, *A. musculus* (the latter two being arboreal mice), *Chrotomys mindorensis*, *Rattus everetti*, and *Rhynchomys tapulao*. They showed a slight preference for fried coconut bait rather than live earthworms, and they probably are omnivorous. Nearly all were captured at night, and nearly all were on the ground surface (not in trees). *Apomys zambalensis* occur in montane forest at adjacent lower elevations; a study of interactions and possible competitive exclusion would be of great interest. Two adult females each carried a single embryo, and each had two pairs of inguinal mammae.

STATUS: Recently discovered and generally poorly known. Apparently abundant and possibly widespread at high elevation in the Zambales Mountains, and able to tolerate some habitat disturbance.

REFERENCES: Balete et al., 2009; Heaney et al., 2011, 2014a; Justiniano et al., 2015.

Apomys datae
Cordillera forest mouse

(Meyer, 1899). Abh. Mus. Dresden, ser. 7, 7:25. Type locality: Mt. Data, Lepanto [now in Mountain Province], Northern Luzon Island.

DESCRIPTION: Total length 260–292 mm; tail 125–143 mm; hind foot 36–39 mm; ear 20–22 mm; weight 67–105 mm. A fairly large, attractive mouse, with large eyes and ears. The fur is soft, dark brown dorsally and nearly white ventrally (sometimes with a pale orange tint); dark fur extends from the body onto the upper sides of the forefeet and hind feet. The tail is usually slightly shorter (89%–95%) than the length of the head and body. *Apomys abrae* are slightly smaller (weight 59–79 g); have a tail that is about equal (97%–101%) to the length of the head and body; and the tops of the feet are nearly entirely white.

DISTRIBUTION: Occur only within the Central Cordillera of northern Luzon, but are widespread within that region. Previous reports of large *Apomys* from 760 m to 1650 m elevation in the Sierra Madre represent *A. sier-*

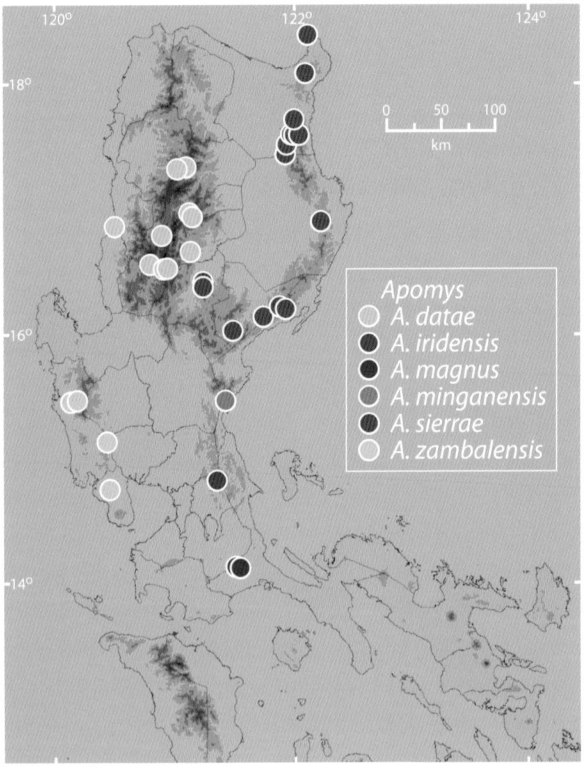

rae, and those from Mt. Banahaw represent *A. magnus* and *A. banahao*.

EVOLUTION AND ECOLOGY: Most closely related to *Apomys abrae*; they last shared a common ancestor about 400,000 years ago (fig. 5.4). There has been some mitochondrial introgression between the two species, but nuclear genes show no evidence of hybridization. *Apomys datae* are abundant in primary and secondary montane and mossy forest, as well as in shrubby habitats adjacent to agricultural fields, from 1600 m to 2800 m elevation in the Central Cordillera; they apparently are absent from agricultural fields and from pine forest, except when broad-leafed forest is regenerating beneath the pine canopy. They are nocturnal, foraging on the ground for earthworms, insects, and seeds, and only rarely climb up onto tree trunks and fallen logs. They occur sympatrically with *A. abrae* in many areas between 1600 m and ca. 2200 m elevation; in such areas, *A. datae* occurs in the more moist, heavily vegetated areas and *A. abrae* in more open and drier habitat (often but not always in pine forest). Females have two pairs of inguinal mammae. Twenty-one pregnant females had an average of 1.66 embryos (range = 1–3). The karyotype is $2n = 44$, FN = 54, similar to that of *A. abrae*.

STATUS: Abundant and widespread; tolerant of disturbed forest.

REFERENCES: Danielsen et al., 1994; Feiler, 1999; Heaney et al., 2004, 2011, 2013b; Justiniano et al., 2015; Mallari and Jensen, 1993; Musser, 1982b; Rabor, 1955; Reginaldo and de Guia, 2014; Rickart and Heaney, 2002; Rickart et al., 2011a, 2011b; Sanborn, 1952; Stuart et al., 2007.

Apomys iridensis
Southern Sierra Madre forest mouse

Heaney et al., 2014. Proc. Biol. Soc. Washington, 126:408. Type locality: 1.25 km S, 0.5 km W Mt. Irid peak, 920 m elevation, Rodriguez Municipality, Rizal Province, Luzon Island, 14.78000° N, 121.32116° E.

DESCRIPTION: Total length 274–310 mm; tail 132–157 mm; hind foot 35–40 mm; ear 19–21 mm; weight 88–104 g. This attractive mouse is one of the largest *Apomys*, with large eyes and ears. The fur is soft, dark brown with deep reddish tones dorsally and pale gray ventrally (sometimes with a pale orange wash). The tail is sharply bicolored (dark dorsally and nearly white ventrally) and long, roughly equal to the length of the head and body. The hind feet are large, with white hair dorsally, except for a few scattered dark hairs. *Apomys aurorae*, which are closely related and live nearby, usually are smaller (total length 262–295 mm; weight 58–92 g); have slightly paler fur, with rusty-orange tones dorsally; and have many scattered dark hairs on the dorsal surface of the hind feet. *Apomys microdon* are much smaller (total length 256 mm or less; weight 42 g or less).

DISTRIBUTION: Currently recorded only from Mt. Irid, but may be more widespread in the mountains of the Southern Sierra Madre. (For map, see *Apomys datae* species account.)

EVOLUTION AND ECOLOGY: Diverged about 400,000 years ago from its common ancestor with *Apomys zambalensis*; it is somewhat more distantly related to *A. aurorae* and *A. magnus* (fig. 5.4). *Apomys iridensis* have been documented from 700 m to 1330 m elevation in regenerating lowland and primary montane forest, but they were abundant at our lowest survey area and probably also occur at lower elevations. They were the most abundant species at all of our survey sites, where the native species *A. microdon*, *Bullimus luzonicus*, *Chrotomys mindorensis*, and *Rattus everetti*, and the non-native species *Suncus murinus*, *R. exulans*, and *R. tanezumi* were also present. *Apomys iridensis* showed a weak preference for live earthworm bait rather than fried coconut but probably are best considered to be omnivorous. Nearly all were captured at night, and all but one were captured on the ground surface. Three pregnant females each carried one embryo, and one had two embryos.

STATUS: Apparently abundant and tolerant of some habitat disturbance. We suspect that they are widespread in the Southern Sierra Madre above about 500 m elevation.

REFERENCES: Balete et al., 2013a; Heaney et al., 2011, 2013b; Justiniano et al., 2015.

Apomys magnus
large Banahao forest mouse

Heaney et al., 2011. Fieldiana: Life and Earth Sci. 2:45. Type locality: Mt. Banahaw, 1250 m elevation, Hasa-an, Barangay Lalo, Tayabas Municipality, Quezon Province, Luzon Island, 14°03′44″ N, 121°31′08″ E.

DESCRIPTION: Total length 272–305 mm; tail 133–154 mm; hind foot 37–41 mm; ear 21–23 mm; weight 92–128 g. This attractive mouse is the largest *Apomys*, with large eyes and ears. The fur is soft, dark brown dorsally and white or nearly white ventrally. The tail is sharply bicolored (dark dorsally and nearly white ventrally) and long, roughly equal to the length of the head and body. The hind feet are large, covered with nearly white hair dorsally. *Apomys banahao* are smaller (total length 250–287 mm); have a tail shorter than the length of the head and body; have darker fur dorsally; and have dark hair on the dorsal surface of the hind feet. Other *Apomys* on Mt. Banahaw are much smaller (total length

256 mm or less; weight 42 g or less). *Rhynchomys banahao* are larger (total length 305–320 mm; weight 135–155 g) and have a long, slender snout, with delicate incisors.

DISTRIBUTION: Recorded only from Mt. Banahaw and adjacent Mts. Banahaw de Lucban and San Cristobal; they should be sought on other mountains in the vicinity. (For map, see *Apomys datae* species account.)

EVOLUTION AND ECOLOGY: Diverged about 1.0 million years ago from the common ancestor of the clade that includes *Apomys aurorae, A. iridensis,* and *A. zambalensis* (fig. 5.4). *Apomys magnus* are quite distantly related to *A. banahao,* which also occur on Mt. Banahaw, last sharing a common ancestor about 2.0 million years ago. *Apomys magnus* have been documented from 765 m to 1465 m elevation, in regenerating lowland and primary lowland and montane forest. In heavily disturbed lowland forest at 765 m to 900 m elevation, *Rattus everetti* were more abundant; in mature or old-growth montane forest from 1100 m to 1250 m elevation, *A. magnus* were more common; and *A. banahao* predominated at 1465 m and above. A study of interactions and possible competitive exclusion would be of great interest. *Apomys magnus* showed a weak preference for live earthworm bait rather than fried coconut but probably are best considered omnivorous. All were captured at night, and all were captured on the ground surface (not in trees). One pregnant female carried three embryos.

STATUS: Apparently abundant and tolerant of some habitat disturbance on Mt. Banahaw and adjacent peaks; most of the distribution lies within Mts. Banahaw–San Cristobal National Park.

REFERENCES: Heaney et al., 2011, 2013b; Justiniano et al., 2015.

Apomys minganensis
Mingan forest mouse

Heaney et al., 2011. Fieldiana: Life and Earth Sci. 2:30. Type locality: 1.5 km S, 0.5 km W Mt. Mingan peak, 1681 m elevation, Dingalan Municipality, Aurora Province, Luzon Island, 15.46802° N, 121.40039° E.

DESCRIPTION: Total length 246–279 mm; tail 116–138 mm; hind foot 31–35 mm; ear 18–19 mm; weight 66–92 g. A large, attractive mouse, with large eyes and ears. The fur is soft, dark brown dorsally and gray (with a slight orange wash) ventrally. The tail is sharply bicolored (dark dorsally and white with many scattered dark hairs and scales ventrally) and less than or equal to the length of the head and body. The hind feet are covered by brown hairs, with only scattered pale hairs. *Apomys aurorae* usually are somewhat larger (total length 262–295 mm); have a tail about equal to the length of the head and body and almost entirely white ventrally; have less-dark, more rusty-red fur dorsally; and have almost entirely white hair on the dorsal surface of the hind feet. Other *Apomys* in the Mingan Mountains are much smaller (total length 256 mm or less; weight 42 g or less). *Rhynchomys* sp. are much larger (total length 312–328 mm; weight 158–176 g).

DISTRIBUTION: Recorded only from Mt. Mingan, but possibly widespread on the high peaks of the Mingan Mountains. (For map, see *Apomys datae* species account.)

EVOLUTION AND ECOLOGY: Diverged about 1.4 million years ago from the common ancestor of the clade that includes *Apomys banahao*, *A. brownorum*, *A. lubangensis*, and *A. sacobianus*. *Apomys minganensis* are distantly related to *A. aurorae*, which also occur in the Mingan Mountains, last sharing a common ancestor about 2.0 million years ago (fig. 5.4). *Apomys minganensis* have been documented from 1540 m to 1785 m elevation, in mossy forest and the uppermost edge of montane forest. They were the most abundant small mammal species in mossy forest from 1540 m elevation up to our highest sampling area (1785 m) and less common in transitional montane-mossy forest at 1540 m. Much of the mossy forest had been naturally but severely disturbed by typhoons, leading to a widespread abundance of climbing pandans and bamboos. They show a strong preference (even greater than that of *A. aurorae*) for live earthworm bait rather than fried coconut but probably are at least somewhat omnivorous. Nearly all were captured at night, and all were captured on the ground surface (not in trees). They appear to be replaced by *A. minganensis* in montane forest; a study of interactions and possible competitive exclusion would be of great interest. Four adult females carried one embryo each, and one had three.

STATUS: Recently discovered and poorly known, but apparently abundant in mossy forest that is not subject to human disturbance.

REFERENCES: Balete et al., 2011; Heaney et al., 2011; Justiniano et al., 2015.

Apomys sacobianus
Pinatubo volcano mouse

Johnson, 1962. Proc. Biol. Soc. Washington, 75:317. Type locality: "A narrow forested canyon of the Sacobia River . . . where it emerges from the foothills of the Zambales Mountains," Clark Air Force Base, Pampanga Province, Luzon Island.

DESCRIPTION: Total length 277–315 mm; tail 132–153 mm; hind foot 35–40 mm; ear 21–25 mm; weight 79–105 g. A large, attractive mouse, with large eyes and ears. The fur is grayish-brown dorsally and grayish-white ventrally. The tail is unusually thick; sharply bicolored (dark dorsally and nearly white ventrally); and long (95%–100% of the length of the head and body). The fur on the dorsal surface of the hind feet is nearly white, with only scattered dark hairs. *Apomys brownorum* are smaller (total length 230–255 mm; weight 60–84 g); have a tail shorter than the length of the head and body; have dark brown fur dorsally; and have dark hair on the dorsal surface of the hind feet. *Apomys zambalensis* are similar in appearance but are slightly larger (total length 294–311 mm; weight 74–125 g); have dorsal fur that is slightly darker and more rusty; and have a tail that is more slender and less dark (brown rather

than nearly black) dorsally. Other *Apomys* in the Zambales Mountains are much smaller (total length 256 mm or less; weight 42 g or less).

DISTRIBUTION: Currently known only from Mt. Pinatubo, but possibly more widespread in the lowlands of the Zambales Mountains. (For map, see *Apomys abrae* species account.)

EVOLUTION AND ECOLOGY: Most closely related to *Apomys lubangensis*, having last shared a common ancestor about 500,000 years ago. These two species, in turn, are related to a pair of closely related species, *A. brownorum* (which also occur in the Zambales Mountains) and *A. banahao*, last sharing a common ancestor with them about 700,000 years ago (fig. 5.4). *Apomys sacobianus* have been documented from 365 m to 1080 m elevation; they may also occur higher, but our sampling did not extend above that elevation. They are highly tolerant of disturbance, occurring entirely in sparsely regenerating vegetation on Mt. Pinatubo, which underwent a massive eruption in June 1991; indeed, they may specialize in this type of habitat. They were the most abundant small mammal species on Mt. Pinatubo during our surveys, with the native species *A. zambalensis*, *Bullimus luzonicus*, *Chrotomys mindorensis*, and *Rattus everetti*, and the non-native *R. exulans* and *R. tanezumi* also present, but at much lower densities. They avidly consumed both live earthworm and fried coconut baits and are best considered to be omnivorous. They are active almost exclusively at night and almost never climb above the ground surface (and never more than 1 m above the ground). *Apomys sacobianus* are externally very similar to *A. zambalensis*; a study of ecological interactions and possible competitive exclusion would be of great interest. Females have two pairs of inguinal mammae; five pregnant females each had two embryos. Males have testes up to 13 × 6 mm.

STATUS: Abundant and widespread on Mt. Pinatubo, and able to tolerate (and may prefer) a great deal of habitat disturbance. Further surveys elsewhere in the Zambales area are needed to determine the full distribution.

REFERENCES: Heaney et al., 2011, 2014a; Johnson, 1962; Justiniano et al., 2015.

Apomys sierrae
Northern Sierra Madre forest mouse

Heaney et al., 2011. Fieldiana: Life and Earth Sci. 2:34. Type locality: 3.5 km SW Mt. Cetaceo peak, 1400 m elevation, Cagayan Province, Luzon Island.

DESCRIPTION: Total length 250–304 mm; tail 118–154 mm; hind foot 34–39 mm; ear 18–21 mm; weight 73–111 g. A large, attractive mouse, with large eyes and ears. Pelage color is variable: in the Sierra Madre, the fur is dark brown, with rusty tints dorsally; in the Caraballos Mountains, the dorsal fur is medium brown, with yellow tints; in both areas, the ventral fur is white or nearly white with a slight orange wash. The tail is sharply bicolored (dark dorsally and nearly white ventrally) and long, roughly equal to the length of the head and body; a few individuals have a small white tip on the tail. The hind feet are nearly white, with only scattered dark hairs. *Apomys aurorae*, which occur nearby, are similar. They usually are somewhat smaller (total

Sierra Madre morph.

Caraballos morph.

length 262–295 mm; weight 58–92 g); have dark fur, with a yellowish tint dorsally; and have slightly shorter hind feet (33–37 mm). Other *Apomys* in the Northern Sierra Madre are much smaller (total length 256 mm or less; weight 42 g or less).

DISTRIBUTION: Widespread in the Sierra Madre and probably in the Caraballos Mountains, including Palaui Island, a small island in shallow water off the NE tip of Luzon. (For map, see *Apomys datae* species account.)

EVOLUTION AND ECOLOGY: Diverged about 1.4 million years ago from the common ancestor of the clade that includes *Apomys aurorae*, *A. iridensis*, *A. magnus*, and *A. zambalensis* (fig. 5.4). *Apomys sierrae* have been documented on Luzon, from 475 m to 1700 m elevation, in heavily to lightly disturbed and primary lowland and montane forest, and in the lower portions of mossy forest. On Palaui Island, they have been found at 153 m elevation; the island has a maximum elevation of about 300 m. They were the most abundant small mammal species from 780 m to 1707 m elevation in lowland, montane, and mossy forest on Mt. Palali, and from 1300 m to 1550 m elevation in montane and mossy forest on Mt. Cetaceo. They showed a strong preference for live earthworm bait rather than fried coconut on Mt. Cetaceo, but a roughly equal preference on Mt. Palali; they appear to be omnivorous overall. Nearly all were captured at night, and nearly all were captured on the ground surface. Seven pregnant females had an average of 1.7 embryos (range = 1–2).

STATUS: One of the most widespread and abundant small mammals on Luzon, able to tolerate a moderate amount of habitat disturbance.

REFERENCES: Alviola et al., 2011; Danielsen et al., 1994; Duya et al., 2007, 2011; Heaney et al., 2011, 2013a; Justiniano et al., 2015.

Apomys zambalensis
Zambales forest mouse

Heaney et al., 2011. Fieldiana: Life and Earth Sci. 2:45. Type locality: 0.1 km N Mt. Natib peak, 1150 m elevation, Bataan Province, Luzon Island, 14.71513° N, 120.39892° E.

DESCRIPTION: Total length 294–311 mm; tail 123–158 mm; hind foot 35–40 mm; ear 20–23 mm; weight 74–125 g. A large, attractive mouse, with large eyes and ears. The fur is soft, bright rusty-orange dorsally and white or nearly white with a pale orange wash ventrally. The tail is sharply bicolored (dark dorsally and nearly white ventrally) and long (94%–100% of the length of the head and body). The hind feet are nearly white, with only scattered dark hairs. *Apomys brownorum* are smaller (total length 230–255 mm; weight 60–84 g); have a tail shorter than the length of the head and body; have dark brown fur dorsally; and have dark hair on the dorsal surface of the hind feet. *Apomys sacobianus* are similar in appearance but are slightly smaller (total length 277–315 mm; weight 79–105 g); have dorsal fur that is slightly paler, grayer, and somewhat variegated (i.e., salt-and-pepper); and have a tail that is thicker and darker (nearly black rather than brown) dorsally. Other *Apomys* in the Zambales Mountains are

much smaller (total length 256 mm or less; weight 42 g or less).

DISTRIBUTION: Widespread in the Zambales Mountains, including Mt. Natib. (For map, see *Apomys datae* species account.)

EVOLUTION AND ECOLOGY: Most closely related to *Apomys iridensis*, having last shared a common ancestor about 700,000 years ago. They are distantly related to *A. brownorum* and *A. sacobianus*, which also occur in the Zambales Mountains, last sharing a common ancestor with them about 2.0 million years ago. *Apomys zambalensis* have been documented from 365 m to 1690 m elevation, which is the greatest documented elevational range of any *Apomys*. They are highly tolerant of disturbance, occurring in sparsely regenerating vegetation on Mt. Pinatubo; in secondary and primary lowland and montane forest on Mts. Natib and Tapulao; and in the lowermost portions of mossy forest on Mt. Tapulao. They were the most abundant small mammal species from 925 m to 1690 m elevation on Mt. Tapulao, and from 900 m to 1250 m on Mt. Natib; *Rattus everetti* were more abundant at lower elevations on Mt. Natib, and *A. sacobianus* were the most abundant species at all elevations on Mt. Pinatubo. *Apomys zambalensis* avidly consume both live earthworm bait and fried coconut and are best considered to be omnivorous. They are active almost exclusively at night and almost never climb above the ground surface (and never more than 1 m above the ground). They appear to be replaced by *A. brownorum* in mossy forest, and they overlap extensively with the externally similar *A. sacobianus* on Mt. Pinatubo; a study of interactions and possible competitive exclusion would be of great interest. Females have two pairs of inguinal mammae; one female carried two embryos. This species has a unique karyotype of $2n = 44$, FN = 58.

STATUS: Abundant and widespread in the Zambales Mountains, and able to tolerate a great deal of habitat disturbance.

REFERENCES: Balete et al., 2009; Heaney et al., 2011, 2014a; Justiniano et al., 2015; Rickart et al., 2013.

Archboldomys luzonensis
Mt. Isarog shrew-mouse

Musser, 1982. Bull. Am. Mus. Nat. Hist., 174:30. Type locality: Mt. Isarog, 2400 feet elevation, Camarines Sur Province, Luzon.

DESCRIPTION: Total length 167–190 mm; tail 60–80 mm; hind foot 26–29 mm; ear 16–17 mm; weight 31–37 g. A small mouse, with small eyes and ears. The tail is shorter (65%–75%) than the length of the head and body. The pelage is a dark reddish-brown over the entire body but slightly paler on the abdomen. *Apomys microdon* are similar in weight (30–42 g) but longer (total length 227–256 mm), including a relatively longer tail (125–145 mm). *Apomys musculus* usually weigh less (16–24 g) but have a much longer tail (97–121 mm); both of these *Apomys* species have ventral pelage that is much paler than the dorsum. *Crocidura grayi* are superficially similar but smaller (total length 125–149 mm; weight 8.5–12 g) and have tiny eyes; a tapering, flexible snout; a continuous row of teeth without a gap between the incisors and molars; dark gray or nearly black fur; and a tail with scattered long hairs.

DISTRIBUTION: Known only from Mt. Isarog, Camarines Sur Province, Luzon.

EVOLUTION AND ECOLOGY: Only two species of *Archboldomys* are known; the other occurs in the Central Cordillera. *Archboldomys luzonensis* are similar to *Soricomys*, but that genus is most closely related to *Chrotomys*. *Archboldomys luzonensis* are moderately common in primary montane and mossy forest on Mt. Isarog, from 1350 m to 1750 m elevation. They are active during the day and are principally, if not exclusively, diurnal. They forage in the leaf-litter and moss on the forest floor, often in well-defined runways beside fallen logs or in root tangles. Limited evidence indicates that they feed on earthworms, insects, amphipods, and other invertebrates. They occur together with *Crocidura grayi*, *Chrotomys gonzalesi*, and *Rhynchomys isarogensis*, all of which also feed on earthworms and other invertebrates. Their density in upper mossy forest was estimated at 4.5 individuals/ha, and they were second in abundance only to *Apomys musculus*, but because of their small mass, *A. luzonensis* had the lowest biomass of any small mammal at the study site. Two females each had a single embryo; females have two pairs of inguinal mammae. Adult males have large testes, up to 11 × 6 mm. Among the earthworm mice, *A. luzonensis* have one of the most distinctive karyotypes: $2n = 26$, FN = 43, with heteromorphic sex chromosomes in both sexes.

STATUS: Occur within a currently stable national park, but are quite restricted in distribution. Recent surveys on Mt. Labo and Mt. Malinao did not find this species, but additional, focused surveys at high elevation are needed.

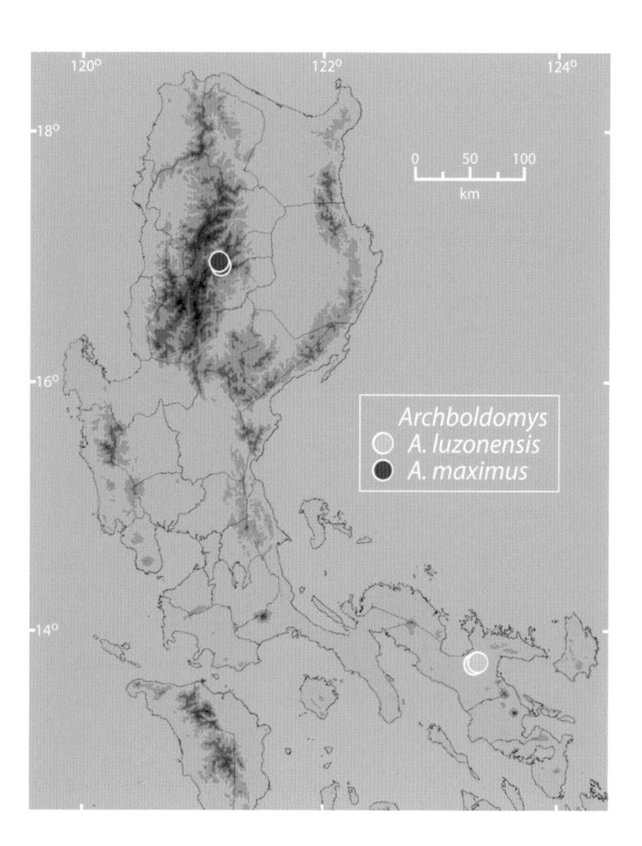

REFERENCES: Balete & Heaney, 1997; Balete et al., 2012, 2013b; Heaney and Regalado, 1998; Heaney and Utzurrum, 1991; Heaney et al., 1999; Musser, 1982c; Rickart and Musser, 1993; Rickart et al., 1991, 1998.

Archboldomys maximus
large Cordillera shrew-mouse

Balete et al., 2012. Am. Mus. Novitates, 3754:33.
Type locality: 1.75 km N, 0.4 km W Mt. Amuyao peak, 1885 m elevation, Barlig Municipality, Mountain Province, Luzon Island, 17.02929° N, 121.12466° E.

DESCRIPTION: Total length 200–232 mm; tail 90–108 mm; hind foot 29–33 mm; ear 18–21 mm; weight 40–55 g. A mouse of moderate size, with a stout body; dark brown pelage both dorsally and ventrally; a tail nearly

as long as the head and body; and relatively long hind feet. *Soricomys montanus* are similar in general appearance but are smaller (head and body 93–103 mm), and have slightly less dark pelage with a rusty tint; a lighter body (23–31 g); a proportionately longer tail (87–98 mm); and a shorter hind foot (23–25 mm). *Crocidura grayi* are much smaller (total length 125–149 mm; weight 8.5–12 g). *Batomys granti* (total length 342–355 mm; weight 182–226 g) and *Chrotomys silaceus* (total length 239–299 mm; weight 71–160 g) are both dark-colored, stout-bodied rodents, but they are much larger. *Apomys datae* have white or nearly white ventral pelage; a long (125–143 mm) tail that is dark on top and nearly white ventrally; and are heavier (67–105 g).

DISTRIBUTION: Currently known only from Mt. Amuyao, Mountain Province, but probably more widespread on nearby peaks in the southern Central Cordillera. (For map, see *Archboldomys luzonensis* species account.)

EVOLUTION AND ECOLOGY: *Archboldomys* was one of the first of the genera of earthworm mice to diverge from the other members of the clade, probably initially appearing about 4 million years ago (fig. 5.2). The event that led to the development of the two current species probably took place about 2 million years ago and most likely involved colonization from the Central Cordillera to the Bicol Peninsula (where they now occur on Mt. Isarog), which was then an island separate from the main body of Luzon. Both species of *Archboldomys* are restricted to high elevations, occurring in montane and especially mossy forest; on Mt. Amuyao, *A. maximus* were found in old-growth and high-quality regenerating forest between 1650 m and 2690 m elevation. They are the largest of Luzon's shrew-mice. They feed on earthworms, centipedes, and insects that live in leaf-litter on the forest floor. They occur together with *Crocidura grayi*, *Chrotomys silaceus*, *Ch. whiteheadi*, *Rhynchomys soricoides*, and *Soricomys montanus*, all of which prey on earthworms, and with *Apomys datae*, which feed opportunistically on earthworms (as well as on seeds and fruits). Females have two pairs of mammae, and limited data imply a litter size of one. Adult males have large testes, up to 10 × 7 mm.

STATUS: Stable and moderately common on Mt. Amuyao; further studies of their distribution and abundance are needed.

REFERENCES: Balete et al., 2012; Jansa et al., 2006; Rickart et al., 2011b, submitted; Schenk et al., 2013.

Chrotomys gonzalesi
Isarog striped rat, Isarog chrotomys

Rickart and Heaney, 1991. Proc. Biol. Soc. Washington, 104:389. Type locality: Western slope of Mt. Isarog, 1350 m elevation, 4 km N, 21 km E Naga, Camarines Sur Province, Luzon Island, 13°40′ N, 123°22′ E.

DESCRIPTION: Total length 232–293 mm; tail 83–105 mm; hind foot 34–39 mm; ear 20–23 mm; weight 70–190 g. A stout-bodied animal, with a short, thick tail; slightly enlarged forefeet; small eyes; and a slightly elongated snout. The fur is soft, dense, and long; very dark on the back, with a narrow (sometimes intermittent) pale stripe extending from the crown of the head to the base of the tail; and gray/brown on the sides and abdomen. Adult males are probably larger than females, on average. *Chrotomys mindorensis* are larger (total length 260–309 mm); have fur that is shorter and coarser textured; and are more brightly colored.

DISTRIBUTION: Known only from Mt. Isarog, Camarines Sur Province, southern Luzon. Should be sought elsewhere on the Bicol Peninsula of southern Luzon. A specimen of *Chrotomys* recently obtained near Saddle Peak in southern Luzon is provisionally assigned to this species but may prove to be distinct, pending further study.

EVOLUTION AND ECOLOGY: There are five recognized species of *Chrotomys*: *C. sibuyanensis* are restricted to Sibuyan Island, *C. mindorensis* occur on Mindoro Island and in the lowlands of central Luzon, *C. silaceus* and *C. whiteheadi* occur in the Central Cordillera of northern Luzon, and *C. gonzalesi* are known only from Mt. Isarog in southern Luzon. *Chrotomys gonzalesi* live in montane and mossy forest, from ca. 1350 m to at least 1800 m elevation. They are active both day and night, burrowing through humus and soil and feeding on earthworms, insects, and other invertebrates. They co-occur with *Apomys microdon*, *A. musculus*, *Archboldomys luzonensis*, *Batomys uragon*, *Crocidura grayi*, *Rattus everetti*, and *Rhynchomys isarogensis*. Females have two pairs of inguinal mammae; one adult female had two embryos. The standard karyotype is $2n = 44$, FN = 52.

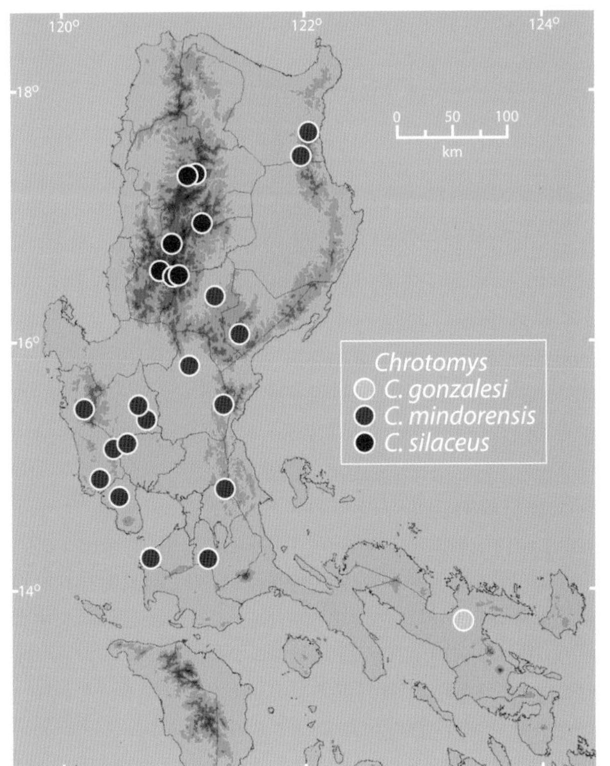

STATUS: Population stable but geographically restricted. Habitat destruction on Mt. Isarog (a national park) in the 1980s has largely ceased, and there seem to be no current threats there.

REFERENCES: Balete and Heaney, 1997; Heaney et al., 1999; Rickart and Heaney, 1991; Rickart and Musser, 1993; Rickart et al., 1991.

Chrotomys mindorensis
lowland striped rat, lowland chrotomys

Kellogg, 1945. Proc. Biol. Soc. Washington, 58:123.
Type locality: 3 miles SSE San Jose, 200 feet elevation, Mindoro Island.

DESCRIPTION: Total length 260–309 mm; tail 99–123 mm; hind foot 36–40 mm; ear 19–23 mm; weight 152–199 g. A stout-bodied animal, with a short, thick tail; slightly enlarged forefeet; small eyes; and a slightly elongated snout. The fur is moderately long and coarse-textured. Dorsal coloration is very striking, with a central pale yellowish stripe extending down the midline from the crown of the head to the base of the tail, bordered by two black stripes. The sides of the body are grayish-brown; the abdomen is slightly paler. Adult males are larger than females, on average. *Chrotomys*

whiteheadi have longer, softer, denser fur; a somewhat less distinct dorsal color pattern; and a slightly longer ear (23–27 mm). No other rodent within the range of this species has dorsal stripes.

DISTRIBUTION: Occur in lowlands throughout central Luzon and the Cagayan River Valley, and also at higher elevations in Bataan and Zambales Provinces. Also widespread on Mindoro. (For map, see *Chrotomys gonzalesi* species account.)

EVOLUTION AND ECOLOGY: The genus *Chrotomys* belongs to the Old Endemic clade of Philippine earthworm mice. It is most closely related to the shrew-mouse genus *Soricomys*; the two genera probably arose from a common ancestor within the Central Cordillera roughly 3.0 million years ago (fig. 5.2). This indicates that the specialized adaptations for burrowing seen in *Chrotomys* arose after separation from the more generalized *Soricomys*. Preliminary genetic data imply that *C. mindorensis* are most closely related to *C. gonzalesi*. Additionally, the genetic similarity of *C. mindorensis* from central Luzon and Mindoro Island indicate relatively recent overwater dispersal from Luzon to Mindoro. *Chrotomys mindorensis* occur in primary and secondary forest and adjacent agricultural areas, from near sea level to about 500 m elevation, except in the Zambales Mountains, where they have been documented up to 2025 m elevation. They are active both day and night. They dig extensively through humus and soil, feeding on earthworms, insects, and other invertebrates. Their presence is often conspicuous, due to the abundance of shallow pits and small piles of dirt where they have dug for food. They are extremely tolerant of habitat disturbance and are often common in agricultural areas, where they feed on non-native invertebrates, including significant crop pests. In agricultural areas, they are preyed upon by domestic cats and dogs. Rural farmers are often familiar with the species because of its distinctive coloration and daytime activity. Females have two pairs of inguinal mammae. The standard karyotype of $2n = 44$, FN = 52 is indistinguishable from those of *C. gonzalesi* and *C. silaceus* but differs from that of *C. whiteheadi*.

STATUS: Widespread; occur in any type of forest within their geographic range, as well as in second growth and agricultural areas.

REFERENCES: Alviola et al., 2011; Balete et al., 2009, 2011, 2013a; Barbehenn et al., 1973; Heaney et al., 2013b; Jansa et al., 2006; Kellogg, 1945; Musser et al., 1981; Rickart and Heaney, 2002; Rickart and Musser, 1993; Rickart et al., 2005, 2011b, 2013; Schenk et al., 2013; Stuart et al., 2008.

Chrotomys silaceus
silver earth-rat, silver chrotomys

(Thomas, 1895). Ann. Mag. Nat. Hist., ser. 6, 16:161. Type locality: "Highlands of Northern Luzon," subsequently defined by Thomas (1898) as the plateau of Mt. Data, 7000–8000 feet elevation, Lepanto [now in Mountain Province], Luzon Island.

DESCRIPTION: Total length 239–299 mm; tail 97–121 mm; hind foot 34–39 mm; ear 19–23 mm; weight 71–160 g. A medium-sized animal, with a stout body; a thick tail that is shorter than the head and body; slightly enlarged forefeet; small eyes; and a slightly elongated snout. The fur is soft, dense, and moderately long; it is dark gray or silver-gray over the back and sides, and a paler shade of gray or silver-gray on the ab-

domen. Some individuals have a trace of a short white stripe (or blaze) on the forehead. *Chrotomys whiteheadi* have a similar body form and size but have longer fur and a distinctive pale stripe extending from the head to the base of the tail. *Rhynchomys soricoides* are similar in size but have longer hind limbs; a slender head, with an elongated snout; and shorter pelage that is blackish dorsally rather than gray.

DISTRIBUTION: Central Cordillera of northern Luzon only. (For map, see *Chrotomys gonzalesi* species account.)

EVOLUTION AND ECOLOGY: Genetic data place *Chrotomys silaceus* as the sister-species to all other members of the genus (fig. 5.2). Compared with other *Chrotomys*, *C. silaceus* is distinctive in lacking a dorsal stripe and having a more delicate skull and smaller molars; all of these features are shared with *Soricomys*, to which *Chrotomys* is closely related. *Chrotomys silaceus* have been documented at 1,800–2,800 m elevation in primary and secondary transitional montane/mossy and mossy forest, and occasionally in shrubby vegetation near agricultural fields; they are more abundant at the higher portion of their elevational range. They are primarily active during the day but also at night, foraging for earthworms and other soil invertebrates by digging in leaf-litter and humus. Individuals that we kept briefly in captivity were docile and readily ate earthworms. *Chrotomys silaceus* and *C. whiteheadi* have overlapping elevational distributions, but where they co-occur, *C. silaceus* are more abundant, and *C. whiteheadi* tend to be present in relatively drier habitat. In addition, *C. silaceus* occur with *Apomys datae*, *Archboldomys maximus*, *Crocidura grayi*, *Rhynchomys soricoides*, and *Soricomys montanus*, all of which also consume invertebrates and forage on the ground. Females have two pairs of mammae; five adult females each carried two embryos. The standard karyotype is $2n = 44$, FN = 52.

STATUS: Widespread and moderately common at high elevations in the Central Cordillera. A 2006 resurvey failed to document this species on Mt. Data (the type locality), which has largely been converted to intensive agriculture; they are now presumed to be extinct there.

REFERENCES: Heaney et al., 2004, 2006b; Musser and Carleton, 2005; Rickart and Heaney, 2002; Rickart et al., 2005, 2011a, 2011b; Sanborn, 1952; Thomas, 1895.

Chrotomys whiteheadi
Cordillera striped rat, Cordillera chrotomys

Thomas, 1895. Ann. Mag. Nat. Hist., ser. 6, 16:161. Type locality: "Highlands of Northern Luzon," subsequently defined by Thomas (1898) as the plateau of Mt. Data, 7000–8000 feet elevation, Lepanto [now in Mountain Province], Luzon Island.

DESCRIPTION: Total length 254–310 mm; tail 95–135 mm; hind foot 36–40 mm; ear 23–27 mm; weight 105–190 g. A medium-sized animal, with a stout body; a thick tail shorter than the length of the head and body; slightly enlarged forefeet; small eyes; and a slightly elongated snout. The fur is soft, dense, and long; it is brownish-black on the back, with a distinctive pale stripe extending from the crown of the head to the base of the tail. The width of the dorsal stripe varies with age and is most pronounced in adults. The sides of the body are brown; the abdomen is grayish-brown. Adult males are larger than females, on average. *Chrotomys mindorensis* are slightly larger; have shorter fur; and are generally more brightly colored. *Chrotomys silaceus* are

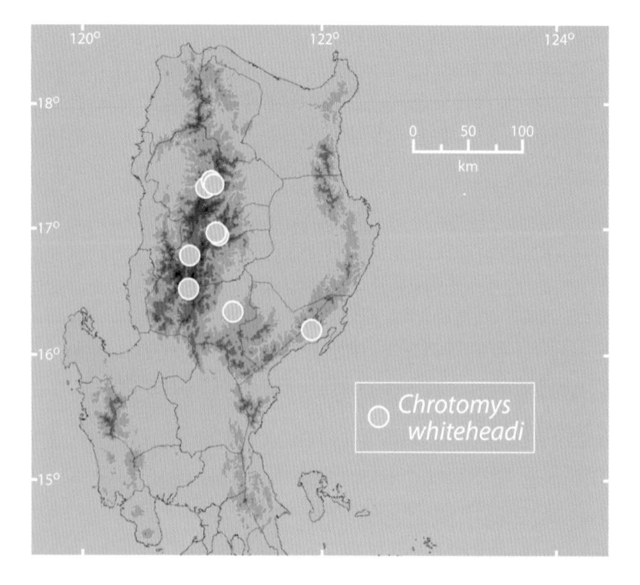

similar in size and shape but have uniformly gray or silver dorsal pelage, without a dorsal stripe, and a shorter ear (19–23 mm).

DISTRIBUTION: Widespread within the Central Cordillera of northern Luzon, with records also from Mt. Palalli in the Caraballos Mountains and Mt. Anacuao in the Northern Sierra Madre.

EVOLUTION AND ECOLOGY: Common in primary and secondary montane and mossy forest, over a broad elevational range (from 850 m to 2690 m). They are highly tolerant of habitat disturbance, often occurring in shrubby second growth and agricultural areas. They occur together with *Apomys datae*, *Archboldomys maximus*, *Crocidura grayi*, *Chrotomys silaceus*, *Rhynchomys soricoides*, and *Soricomys montanus*, all of which feed on earthworms and are principally ground dwelling. *Chrotomys whiteheadi* are active both day and night, burrowing through humus and soil and feeding on earthworms, snails, and other invertebrates, including golden apple snails (*Pomacea canaliculata*), which are a pest in ricefields. Their presence is often conspicuous, due to the abundance of shallow pits and small piles of dirt where they have dug for food. Limited data suggest that they may move more than 100 m in a 24-hour period. Individuals sometimes plug the entrance to a burrow by piling up dirt from inside the burrow. Females have two pairs of inguinal mammae. Four pregnant females had two, two, two, and one embryos. Compared with other groups of native Philippine murid rodents, adult male *Chrotomys* have testes that are large relative to their body size. In contrast to other species of *Chrotomys*, *Rhynchomys*, and *Apomys* (subgenus *Megapomys*) that share a standard karyotype of $2n = 44$, FN = 52, *C. whiteheadi* has a unique derived chromosomal arrangement of $2n = 38$, FN = 52.

STATUS: Common and widespread in northern Luzon in both old-growth and heavily disturbed habitats.

REFERENCES: Alviola et al., 2011; Heaney et al., 2004, 2013a; Largen, 1985; Rabor, 1955; Reginaldo and de Guia, 2014; Rickart and Heaney, 2002; Rickart and Musser, 1993; Rickart et al., 2005, 2011a, 2011b; Sanborn, 1952; Stuart et al., 2007.

Rhynchomys banahao
Banahao shrew-rat, Banahao rhynchomys

Balete et al., 2007. J. Mammal., 88:289. Type locality: Mt. Banahao [= Banahaw], 1465 m elevation, Tayabas Municipality, Quezon Province, Luzon, 14°03′59.4″ N, 121°30′30.9″ E.

DESCRIPTION: Total length 305–320 mm; tail 126–130 mm; hind foot 39–40 mm; ear 24–25 mm; weight 135–

155 g. A highly distinctive animal, with a slender head; long, pointed snout; a small mouth, with tiny incisors; fairly small eyes; large ears; and long vibrissae. The fur is soft and dense. The upperparts are dark gray; the underparts are only slightly paler. Some individuals have white patches of fur on the abdomen. The body is slender, but the hind legs are large and muscular. The tail is much shorter than the length of the head and body and is uniformly dark above and below. The hind feet are long and narrow. *Apomys banahao* and *A. magnus* have tails that are equal to or longer than the length of the head and body; are distinctly darker dorsally than ventrally; and have a shorter, more robust snout.

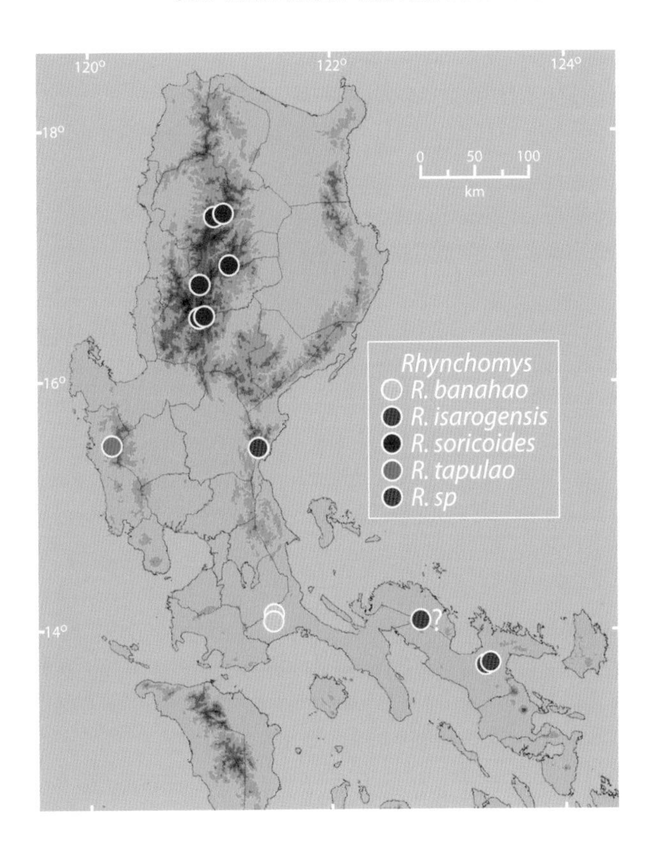

DISTRIBUTION: Currently known only from Mt. Banahaw, Quezon Province, Luzon Island.

EVOLUTION AND ECOLOGY: The genus *Rhynchomys* is the sister-taxon to the common ancestor of *Chrotomys* + *Soricomys* and first appeared about 3.3 million years ago (fig. 5.3). All of the known species in this genus have similar mitochondrial DNA, which may imply that they speciated rather recently. *Rhynchomys banahao* are known from four specimens taken at 1250–1625 m elevation in primary montane forest; none have been caught in intensively sampled mossy forest at higher elevations (1750 m and 2030 m) or in disturbed lowland forest (660 m and 765 m elevation). All were captured on the ground, in traps baited with live earthworms. Trap success for *R. banahao* was low, which indicates that they may be uncommon. They occur together with *Apomys banahao* and *A. magnus*, two other species endemic to Mt. Banahaw, with the latter two taxa even at times being caught in the same traps as the former species. Adult females have two pairs of inguinal mammae.

STATUS: Recently discovered and poorly known. Mt. Banahaw is within Mts. Banahaw–San Cristobal National Park, where, as of 2008, there was no significant threat to habitat.

REFERENCES: Balete et al., 2007; Heaney et al., 2013b.

Rhynchomys isarogensis
Isarog shrew-rat, Isarog rhynchomys

Musser and Freeman, 1981. J. Mammal., 62:154. Type locality: Mt. Isarog, 5000 feet elevation, Camarines Sur Province, SE peninsula of Luzon Island.

DESCRIPTION: Total length 270–305 mm; tail 108–126 mm; hind foot 37–40 mm; ear 21–23 mm; weight 110–156 g. A highly distinctive animal, with a slender head; a long, pointed snout; a small mouth, with tiny incisors; fairly small eyes; large ears; and long vibrissae. The fur on the upperparts is grayish-brown, very dense, and soft; the underparts are nearly white. The body is slender, but the hind legs are large and muscular. The tail is much shorter than the length of the head and body and is somewhat darker on the upper side than on the underside. The hind feet are long and narrow. No similar rodent is found where it occurs.

DISTRIBUTION: Mt. Isarog, Camarines Sur Province, southern Luzon. Specimens of *Rhynchomys* recently obtained on Mt. Labo in southern Luzon are provisionally assigned to this species but may prove to be distinct, pending further study. (For map, see *Rhynchomys banahao* species account.)

EVOLUTION AND ECOLOGY: Occur in primary montane and mossy forest on Mt. Isarog, where they have been recorded at elevations between 1125 m and 1750 m.

They are common in mossy forest and less common in montane forest; at 1650 m elevation, density was estimated at 2.6/ha, and movements were at least 50 m in a day. They construct small trails over the surface of the ground, which they patrol in search of earthworms and other soft-bodied soil invertebrates, including amphipods, insect larvae, and some adult insects; there is no evidence that they eat any plant material. Small earthworms are swallowed whole; large worms are torn into pieces and swallowed after squeezing out the gut contents. They are active primarily, but not exclusively, at night. *Archboldomys luzonensis*, *Chrotomys gonzalesi*, and *Crocidura grayi* are co-occurring species that also forage for invertebrates on or beneath the ground surface. Females have two pairs of inguinal mammae; three females each had a single embryo. Males have large testes, up to 12 × 20 mm. The standard karyotype is $2n = 44$, FN = 52.

STATUS: Stable, but geographically restricted to Mt. Isarog National Park. Although logging has occurred in the lowlands and lower montane forest in the past, as of 2008 Mt. Isarog was not threatened by logging, and most montane and mossy forest there is undisturbed.

REFERENCES: Balete and Heaney, 1997; Balete et al., 2013b; Heaney and Regalado, 1998; Heaney et al., 1999; Musser and Freeman, 1981; Rickart and Heaney, 2002; Rickart and Musser, 1993; Rickart et al., 1991.

Rhynchomys soricoides
Cordillera shrew-rat, Cordillera rhynchomys

Thomas, 1895. Ann. Mag. Nat. Hist., ser. 6, 16:160. Type locality: "Highlands of Northern Luzon," subsequently defined by Thomas (1898) as the plateau of Mt. Data, 7000–8000 feet elevation, Lepanto [now in Mountain Province], Luzon Island.

DESCRIPTION: Total length 307–354 mm; tail 132–162 mm; hind foot 39–42 mm; ear 23–25 mm; weight 133–

225 g. A highly distinctive animal, with a slender head; a long, pointed snout; a small mouth, with tiny incisors; fairly small eyes; large ears; and long vibrissae. The fur is short, soft and dense and is blackish-brown dorsally, with pale silvery-gray underparts. Some individuals have patches of pure white fur on the abdomen. The body is slender, but the hind legs are large and muscular. The tail is much shorter than the length of the head and body; it is dark on the upper side and paler on the underside. The hind feet are long and narrow. *Chrotomys silaceus* are similar in size (weight 71–160 g) but have silvery-gray pelage; a more robust body; a shorter hind foot (34–39 mm); and smaller hind limbs.

DISTRIBUTION: Central Cordillera of northern Luzon only. Records from Benguet, Kalinga, and Mountain Provinces. (For map, see *Rhynchomys banahao* species account.)

EVOLUTION AND ECOLOGY: Occur throughout the Cordillera region in montane and mossy forest, from 1600 m to 2695 m elevation. They apparently have a low tolerance for habitat disturbance; they seem to occur only in old-growth or old secondary forest, but not in heavily disturbed forest, early second growth, or agricultural areas. They are active during both day and night and feed exclusively on earthworms and other soft-bodied soil invertebrates; there is no evidence that they eat any plant matter. They occur together with species of *Crocidura*, *Archboldomys*, *Chrotomys*, and *Soricomys*, which also consume invertebrates. Unlike *Chrotomys* (burrowing animals that principally forage underground), *Rhynchomys* construct small trails over the ground surface, which they patrol in search of earthworms and other invertebrate prey. Females have two pairs of inguinal mammae. Eight females had an average of 1.5 embryos (range = 1–2). Adult males have large testes, up to 13 × 25 mm. The standard karyotype is $2n = 44$, FN = 52(?).

STATUS: Widespread at moderately low density in high elevation montane and mossy forest in the Central Cordillera. A 2006 resurvey failed to document this species on Mt. Data (the type locality), which has largely been converted to intensive agriculture; this species is now presumed to be extinct there.

REFERENCES: Heaney et al., 2004, 2006c; Rabor, 1955; Rickart et al., 2011a, 2011b, submitted; Sanborn, 1952; Thomas, 1898.

Rhynchomys tapulao
Zambales shrew-rat, Zambales rhynchomys

Balete et al., 2007. J. Mammal. 88:296. Type locality: Mt. Tapulao, 2024 m elevation, Palauig Municipality, Zambales Province, Luzon Island, 15°28′54.8″ N, 120°07′10.4″ E.

DESCRIPTION: Total length 290–308 mm; tail 120–128 mm; hind foot 38–40 mm; ear 24–25 mm; weight 129–

156 g. A highly distinctive animal, with a long, slender snout; a small mouth, with tiny incisors; fairly small eyes; large ears; and long vibrissae. The fur is golden-brown and dense, with the underparts being nearly white. Some individuals have patches of gray fur on the abdomen. The body is slender, but the hind legs are large and muscular. The tail is much shorter than the length of the head and body; nearly white on the posterior one-third of its length; and the dark basal portion is paler on the underside than on the upper side. The hind feet are long and narrow. *Apomys zambalensis* and *A. brownorum* are smaller (weight 60–125 g); have tails that are equal to or longer than the head and body; and are sharply bicolored dorsoventrally. *Chrotomys mindorensis* are similar in size but have a pale stripe running down the back, from the base of the neck to the base of the tail; a stout body; a broad, sturdy snout; and hind legs that are not especially large.

DISTRIBUTION: Known only from Mt. Tapulao, Zambales Province, Luzon Island, but might occur on the adjacent peaks of the Zambales Mountains. (For map, see *Rhynchomys banahao* species account.)

EVOLUTION AND ECOLOGY: Known from three specimens trapped in primary mossy forest at 2024 m elevation. They were captured during the daytime, in traps placed on the ground; two were taken with earthworm bait, and one with coconut. As with other *Rhynchomys*, they probably forage mostly in trails they construct on the surface of the ground, feeding on earthworms and soft-bodied invertebrates, although *R. tapulao* are active primarily during daylight. They occurred together with *Apomys brownorum*, *Chrotomys mindorensis*, and *Crocidura grayi*, which also are ground-dwelling species that feed on earthworms and other invertebrates. A single female had two pairs of inguinal mammae.

STATUS: Recently discovered and poorly known. Mining is active (as of early 2015) on Mt. Tapulao and other peaks in the Zambales Mountains, but the impact of that disturbance on this species is unknown. A large section of the Zambales Mountains is currently under consideration for designation as a protected area, but much of the high elevation where mining takes place is excluded.

REFERENCES: Balete et al., 2007, 2009.

Rhynchomys sp.
Mingan shrew-rat, Mingan rhynchomys

DESCRIPTION: Total length 312–328 mm; tail 126–132 mm; hind foot 39–41 mm; ear 21–24 mm; weight 158–176 g. A highly distinctive animal, with a slender head; a long, pointed snout; a small mouth, with tiny incisors; fairly small eyes; large ears; long vibrissae; and a tail that is much shorter than the length of the head and body. *Apomys aurorae* and *A. minganensis* are smaller (weight 58–92 g) and have a tail that is about equal to or longer than the length of head and body and is sharply bicolored dorsoventrally.

DISTRIBUTION: Known only from Mt. Mingan, Aurora Province, Luzon, but might occur on nearby peaks in the Mingan Mountains. (For map, see *Rhynchomys banahao* species account.)

EVOLUTION AND ECOLOGY: A recently discovered and still unnamed species, recorded from localities between

1475 m and 1785 m elevation on Mt. Mingan, in habitat ranging from lower montane forest to mossy forest; relative abundance was greatest in mossy forest. Of 23 individuals, all were captured on the ground, in traps baited with live earthworms, and all but one were captured during the night. They occurred together with *Apomys aurorae*, *A. minganensis*, *Crocidura grayi*, *Rattus everetti*, and *Soricomys leonardocoi*; these other species were also captured exclusively on the ground and consume earthworms.

STATUS: Recently discovered and poorly known, but seemingly stable in habitat that is under no current threat.

REFERENCES: Balete et al., 2011.

Soricomys kalinga
Kalinga shrew-mouse

(Balete et al., 2006). Syst. Biodiv., 4:491. Type locality: Mt. Bali-it, 1950 m elevation, Balbalan Municipality, Kalinga Province, Luzon Island, 17°25.8′ N, 121°00.1′ E.

DESCRIPTION: Total length 183–207 mm; tail 85–101 mm; hind foot 23–25 mm; ear 13–15 mm; weight 21–29 g. A small mouse, with a slender but sturdy snout; small eyes and ears; and a tail shorter than the length of the head and body. The fur is dark brown, with a slight rusty tint over the entire animal. *Crocidura grayi* are superficially similar but smaller (total length 125–149 mm; weight 8.5–12 g), with tiny eyes and a tapering, flexible snout. All species of *Apomys* have tails that are dark dorsally, nearly white ventrally, and longer than the length of the head and body. *Chrotomys silaceus* are much larger (total length 239–299 mm; hind foot 34–39 mm) and have silver-gray pelage.

DISTRIBUTION: Currently known only from the vicinity of Mt. Bali-it, Kalinga Province, but possibly widespread within the northern Central Cordillera. Specimens from Benguet and Mountain Provinces formerly referred to this species are now recognized as a separate species, *Soricomys montanus*.

EVOLUTION AND ECOLOGY: The genus *Soricomys* is superficially similar to *Archboldomys*, and *S. kalinga* and *S. musseri* were placed in that genus when they were first described. The four species of *Soricomys* are actually most closely related to *Chrotomys*, with which *Soricomys* shared a common ancestor as recently as ca. 3 million years ago (fig. 5.3). *Soricomys kalinga* is most closely related to *S. montanus*, and the two species are so similar externally that when they were first discovered, they were considered to represent a single species. Kalinga shrew-mice have been documented from 1600 m

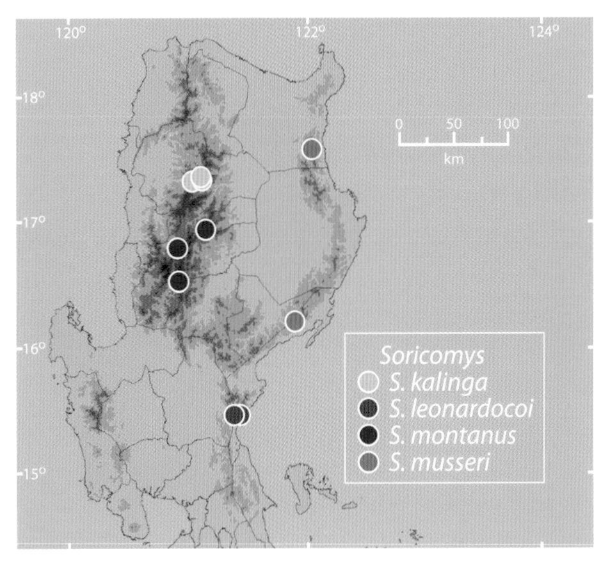

to 2150 m elevation, in montane and mossy forest that is dominated by oaks, podocarps, laurels, and myrtles, with rhododendrons as a common understory plant. These mice forage in the leaf-litter on the surface of the ground, often in runways beside fallen logs and tree trunks and in root tangles, but occasionally they climb up to 2 m above the ground on the leaning trunks of trees. They feed on earthworms, insects, and other invertebrates. They are active primarily during daylight and are fairly common; we occasionally saw them during the day as they ran across trails. They occur together with *Crocidura grayi*, *Apomys abrae*, *A. datae*, *A. musculus*, *Chrotomys silaceus*, *Ch. whiteheadi*, and *Rhynchomys soricoides*, all of which also consume earthworms and other invertebrates. Females have two pairs of inguinal mammae; two individuals captured in April each had two embryos, and one had a single placental scar. Adult males have large scrota and testes (up to 16 × 10 mm). Both sexes appear to reach reproductive maturity at ca. 24–26 g. The karyotype is $2n = 44$.

STATUS: Currently known from a single mountain in Kalinga Province, where they are common, but probably more widespread in the northern Central Cordillera.

REFERENCES: Balete et al., 2006, 2012; Heaney et al., 2003; Rickart and Heaney, 2002; Rickart et al., 2005, 2011a, 2011b; Schenk et al., 2013.

Soricomys leonardocoi Mingan shrew-mouse

Balete et al., 2012. Am. Mus. Novitates, 3754:44. Type locality: 0.9 km S, 0.3 km W Mt. Mingan peak, 1785 m elevation, Dingalan Municipality, Aurora Province, Luzon Island, 15.47390° N, 121.40066° E.

DESCRIPTION: Total length 180–208 mm; tail 82–95 mm; hind foot 21–25 mm; ear 13–15 mm; weight 26–36 g. A small mouse, with a slender but sturdy snout; small eyes and ears; and a tail shorter than the length of the

head and body. The fur is dark chestnut-brown dorsally and slightly paler ventrally. Adult males have large scrota. *Crocidura grayi* are superficially similar but smaller (total length 125–149 mm; weight 8.5–12 g), with tiny eyes and a tapering, flexible snout. All species of *Apomys* have tails that are dark dorsally, nearly white ventrally, and longer than the length of the head and body.

DISTRIBUTION: Known only from the Mingan Mountains, Aurora Province. (For map, see *Soricomys kalinga* species account.)

EVOLUTION AND ECOLOGY: Closely related to *Soricomys musseri* (from the Northern Sierra Madre) and the species from the Central Cordillera (*S. kalinga* and *S. montanus*). *Soricomys leonardocoi* occur in montane and mossy forest, from 1476 m to 1785 m elevation. They are active during both day and night, feeding on earthworms and other soil invertebrates. *Crocidura grayi* and *Rhynchomys* sp., which also consume earthworms and other soil invertebrates, occur together with this species. Females have two pairs of inguinal mammae.

STATUS: Apparently common on Mt. Mingan, which is largely undisturbed by human activities.

REFERENCES: Balete et al., 2011, 2012.

Soricomys montanus
Southern Cordillera shrew-mouse

Balete et al., 2012. Am. Mus. Novitates, 3754:44.
Type locality: 0.75 km N, 0.6 km E of S peak Mt. Data, 2241 m elevation, Mountain Province, Luzon Island, 16.86287° N, 120.86108° E.

DESCRIPTION: Total length 182–199 mm; tail 87–98 mm; hind foot 23–25 mm; ear 13–15 mm; weight 23–31 g. A small mouse, with a slender but sturdy snout; small eyes and ears; and a tail shorter than the length of the head and body. The fur is dark rusty-brown over the entire animal and slightly brighter and shorter on the abdomen. Adult males have large scrota. *Archboldomys maximus* are larger (total length 200–232 mm; weight 40–55 g), with a longer hind foot (29–33 mm). *Crocidura grayi* are superficially similar but smaller (total length 125–149 mm; weight 8.5–12 g), with tiny eyes and a tapering, flexible snout. All species of *Apomys* have tails that are dark dorsally, nearly white ventrally, and longer than the length of the head and body. *Chrotomys silaceus* are much larger (total length 239–299 mm; hind foot 34–39 mm) and have silver-gray pelage.

DISTRIBUTION: Widely distributed over the southern portion of the Central Cordillera, including Mt. Amuyao, Mt. Data, and Mt. Pulag. (For map, see *Soricomys kalinga* species account.)

EVOLUTION AND ECOLOGY: Formerly included within *Soricomys kalinga*, to which *S. montanus* is externally similar, but the latter is distinctive both morphologically and genetically; the two species probably diverged a little over 1 million years ago. Southern Cordillera shrew-mice occur from 1950 m to 2690 m elevation, in old growth and lightly disturbed forest on Mt. Amuyao; they were common in heavily disturbed fragments of mossy forest on Mt. Data and Mt. Pulag, but were absent from pine forest and agricultural fields in all areas. They are predominantly diurnal, and we often saw them foraging in the leaf-litter and running across trails on Mt. Amuyao and Mt. Data. They feed on earthworms and other soil invertebrates; captives avidly consumed earthworms, quickly cleaning off the dirt and then biting off and swallowing small pieces. Females have two pairs of inguinal mammae. The karyotype has $2n = 44$, $FN = 52$, which is similar to that seen for species of *Chrotomys*, *Rhynchomys*, and many of the large-bodied *Apomys* (subgenus *Megapomys*).

STATUS: Widespread in the Central Cordillera; tolerant of forest disturbance.

REFERENCES: Balete et al., 2012; Heaney et al., 2006b; Rickart et al., 2011a, 2011b; Schenk et al., 2013.

Soricomys musseri
Sierra Madre shrew-mouse

(Rickart et al., 1998). Fieldiana Zool., 89:17. Type locality: Mt. Cetaceo, 1650 m, Sierra Madre Range, Callao Municipality, Cagayan Province, Luzon Island, 17°42′ N, 122°02′ E.

DESCRIPTION: Total length 180–198 mm; tail 88–93 mm; hind foot 19–25 mm; ear 14–15 mm; weight 30–32 g. A small mouse, with a rather narrow snout; small eyes and ears; and a dark tail that is shorter than the length of the head and body. The fur is rusty-brown over the entire body and slightly paler on the abdomen. Adult males have large scrota. All species of *Apomys* have tails that are dark dorsally, nearly white ventrally, and longer than the head and body. *Crocidura grayi* are super-

ficially similar but smaller (total length 125–149 mm; weight 8.5–12 g); have tiny eyes; and have a tapering, flexible snout.

DISTRIBUTION: Currently known only from Mt. Cetaceo, Cagayan Province, and Mt. Anacuao, Aurora Province. (For map, see *Soricomys kalinga* species account.)

EVOLUTION AND ECOLOGY: *Soricomys musseri* was initially assigned to the genus *Archboldomys*, but subsequent studies showed that two distinct genera of shrew-mice are present on Luzon (fig. 5.3). *Soricomys musseri* diverged from its relatives roughly 1.5 million years ago; it appears to be about equally closely related to the species from the Mingan Mountains (*S. leonardocoi*) and the ones from the Central Cordillera (*S. kalinga* and *S. montanus*). On Mt. Cetaceo, *S. musseri* occur in mossy forest from at least 1450 m to 1650 m elevation (and probably higher), and at 1760 m elevation on Mt. Anacuao. They are active primarily during the daytime, foraging in the leaf-litter on the surface of the ground, and often in runways beside fallen logs, under low vegetation, and the like. They feed on insects, earthworms, and other invertebrates. Their densities appear to be fairly low at all known localities. Females have two pairs of inguinal mammae.

STATUS: Moderately common in mossy forest on Mt. Cetaceo, and apparently less common on Mt. Anacuao. Further studies of their distribution in the Sierra Madre Range are needed.

COMMENTS: The first specimen was mistakenly identified by Heaney as *Crunomys fallax* in Danielsen et al. (1994). Listed by Heaney et al. (1998) as *Archboldomys* sp. A.

REFERENCES: Balete et al., 2011, 2012; Danielsen et al., 1994; Heaney et al., 1998, 2013a; Rickart et al., 1998.

SECTION 4

New Endemic Rodents

The New Endemic rodents are a miscellaneous lot, presented together here simply for convenience. All five species are native to Luzon, and all are descendents of ancestors that arrived in recent geological time, probably within the past 4 million years or less (fig. 5.3). They probably represent four different colonizations of Luzon; that is, they are the descendants of four different ancestors. Remarkably, they share some similarities, both in terms of what we do and do not know about them.

Two of these species are common and well known. *Bullimus luzonicus* is a member of an endemic Philippine genus that contains three species; it is most closely related to the genus *Sundamys*, which is diverse and widespread on the Sunda Shelf of Southeast Asia. The two genera last shared a common ancestor about 4 million years ago, presumably reflecting the arrival of the ancestor of *Bullimus* in the Philippines. The animal we know as *Rattus everetti* is widespread in the Philippines. It is most closely related to *Limnomys* and *Tarsomys*, both of which occur on Mindanao, and not to the proper members of the genus *Rattus*. Presumably when its phylogeny is known with more certainty, *R. everetti* will acquire a different generic name. Current evidence suggests that the Philippine endemic members of the *Tarsomys* clade last shared a common ancestor about 2 million years ago. *Bullimus luzonicus* and *R. everetti* are most common primarily in moderately to heavily

A: *Rattus everetti.* B: *Bullimus luzonicus.*

disturbed forest, at low to medium elevation, but have broad elevational ranges. They seem to be recent arrivals that landed on the shores of Luzon and have had limited success in invading mature forest high in the mountains, where murid rodent diversity and abundance are greatest (chapters 3 and 6).

The remaining three species are notable for how very little we know about them. *Abditomys latidens*, a Luzon endemic genus and species, is known from only a few specimens obtained in 1946 and 1967. It has a long tail and broad hind feet, which suggest that it is arboreal, but we have been unable to capture even a single specimen during our extensive surveys. *Tryphomys adustus*, another Luzon endemic genus and species, has a rather short tail and narrow hind feet, which suggest that it lives on the surface of the ground. It has only rarely been seen since it was first captured in 1907, and we failed to find a single animal. These enigmatic genera share a number of features in their cranial anatomy and may be the descendents of a single ancestral species that reached the Philippines. When that happened has yet to be estimated, because without tissue samples for genetic studies, we cannot place them into the phylogeny that allows us to understand their evolution. Both species are represented by specimens from the high Central Cordillera and from Luzon's lowland Central Plains, implying that they are ecologically tolerant—and yet we continue to know almost nothing about them.

The final species of the five New Endemics is *Crunomys fallax*, perhaps the most enigmatic of all. The single known specimen was obtained by John Whitehead in 1894, and the species has not been seen since. The genus *Crunomys* is represented in the Philippines by this species and two others on Greater Mindanao, and by a species on the Indonesian island of Sulawesi. Recent molecular evidence shows that *Crunomys* are closely related to *Maxomys*, but the details of their history of diversification remain unclear.

This summary is a powerful reminder of how much remains to be learned about the mammals of Luzon. After 12 years of intensive field surveys, including our best efforts to track down these enigmatic species, we know no more about *Abditomys*, *Crunomys*, and *Tryphomys* than we did when we began. We think it likely that they are still alive on Luzon; we simply do not yet know how to find them, so their place in the story of the evolution of mammalian diversity on Luzon remains to be told.

REFERENCES: Achmadi et al., 2013; Jansa et al., 2006; Musser and Heaney, 1992; Rickart et al., 1998, 2002; Schenk et al., 2013.

Abditomys latidens
Luzon broad-toothed rat

(Sanborn, 1952). Fieldiana Zool., 33:125. Type locality: Mt. Data, Mountain Province, Luzon Island.

DESCRIPTION: Total length 474–487 mm; tail 242–271 mm; hind foot 45–47 mm; ear 21–24 mm; weight unknown. A large rat, with a broad, deep head and unusually broad incisors. The tail is dark brown / black and is

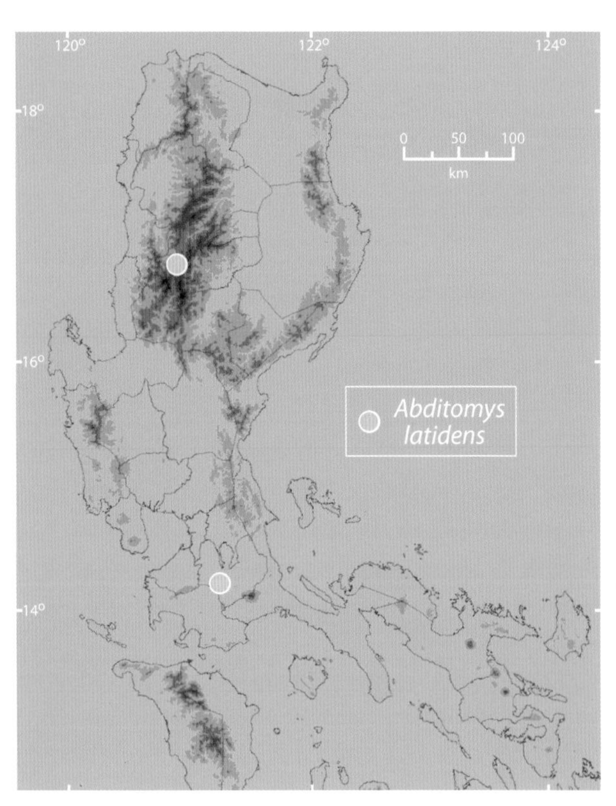

slightly longer than the length of the head and body. The dorsal pelage is brown, with relatively short guard hairs that are black at the base and tipped with white, giving the pelage a salt-and-pepper appearance. The ventral pelage is pale, either yellowish-brown or tan. The hallux (i.e., big toe) has a nail, not a claw. Females have four pairs of mammae. *Rattus everetti* have a tail that is longer than the head and body, with a white distal portion; a claw on the big toe; incisors of normal width; and long black guard hairs dorsally that project above the rest of the fur.

DISTRIBUTION: Poorly known, but apparently endemic to central and northern Luzon Island.

EVOLUTION AND ECOLOGY: One of the most enigmatic mammals on Luzon; very poorly known. The genus *Abditomys* is currently represented by a single species. Morphological evidence suggests that *Abditomys* is most closely related to *Tryphomys*, but no molecular phylogenetic studies have been conducted. The first specimen was captured in a densely vegetated gully in pine forest at about 2250 m elevation on Mt. Data. This specimen was seen and captured in a tree, and the long tail implies that probably it is arboreal. A second specimen was captured in scrubby second growth in the lowlands, at about 75 m elevation. The broad, strong incisors and muzzle shape suggest that they consume hard foods, probably plant material and perhaps seeds. The presence of four pairs of mammae suggests that females may typically give birth to three to four young per litter.

STATUS: Uncertain. Targeted field and phylogenetic studies are needed.

REFERENCES: Barbehenn et al., 1973; Musser, 1982a; Musser and Heaney, 1992; Rabor, 1955, 1986; Sanborn, 1952.

Bullimus luzonicus
large Luzon forest rat, Luzon bullimus

(Thomas, 1895). Ann. Mag. Nat. Hist., ser. 6, 16:163. Type locality: Mt. Data, 8000 feet elevation, Lepanto [now in Mountain Province], Northern Luzon.

DESCRIPTION: Total length 433–500 mm; tail 192–233 mm; hind foot 53–58 mm; ear 31–36 mm; weight 375–

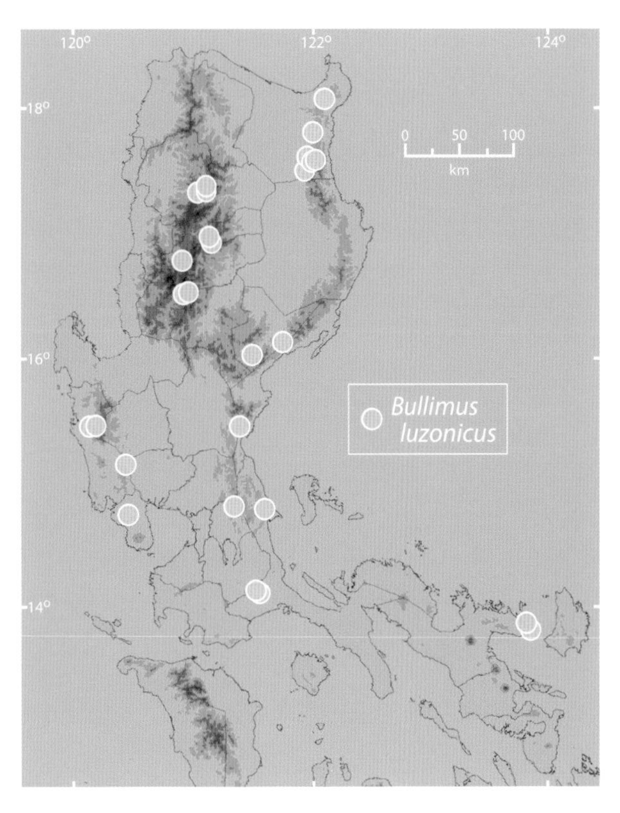

520 g. A large rat, with the tail shorter than the length of the head and body. The eyes and ears are large. The fur on the upperparts is usually dark brown, grading to a silvery-brown on the underparts, but some populations have either gray-brown or golden-brown dorsal fur. The tail is nearly hairless; usually is dark brown near the base and white on the posterior 10%–25%; and paler ventrally than dorsally. *Rattus everetti* have a tail that is longer (181–244 mm) than the length of the head and body, is nearly as dark ventrally as dorsally, and typically has a longer white tip. They also have a proportionately smaller head and eyes; the fur on the abdomen is nearly white (often stained with brown and/or yellow); and the hind foot is shorter (42–50 mm) and usually less heavily pigmented ventrally.

DISTRIBUTION: Widespread on Luzon, though seemingly absent from much of the Bicol Peninsula.

EVOLUTION AND ECOLOGY: Genetic data indicate that *Bullimus* is most closely related to *Sundamys*, which occurs widely on the Sunda Shelf of Southeast Asia, including one species on Palawan. Three species of *Bullimus* are currently recognized, with *B. bagobus* found on Greater Mindanao and *B. gamay* on Camiguin Island. There is considerable geographic variation within *B. luzonicus* that deserves detailed study. *Bullimus luzonicus* occur in a wide range of habitats, from sea level to 2740 m elevation—one of the widest ranges for any Luzon mammal. Their habitat includes heavily disturbed areas with dense grass; regenerating secondary forest; mixed primary and disturbed lowland forest; pine forest with regenerating undergrowth; and primary montane and mossy forest. They seem to be associated with forest over limestone in some areas (e.g., Mt. Irid and Caramoan National Park). Although at times they occur in mature forest, they are most common in secondary forest and in other areas of disturbance (such as along trails) where the canopy is at least partially open. They are usually most common at middle elevations (ca. 1000–1450 m) and at times are the most common small mammal at these elevations, but on some mountains they appear to be entirely absent. They are most active at night, but are also active during the day, foraging on the ground for plant material. They often make large runways through heavy grass or other vegetation and leave pieces of cut grass (ca. 10–20 cm long) in the runways. We suspect that they are omnivorous but feed mostly on green plant material. Five pregnant females had an average of 2.0 embryos (range = 1–3), and had three to four pairs of mammae. Adult males have large testes, up to 16 × 10 mm.

STATUS: Common and widespread in many habitats, and tolerant of moderate to heavy disturbance.

REFERENCES: Balete et al., 2009, 2011, 2013a, 2013b; Heaney et al., 1999, 2004, 2013a, 2013b; Largen, 1985; Musser, 1981; Rabor, 1955; Reginaldo and de Guia, 2014; Reginaldo et al., 2013; Rickart et al., 2013; Sanborn, 1952; Schenk et al., 2013; Stuart et al., 2007; Thomas, 1898.

Crunomys fallax
Northern Luzon shrew-mouse

Thomas, 1898. Trans. Zool. Soc. Lond., 14(6):394. Type locality: "Isabella, Central N. Luzon. Alt. 1000 feet." [= Sierra Madre Range, ca. 300 m elevation, Isabela Province].

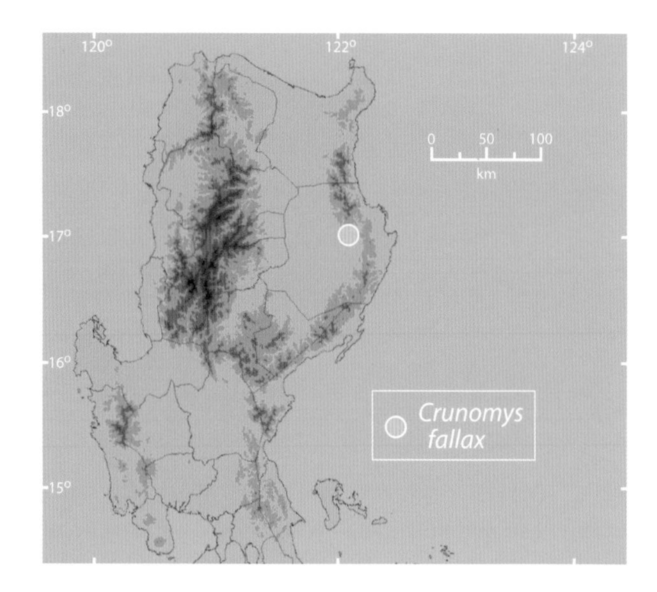

DESCRIPTION: Total length 185 mm; tail 79 mm; hind foot 23 mm; ear 10 mm; weight unknown. A small mouse, with small eyes and a tail shorter than the length of the head and body. The fur on the upperparts is grayish-brown and somewhat spiny; the underparts are a pale gray color. *Soricomys musseri* are of similar size (total length 180–198 mm), but have a longer tail (88–93 mm) and soft fur that is dark brown dorsally and rusty-brown ventrally. All *Apomys* have large eyes, soft fur, and much longer tails. *Mus musculus* are smaller (total length 151–172 mm), with a shorter hind foot (16–17 mm).

DISTRIBUTION: Known only from a single specimen [in BMNH], taken in May 1894 at about 300 m elevation in the Sierra Madre of northern Luzon (the specific location of the dot on our map is an approximation within Isabela Province). A specimen from Mt. Cetaceo, Cagayan Province, erroneously identified by Heaney (in Danielsen et al., 1994) as this species, has been reidentified as *Soricomys musseri*.

EVOLUTION AND ECOLOGY: The genus *Crunomys* occurs on Luzon (as represented by this one specimen), Greater Mindanao (two species), and Sulawesi Island (one species). *Crunomys* is most closely related to *Maxomys*, a genus that is diverse and widespread on the Sunda Shelf, Palawan, and Sulawesi. The ecology of *C. fallax* is unknown; the single specimen was taken in disturbed lowland forest.

STATUS: Unknown. Possibly dependent on lowland forest in the western foothills of the Northern Sierra Madre, a habitat that has diminished greatly due to logging and agriculture. Field studies are badly needed.

REFERENCES: Achmadi et al., 2013; Balete et al., 2012; Danielsen et al., 1994; Heaney et al., 2014a; Musser, 1982c; Rickart et al., 1998b; Schenk et al., 2013; Thomas, 1898.

Rattus everetti
common Philippine forest rat

(Gunther, 1879). Proc. Zool. Soc. Lond., p. 75. Type locality: northern Mindanao.

DESCRIPTION: Total length 388–481 mm; tail 181–244 mm; hind foot 42–50 mm; ear 25–29 mm; weight 320–390 g. There is substantial geographic variation in size and pelage color; the description here refers to Luzon populations only. A large rat, with rough, dark brown dorsal fur that has long black guard hairs easily visible above the rest of the fur. The ventral fur is usually white, but often stained with orange or yellow. The tail is longer than the length of the head and body; the base and most of the rest of the tail is dark, but roughly 20%–50% of the tip of the tail is white (though often stained by dirt). The hind feet are large and broad; the ventral surface is dark; the pads are large and usually pale. *Bullimus luzonicus* also have a white tip on the tail, but the tail is shorter than the length of the head and body. They also have proportionately larger heads and eyes; longer hind feet (53–58 mm); and abdominal fur that is silvery-brown, gray, or some color other than nearly white. *Rattus tanezumi* are smaller (total length 373–422 mm; weight 160–238 g); have a shorter (36–40 mm) hind foot; and lack a white tip on the tail. *Rattus norvegicus* very rarely occur away from buildings; are similar in size; lack a white tip on the tail; and have shorter ears (20–25 mm).

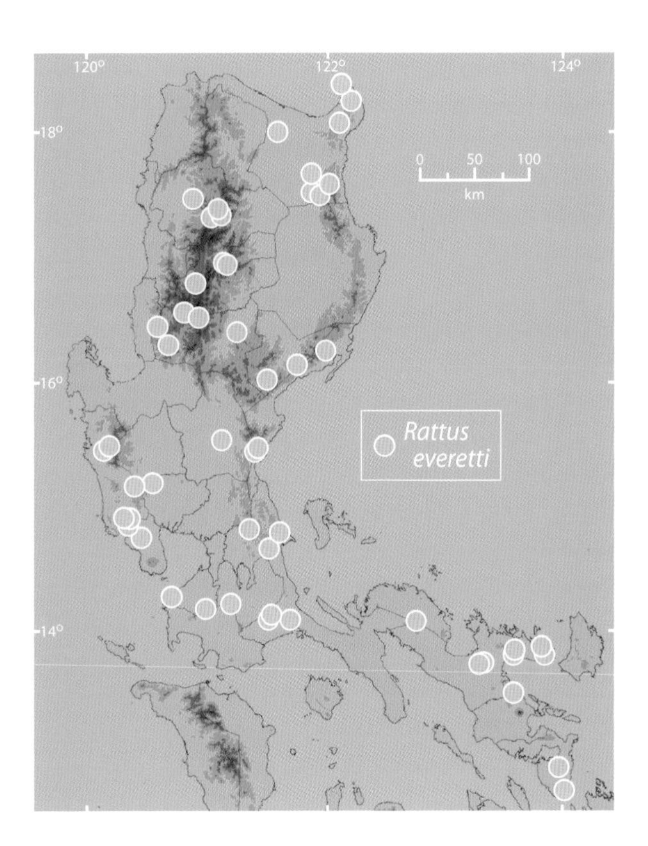

DISTRIBUTION: Endemic but widespread in the Philippines, excluding the Palawan Faunal Region, the Sulu and Batanes/Babuyan island groups, and most or all of Greater Negros-Panay.

EVOLUTION AND ECOLOGY: Recent genetic studies show that *Rattus everetti* are more closely related to *Limnomys* and *Tarsomys* than to other species of *Rattus* and probably should be reassigned to one of those genera; further studies are needed. Found in primary and disturbed lowland, montane, and mossy forest, from sea level to 2530 m elevation. They are sometimes the most abundant native rodent in disturbed lowland and lower montane forest, from sea level to about 1000 m elevation; above that elevation, density typically declines with increasing elevation and decreasing levels of disturbance. They sometimes occur in mixed shrubby and agricultural areas. In mossy forest on Mt. Isarog, their density was estimated at 1.2/ha, with movements of several hundred meters between captures and home ranges up to 1.0 ha. They are active almost exclusively at night, foraging on the ground and in trees for a wide range of foods, including seeds, fruits, and invertebrates, as well as feeding on sweet potatoes in cropland; they are best considered to be broad omnivores. Females have four pairs of mammae; eight pregnant females had an average of 3.75 embryos (range = 2–5). They first become pregnant at weights of about 200 g, before they are fully grown. Adult males have very

large testes, up to 15 × 28 mm. *R. everetti* are sometimes preyed upon by civets (probably *Viverra tangalunga*). The karyotype is $2n = 42$, FN = 64.

STATUS: Abundant and widespread over much of the Philippines.

REFERENCES: Alviola et al., 2011; Balete and Heaney, 1997; Balete et al., 2011, 2013a, 2013b; Danielsen et al., 1994; Duya et al., 2007, 2011; Fabre et al., 2013; Heaney et al., 1991, 1999, 2004, 2006b, 2006c, 2013a, 2013b; Musser and Carleton, 2005; Musser and Heaney, 1992; Rabor, 1955; Reginaldo and de Guia, 2014; Reginaldo et al., 2013; Rickart and Musser, 1993; Rickart et al., 1993, 2011a, 2011b, 2013; Schenk et al., 2013; Stuart et al., 2007.

Tryphomys adustus
Luzon short-nosed rat

Miller, 1910. Proc. U. S. Natl. Mus., 38:39. Type locality: Haights-in-the-Oaks, 7000 feet elevation, Mountain Province [= Atok Municipality, Benguet Province], Luzon.

DESCRIPTION: Total length 324–358 mm; tail 150–181 mm; hind foot 36–42 mm; ear 18–22 mm; weight 198 g. A medium-sized rat, with a tail shorter than or equal to the length of the head and body. The snout is rather short and broad; the ears are small. The dorsal fur is long, rather soft, and medium brown, with a speckled (i.e., salt-and-pepper) appearance. The ventral fur is gray at the base and pale, with a yellow-brown wash at the tips. The tail is uniformly dark brown, with short hairs readily visible along the entire length. The hind feet are long and narrow; the first and fifth digits are shorter than the other digits. *Rattus tanezumi* are slightly larger (total length 373–472 mm); have a longer tail (180–210 mm); and have dorsal pelage that is darker brown, with longer, more conspicuous black guard hairs. *Rattus everetti* are much larger (total length 388–481 mm).

DISTRIBUTION: Luzon Island only; known from the Central Cordillera and the Central Plains.

EVOLUTION AND ECOLOGY: *Tryphomys* is poorly known in all respects, aside from its cranial anatomy. *Tryphomys adustus* is one of the few Luzon mammal species that we have failed to capture during our studies. Many cranial similarities show that *Tryphomys* and *Abditomys* are probably sister-taxa, but no molecular genetic stud-

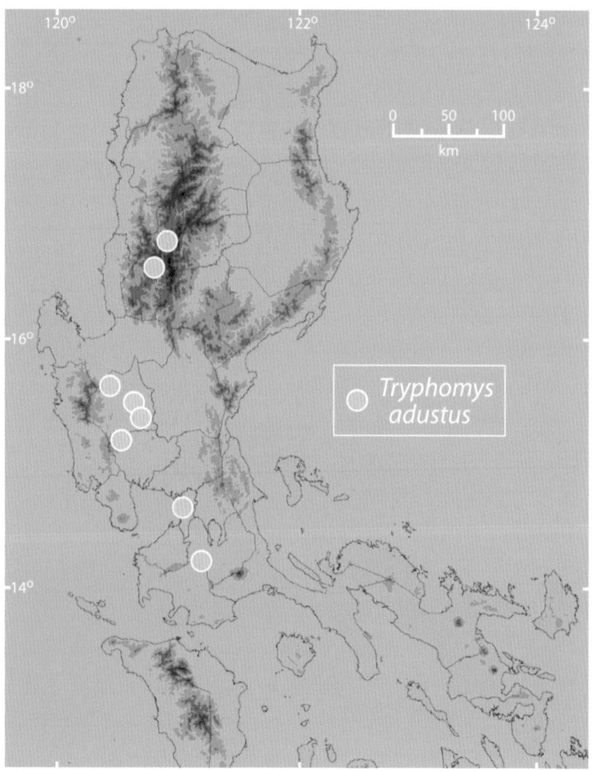

ies have been conducted that place them into a broader context. Records for *T. adustus* are available from about sea level to around 2200 m elevation. Most specimens were taken in secondary or primary forest, but others were captured in grass and shrubs adjacent to ricefields; it seems that most were trapped near water. A specimen was taken in Tondo, Manila, in 1984. Females have five pairs of mammae.

STATUS: Widespread in central Luzon and the southern Central Cordillera (Musser, 1982b), but poorly known.

REFERENCES: Barbehenn et al., 1973; G. Miller, 1910; Musser and Heaney, 1992; Musser and Newcomb, 1983; Rabor, 1955.

SECTION 5

Non-Native Rodents

Given the wonderfully diverse and attractive native small mammals on Luzon, it is terribly unfortunate that most residents of Luzon are familiar only with the small mammals that live in and around their homes, all of which are not native and are best known for the damage they cause and the diseases they carry. As described in chapter 6 and in the following species accounts, there are six species of rats and mice (*Rattus* and *Mus*) that were accidentally brought to the Philippines from the Asian mainland, probably all within the last several thousand years, and are abundant in buildings and agricultural areas. Unlike their distant relatives that are native to the Philippines—which live predominantly in natural forest, actively avoid humans and their habitations, and produce only a few young per litter once each year—the non-native pest rodents survive only in places that have been heavily impacted by humans and reproduce at a remarkably high rate. This is true of the murid rodents included in this section, and also of the non-native house shrew (*Suncus murinus*), included in section 1 of this chapter, though the shrew causes little damage since it subsists mostly on cockroaches and other household insect pests.

Although several of the non-native species are currently widespread and highly successful on Luzon, one (*Rattus argentiventer*) appears to have had very limited success in spreading out on Luzon: specimens were collected only near Los Baños, Laguna Province, in the 1960s, and no specimens have made their way to museum research collections for about 50 years. A second species (*R. nitidus*) was taken fairly commonly in agricultural areas in the Central Cordillera prior to about 1950 but has not been seen by a biologist since then. The latter species is native to mountains in southern China and adjacent regions to the west, and it is perhaps notable that its habitat in the uplands of northern

A: *Rattus tanezumi*. B: *R. exulans*.

Luzon is environmentally similar to that in China. The story of their arrival from China and apparent eventual extinction on Luzon could be an illustrative example of an initially successful invasive species that eventually died out, but we have far too little information to say anything further at this time.

In addition to these non-native rats and mice, recent years have seen the arrival of a very different kind of invasive species. At some time in the 1960s, someone in Forbes Park, a wealthy suburb of Manila where many foreigners live, released some captive squirrels (*Callosciurus finlaysonii*) that came from Thailand. They have an attractive appearance and are often quite visible during the daytime, but they have now become pests that damage mangos, coconuts, and other agricultural crops; they have become infamous for chewing through electrical lines and causing power outages. They appear to have spread over an area at least 40 km long and roughly 10 km wide, and they are now on the verge of moving beyond the urbanized area of Metro Manila. Whether they are able to spread into the agricultural countryside, where they could do great damage, is unknown. Unlike the pest rats and mice that appear to be held in check from invading natural forest by the far more innocuous native mice, squirrels have no direct counterpart on Luzon. We shall eventually see if they will become as widespread and naturalized as the non-native monkeys and civets are today.

Callosciurus finlaysonii
variable squirrel, Finlayson's squirrel

(Horsfield, 1823). Zool. Res. Java. Type locality: Koh Chang Island, Thailand.

DESCRIPTION: Total length 380–400 mm; tail 195–200 mm; hind foot 46–49 mm; ear 19–21 mm; weight 200–252 g. The only squirrel on Luzon. The dorsal pelage on most individuals is pale gray; the ventral pelage is nearly white. The tail is long; the tail hair is bushy, with a band of dark gray hair on the anterior half of the tail and pale yellow hair on the posterior half. Some individuals are nearly white, and others are pale rusty-red dorsally. No other mammal on Luzon is similar.

DISTRIBUTION: Originally from Thailand, Cambodia, Laos, and southern Vietnam, but introduced into Italy, Singapore, and Manila.

EVOLUTION AND ECOLOGY: Introduced into Metro Manila, probably from the vicinity of Bangkok, which is the native range of the subspecies present in the Philippines (*Callosciurus finlaysonii boucourti*). The presence

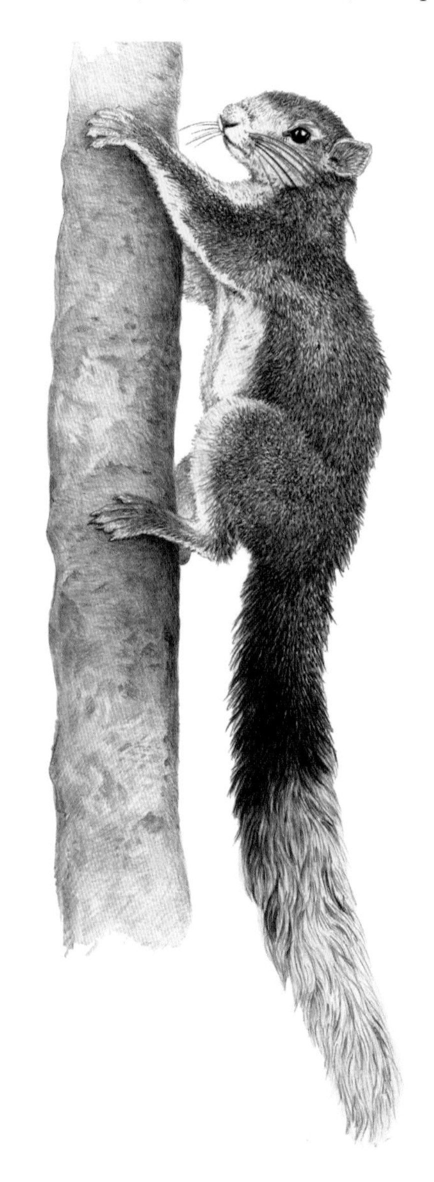

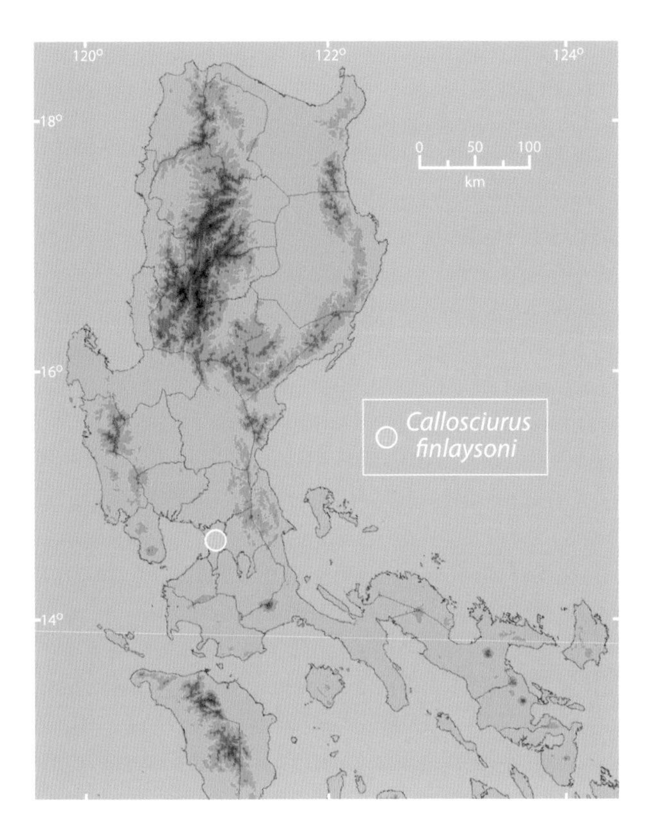

of some rusty-red individuals implies the importation of at least one additional subspecies from Thailand. Unverified reports indicate that the initial population of squirrels was released in the Forbes Park subdivision of Manila in the 1960s, and they soon spread to the adjacent Dasmarinas Village and the American Cemetery at Fort Bonifacio. It is not know if there was only a single introduction. We became aware of them by 2008, when they had reached the campus of the University of the Philippines–Diliman and the neighboring Ninoy Aquino Parks and Wildlife Center. By 2013, they had dispersed as far north as Fairview (at the south end of La Mesa Dam) and as far south as San Pedro, Laguna Province, a distance of about 40 km. They are active during daytime, foraging primarily in trees and rarely descending to the ground. They feed on fruits, seeds, buds, flowers, sap, and occasionally on insects. In Manila, their diet includes mangoes, rambutan, santol, coconuts, the flowers and fruit of macopa (*Syzigium* sp.), and other fruits and seeds. Nests are constructed from small sticks, twigs, and leaves and are found in the high branches of trees. They cause damage to fruit, buildings, and electrical wires, sometimes causing transformers to explode.

STATUS: A non-native pest species, spreading rapidly within Metro Manila.

REFERENCES: Aloise and Bertolino, 2005; Bertolino et al., 2004; Kuramoto et al., 2012; Lekagul and McNeely, 1977; Thorington et al., 2012.

Mus musculus
house mouse

Linnaeus, 1758. Syst. Nat., 10th ed., 1:138. Type locality: Uppsala, Uppsala County, Sweden.

DESCRIPTION: Total length 151–172 mm; tail 79–91 mm; hind foot 16–17 mm; ear 11–14 mm; weight 12–21 g. A small mouse, with small eyes; a rather narrow snout; and a long tail. The fur is gray or grayish-brown, paler on the underparts than on the upperparts. The tail is brown, both dorsally and ventrally. All *Apomys* are larger (total length at least 181 mm; weight at least 17 g). *Musseromys* have reddish pelage; a short, broad snout; and a tail with long hair visible near the tip. *Soricomys* are larger (total length at least 180 mm; weight at least 21 g), with dark fur both dorsally and ventrally.

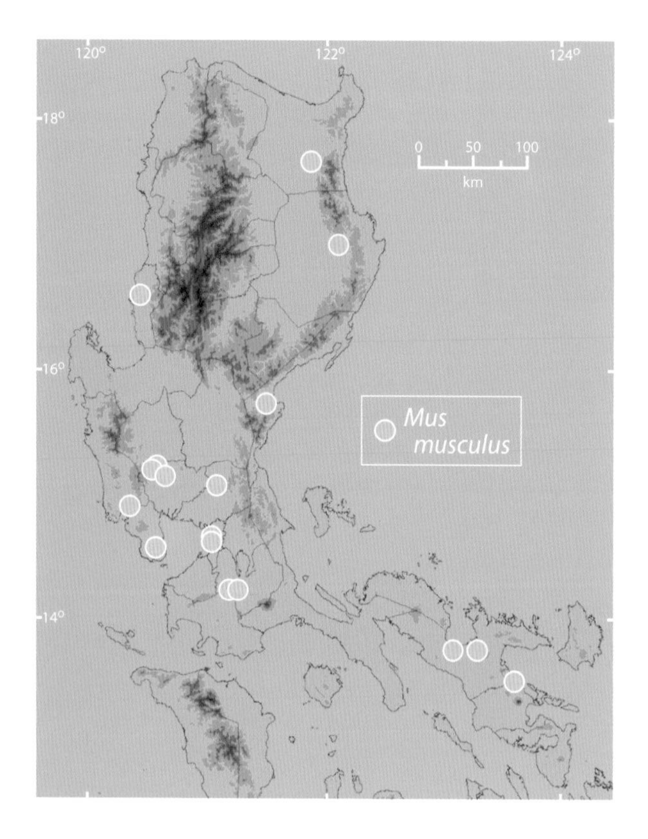

DISTRIBUTION: Nearly worldwide; widespread in Southeast Asia. Found throughout the Philippines.

EVOLUTION AND ECOLOGY: The genus *Mus* is diverse, with about 38 species recognized. Several of the species live with humans as household or crop pests; *M. musculus* is the most common of these species and has been accidentally transported nearly worldwide. On Luzon, they live in or near human residences in urban and rural areas, and sometimes in fields of corn [= maize], from sea level to at least 600 m elevation, but they probably occur at much higher elevations in urban areas. They are not known to occur in forest in the Philippines. They are usually nocturnal. Females have six pairs of mammae. Litter size ranges from four to eight; females produce many litters each year.

STATUS: Non-native and abundant.

REFERENCES: Barbehenn et al., 1973; Esselstyn et al., 2004; Heaney et al., 1999, 2006a, 2006b; Marshall, 1977, 1986; Ong and Rickart, 2008; Rabor, 1986.

Rattus argentiventer ricefield rat

(Robinson and Kloss, 1916). J. Straits Br. Roy. Asiat. Soc., 73:274. Type locality: Pasir Ganting, W coast Sumatra, [Indonesia].

DESCRIPTION: Total length 350–400 mm; tail 160–201 mm; hind foot 35–40 mm; ear 20–24 mm; weight 160–212 g. Similar to *Rattus tanezumi*. In *R. argentiventer*, the fur on the upperparts has a speckled (i.e., salt-and-pepper) appearance and is yellowish-brown; *R. tanezumi* have less salt-and-pepper coloration and are usually dark reddish-brown or dark orange-brown. There is often a tuft of orange hair at the front edge of the ear in *R. argentiventer*, which is lacking in other Philippine *Rattus* species. The underparts usually have a silvery appearance, rather than being mostly white or pale yellow. The pads on the hind feet are smaller than in *R. tanezumi* and are only slightly raised above the surrounding surface; lamellae (i.e., fingerprints) on the pads are nearly absent. The tail is usually slightly shorter than the head and body, rather than longer (as in *R. tanezumi*).

DISTRIBUTION: Thailand to New Guinea. In the Philippines, recorded on Luzon, Mindanao, and Mindoro. On Luzon, they have been recorded only in Laguna Province.

EVOLUTION AND ECOLOGY: A member of the *Rattus rattus* group, and one of many species in that group that are closely associated with areas of intensive human activity. *Rattus argentiventer* usually occur in highly disturbed areas, including ricefields, grasslands, and wastelands in the lowlands, especially when these places are burned frequently. They are occasionally

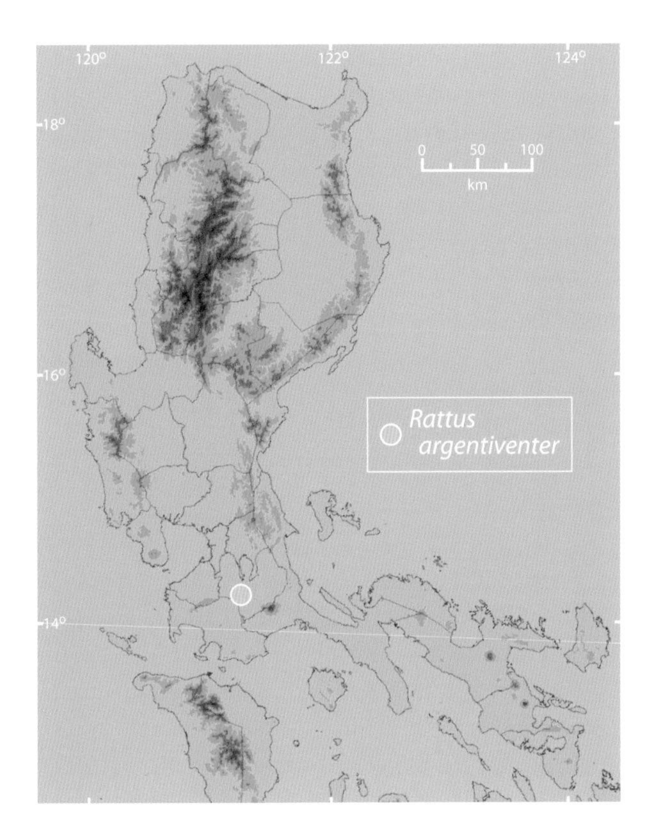

found in coconut groves, and rarely inside houses. They are virtually absent from forest. Females have six pairs of mammae (*R. tanezumi* have five pairs). Litter size in Selangor, Malaysia, was 5–7 (average = 6.0). They sometimes undergo sudden, massive increases in numbers (i.e., irruptions). They are active at night, most often on the surface of the ground, but they are also able to climb shrubs and trees. They are omnivorous, feeding on shoots, roots, seeds, nuts, fruits, insects, and snails.

STATUS: Non-native agricultural pest; may be locally abundant, but seemingly rare on Luzon.

REFERENCES: Barbehenn et al., 1973; Medway, 1969; Musser, 1973; Ong and Rickart, 2008.

Rattus exulans
spiny ricefield rat, Polynesian rat

(Peale, 1848). Mammalia. In Repts. U.S. Expl. Surv., 8:47. Type locality: Tahiti Island, Society Islands.

DESCRIPTION: Total length 217–257 mm; tail 121–136 mm; hind foot 25–28 mm; ear 15–19 mm; weight 45–61 g. A small rat, with a tail slightly longer than the length of the head and body. The fur on the upperparts is dark brown, with many scattered, stiff, somewhat spinous hairs. The underparts range from white to dull gray, often stained a pale yellow. Other *Rattus* are larger (often most easily distinguished by measuring the hind feet) and do not have stiff/spiny hairs in the dorsal fur. *Apomys musculus* and *A. microdon* are smaller (weight 42 g or less); have soft fur without spinous hairs; and have sharply bicolored tails that are much longer than the length of the head and body. Members of the subgenus *Megapomys* always have longer hind feet (31 mm or greater); have soft fur without spinous hairs; and have tails that are sharply bicolored.

DISTRIBUTION: Bangladesh to Easter Island; found throughout the Philippines.

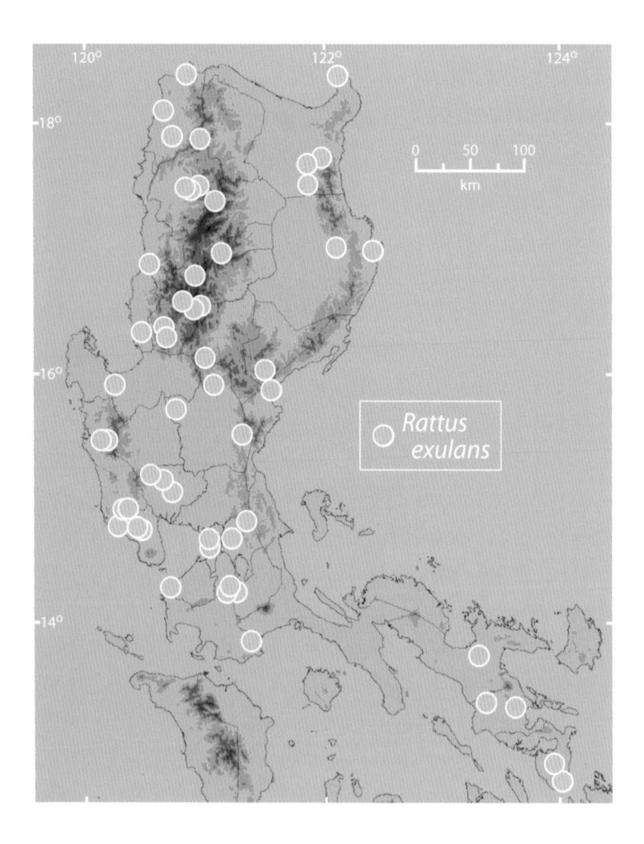

EVOLUTION AND ECOLOGY: The clade that gave rise to *Rattus exulans*, *R. rattus*, and *R. tanezumi* is estimated to have diverged from its common ancestor with *R. norvegicus* about 2.9 million years ago, followed by the divergence of *R. exulans* from the ancestors of *R. tanezumi* and *R. rattus* around 2.2 million years ago. *Rattus exulans* probably were brought into the Philippines within the past 10,000 years, perhaps as recently as 4000 years ago. They are associated with agricultural areas throughout the country at all elevations, often in grassy areas in pine forest, wasteland, and other places that are frequently burned, grazed, or mowed. They tend to be more common in shrubland adjacent to ricefields than within the ricefields, and often occur in grassy areas in cities. On Luzon, they are absent or rare in secondary forest and absent in old-growth forest (chapter 6), but they occur in high-elevation old-growth forest on islands such as Negros and Camiguin, where there are few native rodents. They are active primarily at night, and forage on the ground for seeds, fruits, tender parts of green plants, insects, and snails. Females have four pairs of mammae. They become reproductively active when still quite young and small. The litter size averages about 4.3, ranging up to 8, and females have several litters each year. They have been recorded in several protected areas on Luzon—including Balbalasang-Balbalan, Mts. Banahaw-San Cristobal, Mt. Data, Mt. Isarog, Mt. Natib, and Mt. Pulag—always in areas heavily disturbed by forest fires, logging, *kaingin*, and vegetable farming.

STATUS: Non-native and abundant.

REFERENCES: Balete et al., 2009, 2011, 2013a; Barbehenn et al., 1973; Danielsen et al., 1994; Duya et al., 2007; Esselstyn et al., 2004; Heaney et al., 1989, 1999, 2004, 2006a, 2006b, 2013b; Medway, 1969; Ong and Rickart, 2008; Ong et al., 1999; Rabor, 1986; Reginaldo and de Guia, 2014; Reginaldo et al., 2013; Rickart et al., 1993, 2007, 2011a, 2011b, 2013; Robins et al. 2008; Stuart et al., 2008.

Rattus nitidus
Himalayan field rat, Himalayan white-footed rat

(Hodgson, 1845). Ann. Mag. Nat. Hist., [ser. 1], 15:267. Type locality: Nepal.

DESCRIPTION: Total length 310–386 mm; tail 148–206 mm; hind foot 36–39 mm; ear 15–21 mm; weight ca. 122–136 g. A medium-sized rat, similar to *R. tanezumi*. The dorsal fur is soft, thick, dark brown, and somewhat curly, giving it a woolly appearance. The abdominal fur is gray. The upper surfaces of the forefeet and hind feet are covered by white hairs. The tail is equal to or shorter than the length of the head and body and is dark brown dorsally and ventrally for its entire length. *Rattus tanezumi* have fur that is straight and does not appear woolly; the hind feet usually have some dark

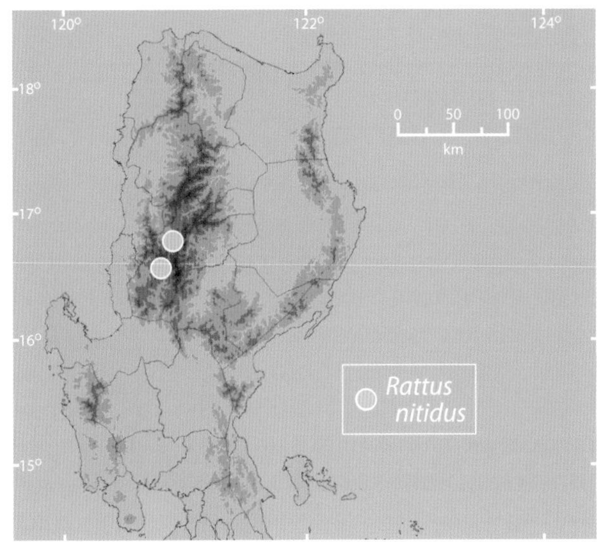

hair on the dorsal surface; and the ears are usually longer (20–24 mm). *R. norvegicus* are substantially larger (total length 423–510 mm; hind foot 40–50 mm).

DISTRIBUTION: Nepal to New Guinea. In the Philippines, known only from Luzon.

EVOLUTION AND ECOLOGY: Closely related to *Rattus norvegicus*. The few Philippine records are all from the Central Cordillera at high elevation. No *R. nitidus* have been seen by biologists since 1946; we captured none during our surveys. It is not known when they first were brought to the Philippines, but given their distribution, they most likely came from southern China. In other parts of their range, they generally are found in houses in hilly areas, often near streams and rivers. They are active primarily at night, often climbing into the upper parts of houses. They have six pairs of mammae; litters range from 4 to 15, with an average of 8. They are pests of rice, corn [= maize], wheat, sweet potatoes (camote), potatoes, and other crops.

STATUS: Non-native pest.

REFERENCES: Barbehenn et al., 1973; Lekagul and McNeely, 1977; Ong and Rickart, 2008; Smith and Xie, 2008.

Rattus norvegicus
brown rat, Norway rat

(Berkenhout, 1769). Outlines Nat. Hist. Great Brit. and Ireland, 1:5. Type locality: Great Britain.

DESCRIPTION: Total length 423–510 mm; tail 200–250 mm; hind foot 40–50 mm; ear 20–25 mm; weight 230–500 g. A large rat, with long, rough brown fur on the upperparts and proportionately small ears. The tail is thick at the base; has large scales; is shorter than the length of the head and body; and is slightly paler on the bottom side than on the top. The hair on the top of the feet is white. The fur on the belly is a uniform gray. *Rattus tanezumi* are smaller (total length 373–422 mm; weight 160–238 g) and rarely have gray fur on the abdomen. *Rattus everetti* usually have a longer ear (25–

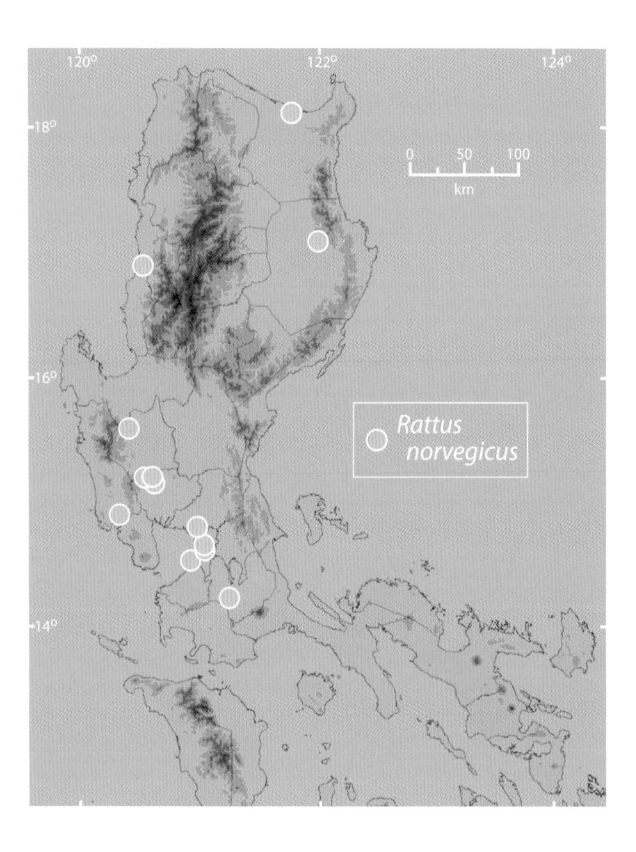

29 mm) and a tail that is longer than the length of the head and body, with a white tip.

DISTRIBUTION: Worldwide. They are widespread in the Philippines, probably in cities throughout Luzon.

EVOLUTION AND ECOLOGY: Estimated to have diverged from the clade that gave rise to *Rattus exulans*, *R. rattus*, and *R. tanezumi* ca. 2.9 million years ago, the earliest within the Asian *Rattus* lineages following their divergence from the New Guinea endemic *R. praetor* clade, ca. 3.5 million years ago. *Rattus norvegicus* are closely related to *R. nitidus*. In the Philippines, *R. norvegicus* are primarily restricted to large cities, places where large ships dock, and other urban areas, but at times they also occurs in small villages, ricefields, and canefields. They are a major pest of stored foods. They are most common in the lowlands, but there are a few records from higher elevations. There are no records on Luzon from any forested areas, and we never encountered them during our surveys. Females have six pairs of mammae; litter size is 3–12, usually 6–8. They are known to host some zoonotic parasites transmissible to humans.

STATUS: An alien pest; abundant in urban areas.

REFERENCES: Barbehenn et al., 1973; Eduardo et al., 2008; Ong et al., 1999; Rabor, 1986; Robins et al., 2008; Smith and Xie, 2008.

Rattus tanezumi
Oriental house rat

Temminck, 1844. In Siebold, Temminck, and Schlegel, Fauna Jap., p. 51. Type locality: Japan.

DESCRIPTION: Total length 373–422 mm; tail 180–210 mm; hind foot 36–40 mm; ear 21–24 mm; weight 160–238 g. A medium-sized rat, with a tail that is longer than the length of the head and body, and dark brown or nearly black for its entire length. The fur on the upperparts is variable in color, from dark brown to gray and reddish; the fur is usually rough, but may be soft. There are prominent, long black guard hairs that project well above the rest of the dorsal pelage. The fur on the belly is usually whitish or pale gray, though often with some pale yellow or orange staining. The age classes of this species are highly variable, often making identification difficult. Other Luzon *Rattus* differ in size, often detected most easily by measuring the hind foot. *Rattus everetti* are larger (total length 388–481 mm; weight

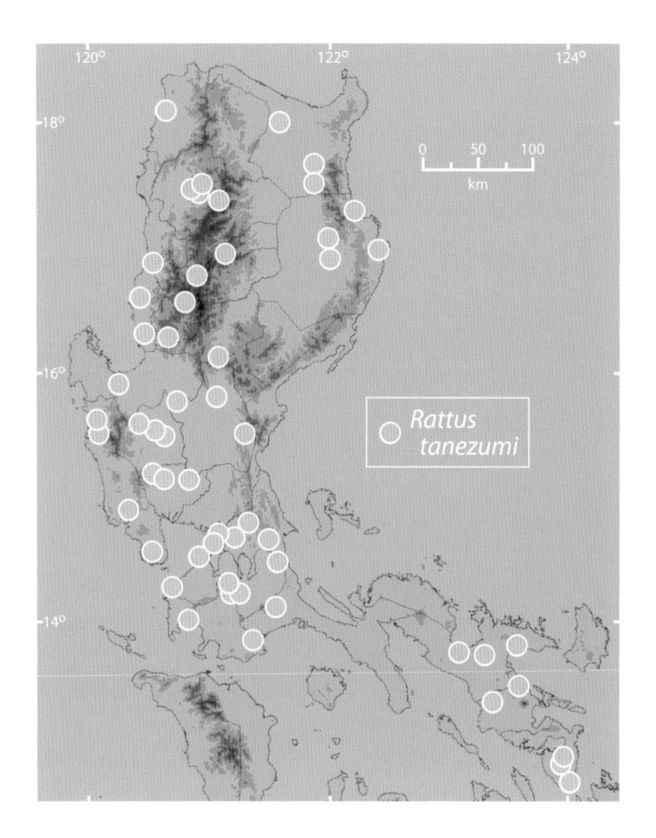

320–390 g); have a longer hind foot (42–50 mm); and usually have a longer tail (181–244 mm) that is white for ca. 20%–40% of its length.

DISTRIBUTION: Afghanistan to New Guinea and Micronesia (except the Samoas); throughout the Philippines.

EVOLUTION AND ECOLOGY: The ancestral lineage of *Rattus tanezumi* and *R. rattus* is estimated to have diverged from that of its sister-species, *R. exulans*, about 2.2 million years ago; *R. tanezumi* subsequently diverged from *R. rattus* around 400,000 years ago. *Rattus tanezumi* was formerly included within *R. rattus*, but it now is recognized as a separate species, based on karyotypic differences (FN=42 vs. FN=38/40, respectively) and subtle morphological differences. As currently defined, *R. tanezumi* includes many populations formerly recognized as distinct species, including *R. mindanensis*. *Rattus tanezumi* are known to host some zoonotic parasites transmissible to humans. They are by far the most widespread and destructive alien pest rats in the Philippines. They are abundant in cities, villages, and agricultural areas at all elevations, and usually absent or uncommon in disturbed lowland and montane forest. They are absent from old-growth forest on Luzon, even when it is adjacent to agricultural fields and houses. In contrast, they occur in primary forest on islands such as Camiguin and Negros where few native rodents are present. They eat a wide range of foods, including seeds, fruits, invertebrates, and occasionally small birds and mice; they consume virtually anything humans will eat. Females have five (rarely six) pairs of mammae; litter sizes of 4–7 embryos have been noted in the Philippines, but up to an average of 8.11 was recorded among females in agricultural areas elsewhere.

STATUS: Non-native pest; abundant. Present in several protected areas on Luzon—including Balbalasang-Balbalan, Mts. Banahaw–San Cristobal, Caramoan, Mt. Data, Mt. Isarog, and Mt. Pulag—in heavily disturbed areas caused by forest fires, logging, *kaingin*, and large-scale vegetable farming.

REFERENCES: Balete et al., 2013b; Danielsen et al., 1994; Eduardo et al., 2008; Esselstyn et al., 2004; Heaney et al., 1989, 1999, 2004, 2006a, 2006b, 2013b; Musser, 1977; Musser and Carleton, 2005; Rabor, 1986; Reginaldo and de Guia, 2014; Rickart and Musser, 1993; Rickart et al., 1993, 2007, 2011a, 2011b; Robins et al., 2008; Sanborn, 1952; Stuart et al., 2007, 2008.

11

A Guide to the Bats

With at least 57 species in 7 families, the Chiroptera is the most diverse order of mammals on Luzon. This diversity is apparent in a great many ways. They range in size from one of the largest bats in the world, golden-crowned flying foxes (*Acerodon jubatus*), which reach nearly 1.0 kg on Luzon and about 1.4 kg elsewhere in the Philippines, to one of the smallest, lesser bamboo bats (*Tylonycteris pachypus*), which weigh only about 2.7 g as adults. Some of these bat species consume fruits, nectar, and blossoms; other feed on large, hard-shelled beetles; some eat small moths; and others primarily feed on spiders. Some species roost in caves, others in hollow trees, some in thick vegetation, and still others in open sunlight in the tops of large trees. The species that feed on insects and other invertebrates perceive their environment primarily through producing and then listening to the echoes of high-pitched clicks and squeaks that are largely inaudible to humans, a system known as echolocation. Those that feed on plants lack this type of sonar and instead rely on their eyes (though they probably also use their excellent sense of smell to find food).

A: *Hipposideros antricola* (Hipposideridae). B: *Cynopterus brachyotis* (Pteropodidae). C: *Myotis rufopictus* (Vespertilionidae).

SECTION 1

Introduction to Bats

Given this remarkable diversity in bat species, it is not surprising that their general appearance also varies greatly, with many features of their external morphology associated with the ability to fly. All of the fruit bats (Pteropodidae) have large eyes, simple ears, and relatively simple noses, though some have elongated nostrils. All of the insectivorous bats have more complex and (usually) larger ears, and usually smaller eyes (though none are blind). Most of them emit their ultrasonic sounds through their mouths; these have simple noses (Emballonuridae, Vespertilionidae, and Molossidae). But some of the bats (Megadermatidae, Hipposideridae, and Rhinolophidae) emit their ultrasonic squeaks through their nostrils; these all have elaborate folds of skin around the nostrils that allow the bats to focus and direct the sounds they produce. Differences in the relative size and placement of the tail and tail membrane vary among the families of bats and are also associated with their means of finding and consuming their food.

Fruit bats are the most thoroughly studied of the Philippine Chiroptera, simply because they lack sonar and are therefore easier to catch (in mist nets) than the insectivorous bats; information on their distribution, habitat preferences, elevational ranges, roosting

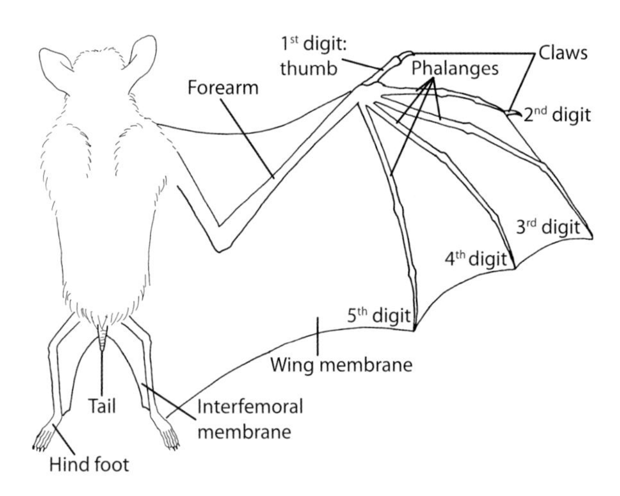

External morphology of a fruit bat, with important structures indicated. Modified from Ingle and Heaney (1992).

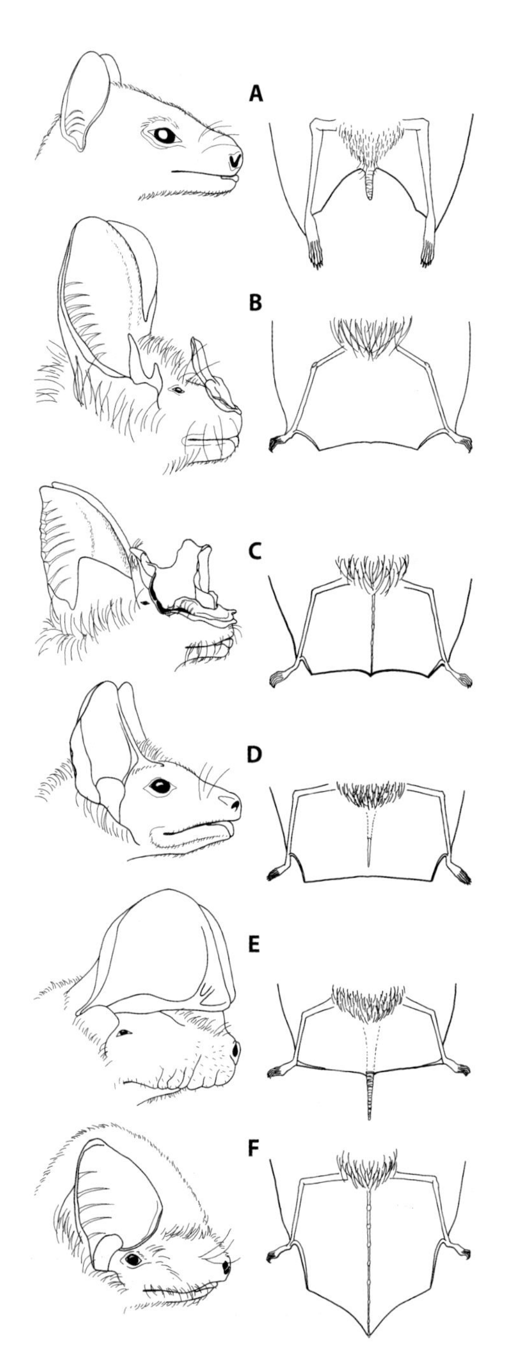

Heads and tail regions of representative Philippine bat families (not to the same scale). A: Pteropodidae (*Rousettus*). B: Megadermatidae (*Megaderma*). C: Rhinolophidae (*Rhinolophus*). D: Emballonuridae (*Taphozous*). E: Molossidae (*Chaerophon*). F: Vespertilionidae (*Miniopterus*). Modified from Ingle and Heaney (1992).

Acoustic Diversity in Philippine Bats

Jodi L. Sedlock, Biology Department, Lawrence University, Appleton, Wisconsin, USA 54911

It is only dusk, but darkness has fallen under the forest canopy. Vines form a latticework obstacle course in front of a small opening in a rock outcrop. A stream of bats begins flowing out of the opening—tiny "fighter jets" expertly negotiating the vines and avoiding collisions with each other—and continues for over an hour. This spectacular scene, made possible by highly adapted echolocation systems, plays out every night on Mt. Makiling, and in countless other places on Luzon, as thousands of bats head out to hunt for insects.

Contrary to what many believe, bats are not blind. In fact, all of the ca. 27 fruit- and nectar-eating bat species in the Philippines use highly sensitive vision to navigate at night. Most (about 55) of the approximately 82 Philippine bat species overall, however, rely not on their eyes but on their voice and ears to expertly avoid obstacles and capture small insects in flight using echolocation. These insect-eating bats squeak out of their mouths or through their noses at frequencies 100–800 times higher than the human voice, and they glean a wealth of information from the returning echoes. And each of the 55 bat species has its own "voice." Just as visually orienting animals that see in diverse wavelengths, such as humans and bees, view the world differently, variation in perception among acoustically orienting bats is influenced by sound-wave attributes. In fact, coupled with flight adaptations, a bat's echolocation call largely determines where it can successfully forage and what it eats.

Using sound to locate tiny, flying, evasive insects in flight does not come without its challenges. For example, just as we squint and strain to see through the glare reflected from a bright, white, sandy beach, bats' prey can be obscured by the echoes from background clutter, such as vegetation or a water surface. Information can also be lost when the echo overlaps with subsequent calls. Thus bats have to distinguish between food echoes and those from background clutter. As a result of these challenges, the design of a bat's echolocation call largely corresponds to its preferred habitat. Forest bats tend to use calls of short duration that are spaced far apart to minimize call-echo and echo-echo overlap. For example, the woolly bat, *Kerivoula whiteheadi*, uses extremely short-duration, broadband calls to flutter slowly among vegetation near the forest floor, hunting for prey at short range. Similarly, *Myotis horsfieldii*, also with a short-duration, broadband call, is commonly seen flying along rivers and creeks scooping up insects as they emerge just above the water surface. In contrast, horseshoe bats and roundleaf bats use almost pure-tone "beeps" emitted through their nostrils, which are surrounded by sound-directing skin flaps and folds called noseleafs. These calls are sent out at a slightly lower frequency than the returning echoes, therefore avoiding call-echo overlap. Moreover, information encoded in the echo distinguishes fluttering insect prey from stationary background clutter. Given this, one might see *Rhinolophus virgo*, hanging from a branch,

Representative frequency sonograms of echolocation calls from species in the bat families found on Luzon Island (kHz = kilohertz; ms = milliseconds). The echolocation calls differ, based in part on the types of habitat in which they forage and on the ways they use those habitats. A: Rhinolophidae, *Rhinolophus virgo*. B–D: Vespertilionidae. B: *Kerivoula whiteheadi*. C: *Myotis horsfieldii*. D: *Scotophilus kuhlii*. E: Emballonuridae, *Emballonura alecto*. F: Hipposideridae, *Hipposideros diadema*. G: Molossidae, *Chaerephon plicata*.

noseleaf twitching, pivoting 360° from one foot, searching for prey in the dense forest understory. Species occupying the night skies, such as the lesser Asian house bat (*Scotophilus kuhlii*) or the wrinkle-lipped bat (*Chaerephon plicatus*) flying over a ricefield, are not concerned about obstacles as much as the long-distance detection of prey during high-speed flight. As a result, these bats use calls that are of relatively long duration, spaced far apart and low in frequency. Low-frequency (long wavelength) calls travel greater distances than high-frequency (short wavelength) calls but are not as precise (i.e., do not detect relatively small prey); think of how we lower our own voice when calling out to someone far away.

Philippine bat calls vary greatly in frequency, which, in turn, may influence what they eat. For example, *Rhinolophus virgo*, with a very high-frequency (short wavelength) call (ca. 85 kHz), may perceive an abundance of available prey types—including very small prey. Whereas in the same forest, the larger *R. rufus*, with a relatively low-frequency (longer wavelength) call (ca. 41 kHz), perceives only a subset: the larger prey types. This is because calls with wavelengths longer than the prey diameter return weak echoes and are therefore difficult to detect. Moreover, many insects (especially moths) have evolved simple ears that allow them to evade a bat attack—when they hear a bat squeak, they dive for cover. Therefore, bats calling at frequencies higher than the hearing range of most insects (less than 60 kHz)—including many of the horseshoe and roundleaf bats—have greater access to more types of prey than bats that can be heard by the insects. One bat species on Luzon Island, *Megaderma spasma*, uses its enormous ears to listen for prey-generated sounds from insects such as cicadas, rather than solely using echolocation to find prey.

The night skies of Luzon are not quiet—in fact, there is a high-frequency cacophony of sounds playing each night over ricefields, in cities, and echoing throughout the forest understory. Each "instrument" is tuned to a particular task, and the bats quietly (to us) provide an invaluable service as they consume millions of insects nightly—many of which may be harmful to crops or human health.

References: Kingston et al., 1999, 2000; Schnitzler and Kalko, 2001; Schnitzler and Ostwald, 1983; Sedlock, 2001; Sedlock et al., 2014a, 2014b.

sites, and other aspects of their general ecology are usually known, though very few studies have gone beyond these most basic attributes. On the other hand, the insectivorous bats, with their sophisticated echolocation systems, are difficult to capture without using labor-intensive techniques (e.g., modified mist nets, specialized traps, etc.), and we often have very incomplete information on even the most basic elements of their ecology. In spite of these difficulties, several detailed studies have been conducted, and an understanding is emerging of how the species with different echolocation calls and morphology use the available habitats (pp. 184–185).

Even small species of bats that roost in large aggregations, whether in caves or in the tops of large trees, are often heavily overhunted; thus some bats are among the most severely threatened mammals on Luzon. Populations of flying foxes have plummeted steadily and are locally extinct in many parts of the country, and numerous caves that once supported large colonies of insectivorous bats that controlled agricultural insect pests are now virtually empty of bats.

REFERENCES: Heaney et al., 1997; Ingle and Heaney, 1992; Mickleburgh et al., 1992; Simmons, 2005; Utzurrum, 1992.

SECTION 2

Family Pteropodidae: Fruit Bats and Flying Foxes

With 15 species now recorded, the Pteropodidae is the third-most species-rich family of mammals on Luzon; only the evening bats (Vespertilionidae) and murid rodents (Muridae) have more. These 15 species vary hugely in size, from ca. 17 g for the smallest to 1 kg for the largest. Their ecology is similarly varied, from roosting habits (some in caves, others at the tops of large trees, and still others hidden in vegetation or hollow trees) and aggregation size (from a few individuals to colonies of tens or hundreds of thousands) to types of habitat (from lowland urban and agricultural areas to old-growth montane forest). All share some features, however: they eat only plant matter (fruits, nectar, blossoms, and/or leaves), and females have only one young at a time. The largest of the species and those that live in caves are among the most heavily hunted and seriously threatened mammal species in the Philippines.

Some of the fruit bats on Luzon are members of entire branches on the tree of life that are restricted to the Philippines. For example, *Alionycteris*, *Haplonycteris* (with one or perhaps two species), and *Otopteropus* form a clade that occurs only within the Philippines, and these genera clearly have undergone their diversification within the archipelago (chapter 5). Similarly, *Ptenochirus* (with two species) has undergone some speciation after its separation from *Cynopterus* (which occurs mostly on the Sunda Shelf), as has the Philippine endemic genus *Desmalopex*. This level of diversification is evidence of their deep evolutionary history within the archipelago. Many other species on Luzon have their closest relatives on the Sunda Shelf (e.g., *Dyacopterus rickarti*, *Eonycteris robusta*, and *Pteropus vampyrus*) or in the New Guinea / Sulawesi area (e.g., *Desmalopex* and *Harpyionycteris whiteheadi*), implying a long history of colonization from both directions.

Unlike the insectivorous bats that use echolocation (i.e., sonar) to perceive their environment, all Philippine fruit bats perceive the world visually (except for *Rousettus*, which has large eyes but also produces clicks that may be used as a very simple form of echolocation). As a result, they all have large eyes, simple ears (without a tragus), and simple noses (without elaborate noseleafs). They have a short tail or no tail, and the interfemoral membrane (i.e., the skin between the hind legs) is short, not reaching to the hind feet or beyond.

REFERENCES: Almeida et al., 2011; Heaney, 1991; Heaney and Rickart, 1990; Heaney and Roberts, 2009.

Fruit bats. A: *Otopteropus cartilagonodus*. B: *Desmalopex leucopterus*. C: *Rousettus amplexicaudatus*. D: *Ptenochirus jagori*.

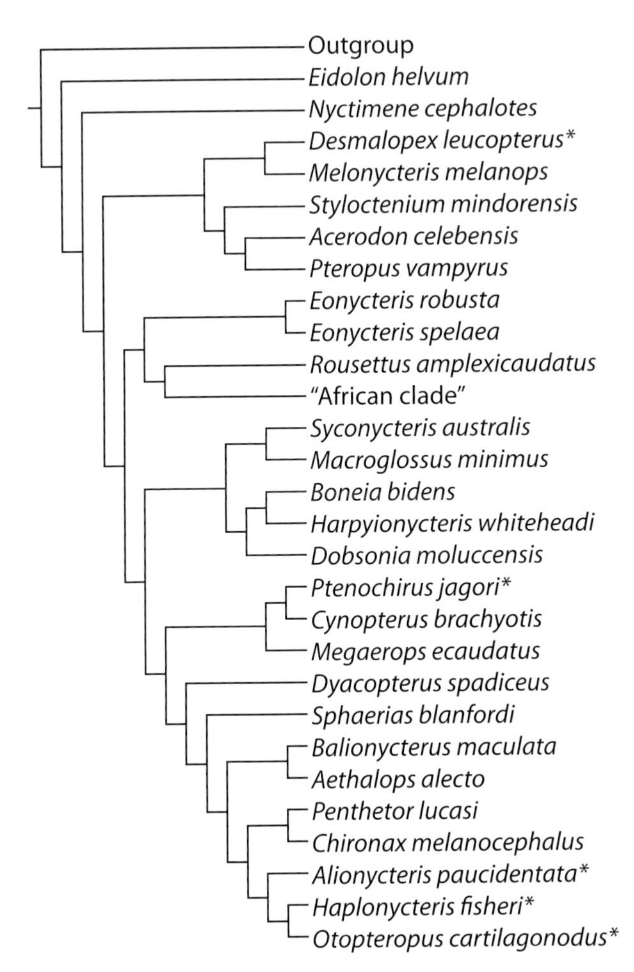

Phylogeny of fruit bats (Pteropodidae). Asterisks indicate genera that are endemic to the Philippines. Modified from Almeida et al. (2011).

Acerodon jubatus golden-crowned flying fox

(Eschscholtz, 1831). Zool. Atlas, part 4:1. Type locality: Manila, Luzon Island.

DESCRIPTION: Total length, 262–300 mm; tail 0 mm (absent); ear 31–34 mm; forearm 180–195 mm; weight 0.8–1.05 kg (reaching ca. 1.4 kg on Mindanao). These spectacular bats are the largest in the Philippines, and by weight they are the largest bats in the world. Golden fur on top of the head extends from between the eyes to either the nape of the neck or the shoulders. The wing membranes are medium to dark brown, with irregular pale and dark blotches. The thumb is unusually thick and long (65–75 mm). *Pteropus vampyrus* are similar in size, but have dark brown (nearly black) wing membranes and either an entirely black or a golden-and-russet mantle that extends from between the ears to over the shoulders. Other large Philippine fruit bats are smaller.

DISTRIBUTION: Endemic to the Philippines; widespread, with the exception of the Palawan Faunal Region and the Batanes/Babuyan group of islands. They formerly occurred throughout Luzon but are now limited to fairly large areas of forest at low elevation, usually in the rugged foothills of mountains.

EVOLUTION AND ECOLOGY: *Acerodon* is most closely related to *Pteropus*. Unlike the very widespread genus

Pteropus, the five species of *Acerodon* occur only in oceanic portions of Indonesia and the Philippines: *A. leucotis* lives only on Palawan and associated islands, *A. jubatus* throughout most of the oceanic Philippines, and the others in eastern Indonesia. *Acerodon jubatus* occur in primary and secondary lowland forest, at up to 1100 m elevation. They feed in secondary and primary forest, especially near rivers, and rarely venture into agricultural areas. Their preferred foods include the fruits (and occasionally flowers) of wild figs (*Ficus*), *kupang* (*Parkia*), santol (*Sandoricum*), *malaruhat* (*Syzygium*), *bangkal* (*Nauclea*), and *talisay* (*Terminalia*). They typically forage up to 5 to 12 km from their roost each night. This species commonly roosts together with *P. vampyrus*, and occasionally with *P. hypomelanus*, in mixed-species colonies in large trees, where they are highly conspicuous. Compared with *P. vampyrus*, they roost in denser vegetation at the edge of the colony. Most roost sites are on trees in lowland forest, but some are in areas with ultrabasic soil. Common reports from the late 1800s and early 1900s of 100,000 individuals in a mixed-species colony contrast with most recent observations of a maximum colony size of 5000, and usually far fewer. The two largest known roosts on Luzon as of October 2014 are in Northern Sierra Madre National Park, along the Pacific coast in Divilacan and Dinapigue Municipalities, estimated as possibly including 60,000 and 25,000 individuals, respectively. There are reports of Philippine eagles (*Pithecophaga jefferyi*), white-bellied sea eagles (*Haliaeetus leucogaster*), reticulated pythons (*Python reticulatus*), and possibly Brahminy kites (*Aliastur indicus*) preying on *A. jubatus* and *P. vampyrus* in the mixed colonies. Colonies move their roosting sites frequently, perhaps to avoid predators (including humans). Females give birth to a single young once each year, or perhaps every other year. A mixed colony of *A. jubatus* and *P. vampyrus* in the Subic Bay Special Export Zone is a major tourist attraction; at this site, a viewer may also see *Tylonycteris pachypus*, one of the tiniest bats in the world.

STATUS: Severely declining as a result of habitat destruction and heavy hunting; the population on Luzon is probably only a small percentage of what it was in 1900. As the size of mixed colonies declines, the proportion of *Acerodon jubatus* decreases faster than that

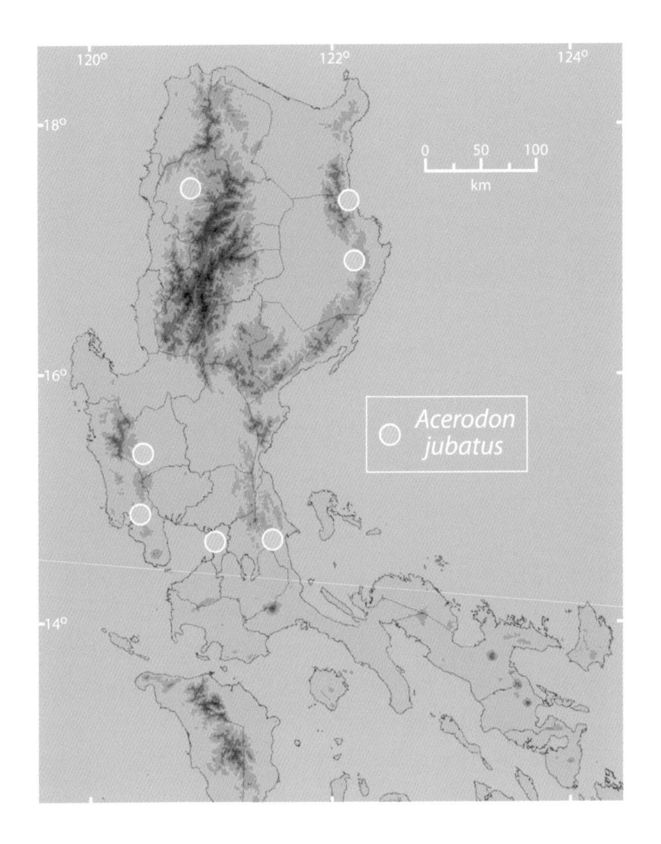

of *Pteropus vampyrus*, which appears to indicate the former's especially high level of vulnerability.

REFERENCES: Heaney and Heideman, 1987; Heaney and Utzurrum, 1991; Hoogstraal, 1951; Lawrence, 1939; Lepiten, 1995; Mickleburgh et al., 1992; Mildenstein et al., 2005; Mudar and Allen, 1986; Rabor, 1986; Rickart et al., 1993; Scheffers et al., 2012; Stier and Mildenstein, 2005; Taylor, 1934; Utzurrum, 1992; van Weerd et al., 2003, 2009, pers. comm. (Oct. 2014).

Cynopterus brachyotis
common short-nosed fruit bat

(Müller, 1838). Tijdschr. Nat. Gesch. Physiol., 5:146. Type locality: Dewei River, Borneo.

DESCRIPTION: Total length 92–108 mm; tail 4–11 mm; hind foot 13–16 mm; ear 16–19 mm; forearm 60–70 mm; weight 26–39 g. A moderately small fruit bat, with large eyes; a broad snout; and four upper and four lower incisors. A tail is always present, though variable in length. In adults, the edge of the ears is pale/white; the skin over the wing phalanges is nearly translucent, causing the white color of the bones to be visible. Adults, especially males, usually have a conspicuous yellow/orange ruff around the shoulders. *Ptenochirus jagori* are larger (forearm 72–88 mm); have pigmented skin over the wing phalanges; and have only two lower incisors. *Haplonycteris fischeri* are smaller (forearm 46–53 mm); have no tail; and have two upper and two lower incisors. *Eonycteris spelaea* are larger (forearm 67–80 mm); have a more elongate muzzle; and lack a claw at the tip of the second digit on the wing.

DISTRIBUTION: Widespread in Southeast Asia; occur throughout the Philippines, including all parts of Luzon.

EVOLUTION AND ECOLOGY: *Cynopterus* is closely related to *Megaerops* and *Ptenochirus*. Genetic evidence suggests that *C. brachyotis* reached Luzon from Southeast Asia within the past 1 million years. There is substantial genetic variation within Philippine populations and high levels of gene flow between populations, al-

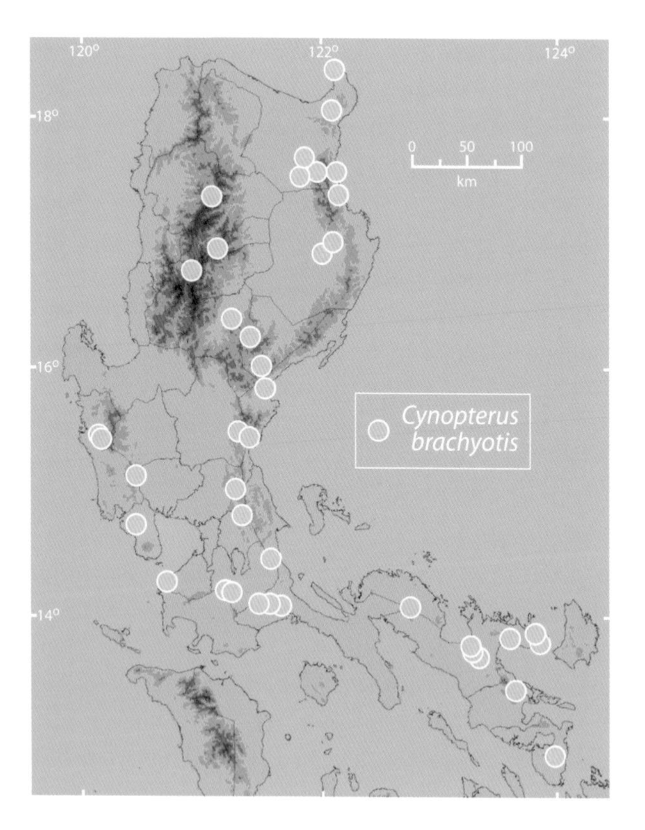

though they show greater genetic similarity within Pleistocene island groups than between the Pleistocene island groups. They are among the most abundant mammals in agricultural areas, are common in secondary forest, and are usually uncommon or absent in primary forest; they persist well in residential and urban areas. They range from sea level to at least 2225 m elevation, are most abundant in lowland areas, generally decrease in abundance up to ca. 800 m elevation, and are scarce above that elevation. They feed only on fruits, including wild figs, wild bananas, and many domestic fruits, and they are important dispersers of seeds into disturbed or deforested areas. Individuals often move up to 1 km over a period of a few weeks. They roost in small groups—usually a single male, several adult females, and some immature individuals—in vegetation, including bananas and palms, and they sometimes modify large leaves to create a "tent"; occasionally they roost in shallow, well-lit caves. Adult males produce an oily orange fluid from the shoulder glands that gives off a distinctive musky odor. Females give birth to a single young twice each year, which they carry with them continuously for several weeks. Gestation is estimated at 120 days, followed by 50 days of lactation. Females may become pregnant at six to eight months of age, and males develop adult-sized testes at about one year of age. Longevity is estimated at four to six years. They produce a loud call when annoyed, often described as similar to that produced by repeatedly squeezing a child's "rubber duckie" toy. Philippine populations have a karyotype of $2n = 34$, FN = 58.

STATUS: Abundant and geographically widespread; stable.

COMMENTS: As traditionally defined, *Cynopterus brachyotis* is composed of several species, two of which occur sympatrically on Borneo. Some authors place Philippine populations in the species *C. luzoniensis*, but the content, distribution, and proper names for the relevant species are unclear, so we retain use of the familiar name until the matter is definitively resolved.

REFERENCES: Alviola et al., 2011; Balete et al., 2011, 2013a; Campbell et al., 2004, 2007; Catibog-Sinha, 1987; Duya et al., 2007; Heaney et al., 1989, 1991, 1998, 2004, 2006a, 2006b; Heaney and Roberts, 2009; Heideman, 1989, 1995; Heideman and Heaney, 1989; Ingle, 1992, 1993, 2003; Kitchener and Maharadatunkamsi, 1991; Lepiten, 1995; Mould, 2012; Ong et al., 1999; Rickart et al., 1989b, 1993, 2013; Roberts 2006a; Sedlock et al., 2008, 2011; Utzurrum, 1995.

Desmalopex leucopterus
mottle-winged flying fox

Temminck, 1853. Esquisses Zool. sur la Côte de Guiné, p. 60. Type locality: "Philippines."

DESCRIPTION: Total length 205–240 mm; tail 0 mm (absent); hind foot 44–45 mm; ear 26–28 mm; forearm 135–150 mm; weight 350–400 g. A fairly large flying fox, with medium to pale brown fur; medium to pale brown wings, with white blotches (especially on the leading edge of the wing and near the wingtip); and pale brown skin on the face. *Pteropus hypomelanus* are similar in size, but have a nearly black face and wings. *Pteropus pumilus* also have pale fur but are smaller (forearm 103–113 mm); the wings and facial skin are medium to dark brown and are not mottled.

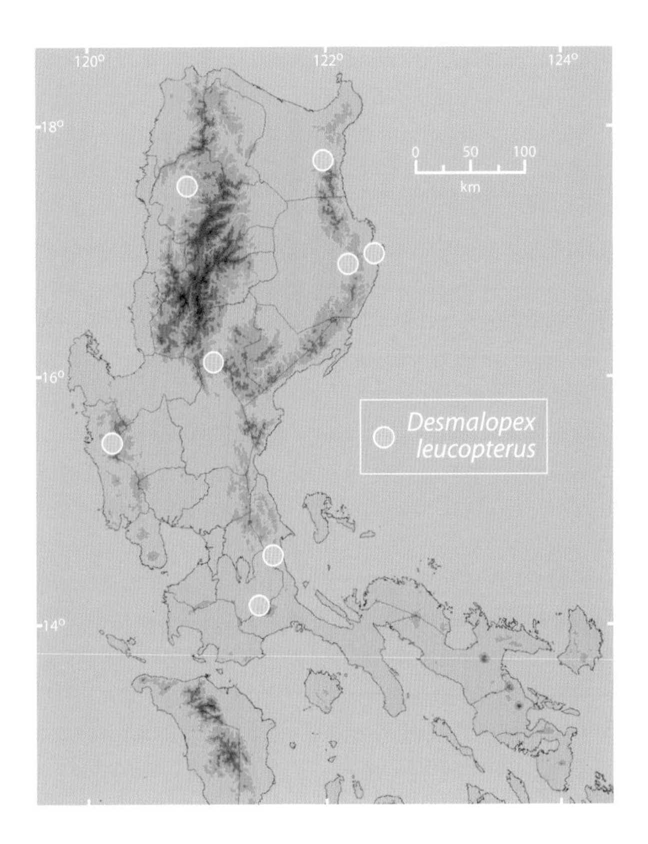

DISTRIBUTION: Endemic to Luzon and Catanduanes Island; probably occur in all mountainous, forested regions of Luzon. A population from Dinagat Island previously referred to this species may be a distinct species (see "Comments").

EVOLUTION AND ECOLOGY: Poorly known overall. Genetic studies show *Desmalopex* to be distantly related to the other Philippine flying foxes, and probably most closely related to *Melonycteris*, which occur in the Solomon and Bismarck Islands. Only two species of *Desmalopex* are currently recognized, and a third may soon be named (see "Comments"); all are restricted to the Philippines, including this species, the recently discovered *D. microleucopterus* on Mindoro, and the possible third species from Dinagat Island. *Desmalopex leucopterus* range from sea level to at least 1200 m elevation. On Catanduanes Island, they are moderately common in primary montane forest and present in lowland forest. Along the Pacific coast in northeastern Luzon, they are present in both lowland forest and forest over ultrabasic soil near the coast. The presence of specimens from Dalton Pass and reports from local people in Balbalasang, Kalinga Province, indicate seasonal movements by this species. Captive females give birth to a single young, not more than once each year. Their roosting habits are not known, but because they have never been seen to form conspicuous roosts in the tops of large trees (as is done by *Acerodon jubatus*, *Pteropus hypomelanus*, and *P. vampyrus*) and have never been found in caves, it is most likely that they roost in inconspicuous places in trees, probably in small groups. They feed exclusively on fruits, including figs. There are no reports of them living or feeding in agricultural areas. They have a karyotype with $2n = 38$.

STATUS: Poorly known; widespread on Luzon, but probably have declined as a result of hunting and extensive lowland-forest habitat destruction.

COMMENTS: The genus *Desmalopex* was described by Gerrit Miller over a century ago to contain this species, on the basis of many distinctive cranial features. Subsequent authors considered *Desmalopex* to be a junior synonym of *Pteropus* and thus placed *leucopterus* in *Pteropus*. Recent studies using DNA-sequence data indicate that *leucopterus* is more closely related to *Pteralopex* than to *Pteropus* and support the recognition of *Desmalopex* as a valid genus. Other recent studies (K. Helgen, pers. comm.) indicate that the population on Dinagat Island is a distinct species, so in this volume we treat *D. leucopterus* as being endemic to Greater Luzon.

REFERENCES: Almeida et al., 2011; Balete et al., 2011; Duya et al., 2007; Esselstyn et al., 2008; Giannini et al., 2008; Heaney and Rabor, 1982; Heaney et al., 1991, 1999; Mickleburgh et al., 1992; G. Miller, 1907; Rickart et al., 1999; Scheffers et al., 2012; Utzurrum, 1992; van Weerd et al., 2003, 2009.

Dyacopterus rickarti
Philippine large-headed fruit bat

Helgen et al., 2007. J. Mammal., 88:311. Type locality: San Isidro, Abra Province, Luzon.

DESCRIPTION: Total length 148–173 mm; tail 18–29 mm; ear 21–25 mm; forearm 91–96 mm; weight 138–148 g. A large, dark fruit bat, with a heavy muzzle. There are four upper and four lower incisors. The head is broad, with prominent masseter muscles; the canines and molars are unusually massive. The fur is rather short and dark on the head and back, and paler on the abdomen. *Ptenochirus jagori* are similar but smaller (forearm 72–88 mm), have a less massive muzzle and teeth; and have only two lower incisors.

DISTRIBUTION: Currently known on Luzon from a single specimen from Abra Province, and from several localities on Mindanao. Related species occur on Sumatra, Borneo, and the Malay Peninsula.

EVOLUTION AND ECOLOGY: *Dyacopterus* is distantly related to *Cynopterus* and other genera that occur primarily on the Sunda Shelf of Southeast Asia. The Philippine species is known only from 1260 m to 1680 m elevation, in regenerating secondary or primary montane and mossy rainforest on Mindanao and Luzon. No ecological information is available for the single specimen from Luzon. On Mindanao, the forest trees at capture sites included *Actinodaphne*, *Agathis*, *Casuarina*, *Cinnamomum*, *Elaeocarpus*, *Ficus*, *Leptospermum*, *Lithocarpus*, *Prunus*, and *Syzygium*, with *Cyathea*, *Freycinetia*, *Medinilla*, and *Musa* in the understory. A female pregnant with a single young was taken on 1 May. Either one or only a few specimens were taken at each locality. We suspect that *D. rickarti* typically fly above the canopy, making them difficult to capture.

STATUS: A very poorly known species; possibly widely distributed in montane forest, which is usually less impacted by human activities than lowland forest. Further research is needed.

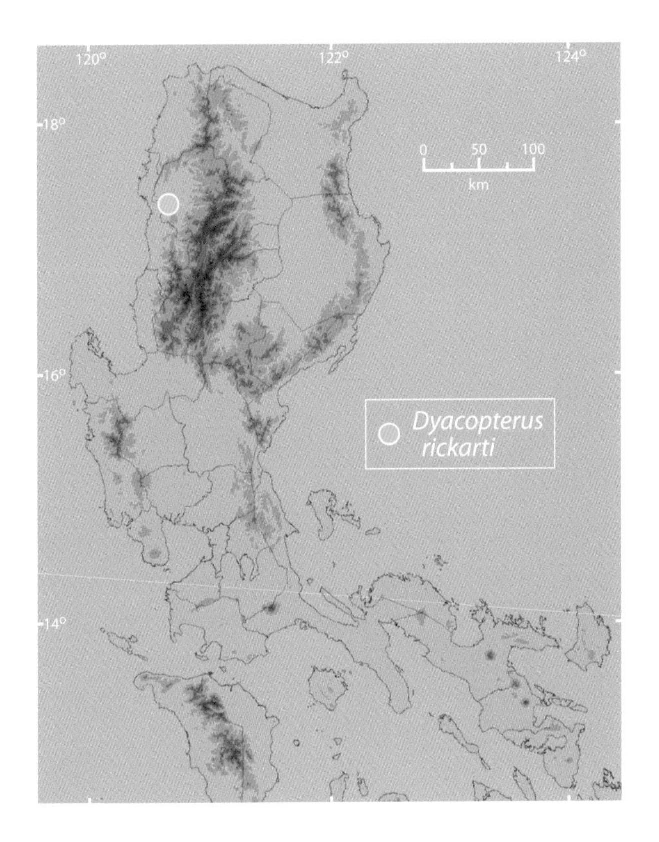

REFERENCES: Gomez et al., 2005; Heaney et al., 2006a; Helgen et al., 2007; Kock, 1969a; Utzurrum, 1992.

Eonycteris robusta
Philippine nectar bat, Philippine dawn bat

Miller, 1913. Proc. Biol. Soc. Washington, 26:73. Type locality: Montalban Caves, Rizal Province, Luzon Island.

DESCRIPTION: Total length 127–155 mm; tail 20–28 mm; hind foot 23–25 mm; ear 19–23 mm; forearm 67–82 mm; weight 56–80 g. A medium-sized fruit bat, with short, dark gray or silver-gray fur and a notably long muzzle and tongue. It lacks a claw at the tip of the second digit (on the leading edge of the wing; chapter 11, "Introduction to Bats") and lacks paired anal glands. The ears are somewhat elongated and bluntly pointed; the tail is moderately long; and the skin over the wing bones is pigmented (not translucent and "white"). Adult males are larger than females and usually have a ruff of elongated hair around the neck, often yellowish in color. *Eonycteris spelaea* are similar but slightly smaller, with a shorter hind foot (15–20 mm); a shorter tail (12–20 mm); and a pair of glands adjacent to the anus. *Rousettus amplexicaudatus* are similar, but are larger (forearm 83–91 mm); usually have a shorter tail (17–23 mm); have a claw on the second digit; and have nearly translucent skin over the bones of the wings that allows the white bone to be seen.

DISTRIBUTION: Endemic to the Philippines, where it is widespread, but absent from the Palawan Faunal Region and Batanes/Babuyan group of islands. Records on Luzon are widespread, with most coming from areas with extensive caves.

EVOLUTION AND ECOLOGY: The genus *Eonycteris* is most closely related to *Rousettus*; *E. robusta* are probably most closely related to *E. major*, which occur on Borneo. The ecology of *E. robusta* is poorly known. We suspect that they feed on nectar, pollen, and soft fruits. The limited information suggests that these bats roost in limestone

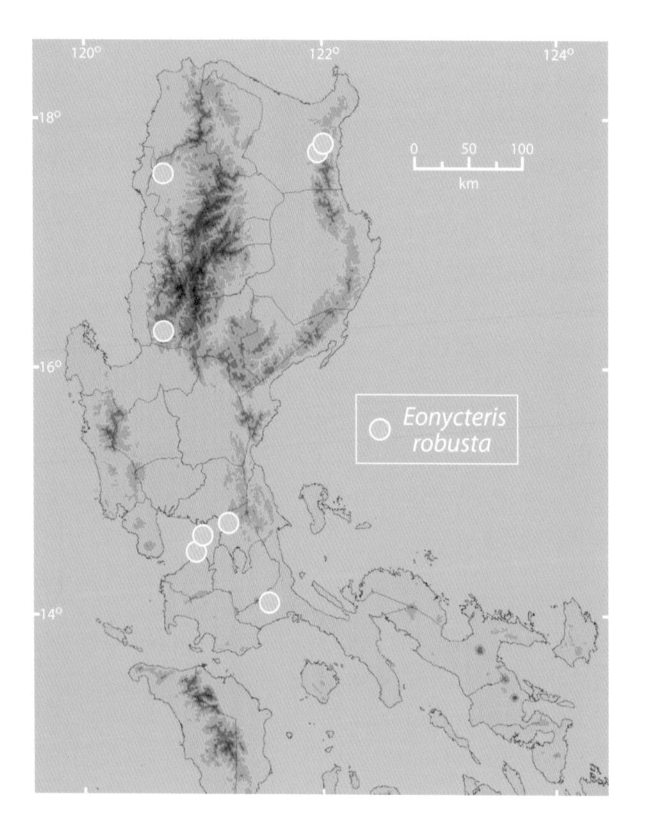

caves (often together with *E. spelaea* and/or *R. amplexicaudatus*) and forage in nearby forest. Until the 1960s, they were commonly found in caves adjacent to forest and commonly netted in and adjacent to forest, from sea level to 1100 m elevation, often in areas with mixed forest and clearings but apparently not in primarily agricultural areas. There are few records from our extensive recent field studies on Luzon, which began in the 1980s. A 68 g female was captured carrying a pup weighing 13 g; females probably give birth to a single young.

STATUS: Captured less frequently in the 1980s to the present than in the 1920s to 1960s, and may now be scarce, perhaps due to hunting and mining in the caves where they roost; more field studies are needed.

COMMENTS: Sometimes included as a subspecies of *Eonycteris major*, but we believe that *E. robusta* is a distinct species, based on cranial features.

REFERENCES: Almeida et al., 2011; Corbet and Hill, 1992; Heaney et al., 1991, 1998; Ingle and Heaney, 1992; Mickleburgh et al., 1992; Rickart et al., 1993; Utzurrum, 1992.

Eonycteris spelaea common nectar bat, common dawn bat

(Dobson, 1871). Proc. Asiat. Soc. Bengal, pp. 105, 106. Type locality: Farm Caves, Moulmein, Tenasserim, Burma [= Myanmar].

DESCRIPTION: Total length 121–132 mm; tail 12–20 mm; hind foot 15–20 mm; ear 18–23 mm; forearm 67–80 mm; weight 55–76 g. A small fruit bat, with short (often sparse), dark brown or gray fur; a fairly long muzzle and tongue; and large eyes. There is no claw at the tip of the second digit (on the leading edge of the wing; chapter 11, "Introduction to Bats"). A pair of

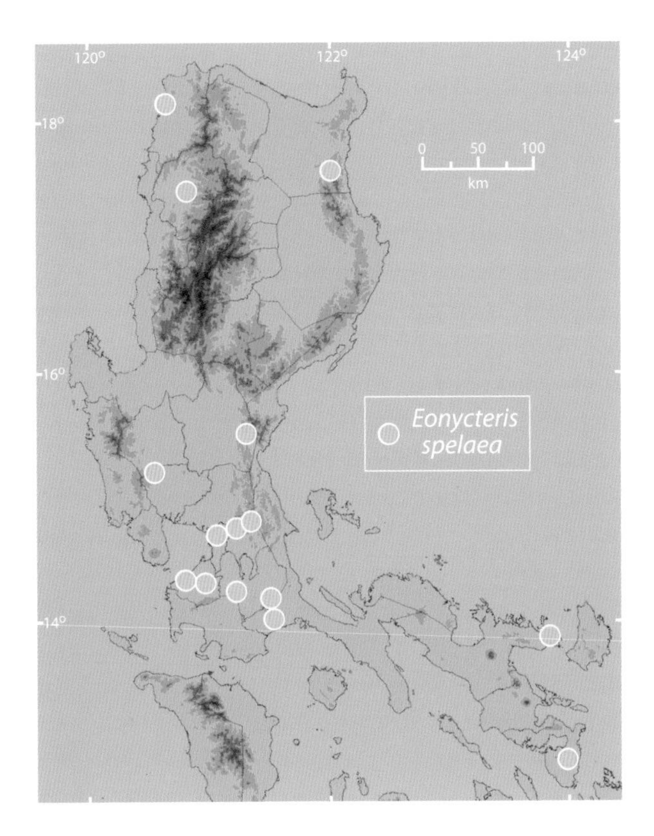

glands at the sides of the anus are always present, but they are small in young animals and largest in breeding males. The ears and tail are moderate in length (the tail is shorter than in *Eonycteris robusta*). Males are larger than females and usually have a ruff of elongated hairs at the sides of the neck. *Rousettus amplexicaudatus* are similar but larger (forearm 83–91 mm); have a claw on the second digit; and lack paired anal glands. *Cynopterus brachyotis* are slightly smaller in size (forearm 60–70 mm); have a short, blunt muzzle; do not have paired anal glands; and have skin over the wing bones that is nearly translucent.

DISTRIBUTION: From India to Timor; occur throughout the Philippines, except in the Batanes/Babuyan Faunal Region, and are widespread on Luzon.

EVOLUTION AND ECOLOGY: *Eonycteris* contains three species: one from Borneo (*E. major*), one from the Philippines (*E. robusta*), and the very widespread *E. spelaea*. Nothing is known of their genetics or evolution within the Philippines. *Eonycteris spelaea* occur from sea level to 700 m elevation on Luzon; they are nearly or entirely absent from montane and mossy forest. In the lowlands, they are often abundant in agricultural areas and secondary forest (including urban areas and orchards) but avoid mature forest. Their abundance generally decreases with increasing elevation. All known Philippine roosting sites have been in caves, where they form colonies of up to thousands of individuals and where they are vulnerable to hunting pressure. Females give birth to a single young, usually twice each year; gestation is estimated to be 120 days. In Malaysia, this bat species feeds on nectar, pollen, and soft (often overripe) fruits, and some individuals forage up to 38 km from their roost. They live for at least five years. They have a karyotype with $2n = 36$, FN = 66.

STATUS: Stable and common in some agricultural areas, but often heavily hunted in the caves where they roost. Esselstyn et al. (2004) found one population on Palawan Island estimated as exceeding 2000 individuals and another that probably exceeded 50,000.

REFERENCES: Balete et al., 2013a; Heaney et al., 1989, 1999; Heideman and Heaney, 1989; Heideman and Utzurrum, 2003; Ingle, 1992; Lepiten, 1995; Ong et al., 1999; Rickart et al., 1989a, 1993; Sedlock et al., 2014b; Start and Marshall, 1975; Utzurrum, 1992.

Haplonycteris fischeri
Philippine pygmy fruit bat

Lawrence, 1939. Bull. Mus. Comp. Zool., 86:33. Type locality: Mt. Halcon, Mindoro Island.

DESCRIPTION: Total length 65–83 mm; tail 0 mm (absent); hind foot 10–13 mm; ear 12–15 mm; forearm 46–53 mm; weight 16–22 g. A small fruit bat, with a short, blunt muzzle; two upper and two lower incisors; and a distinctive band of pale fur along the upper surface of the forearm. Adult males have a gland on each side

of the neck that produces a dark orange oily substance that stains the fur around it. Young *Cynopterus brachyotis* are superficially similar but lack the band of fur on the forearm; have four upper and lower incisors; have a longer forearm (60–70 mm); and have nearly clear skin over the bones of the phalanges. *Otopteropus cartilagonodus* are similar in size but have distinctive bits of white tissue at the front and back of each ear; lack the band of pale fur on the forearm; and have larger eyes.

DISTRIBUTION: Endemic to the Philippines; widespread, but excluding the Camiguin, Palawan, and Batanes/Babuyan Faunal Regions. Widespread on Luzon, with the possible exception of the Central Cordillera, where no records are currently available.

EVOLUTION AND ECOLOGY: *Haplonycteris* is closely related to *Alionycteris* and *Otopteropus*, all three of which are endemic to the oceanic portion of the Philippines. Genetic variation within populations is moderately high. Each of the faunal regions sampled thus far has a genetically distinct population, implying low rates of gene flow among islands. *Haplonycteris fischeri* are uncommon on Luzon, where they occur from sea level to 900 m elevation (rarely to 1200 m); on other islands, they are more abundant and are found from about sea level to 2250 m elevation. They are most common in primary forest, especially at middle elevations; they are less abundant in secondary forest but can persist in degraded regenerating forest at middle elevations. They are usually absent in agricultural areas. They appear to fly most often within the subcanopy, rather than at ground level, and do not often fly far from canopy cover. On Luzon, this species is relatively most common where *O. cartilagonodus* is uncommon or absent, such as on Mt. Natib, suggesting that competition may occur between the two. Density at 860–1100 m elevation on southern Negros was estimated at 3.7/ha; at most places on Luzon they appear to be much less abundant. Longevity of 8–10 years may be typical; limited data suggest that mortality is highest during typhoons. All adult females in a given population give birth to a

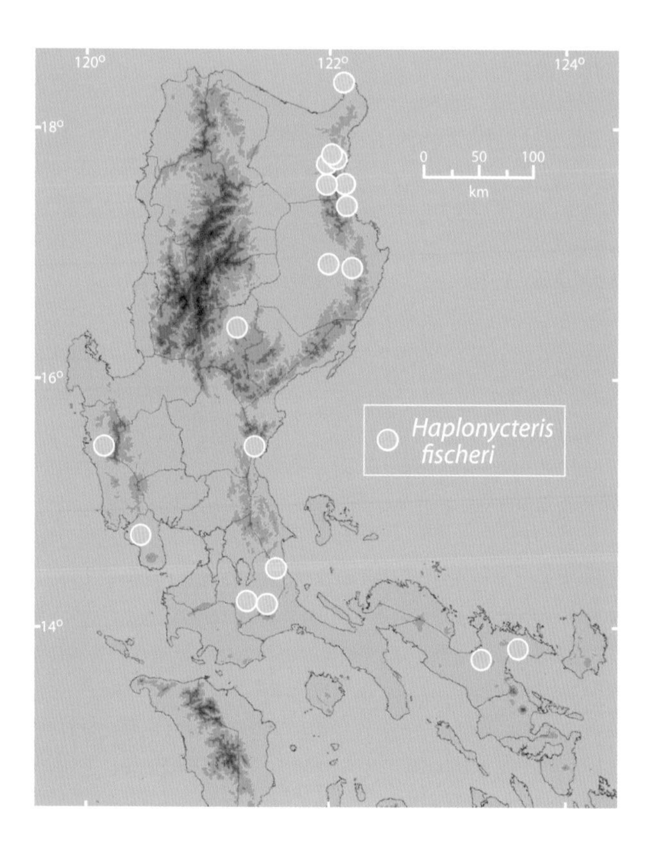

single young at about the same time each year, mate soon afterwards, and become pregnant after an extended period of delayed implantation; they give birth about 50 weeks after mating. Like other fruit bats, they play a significant role in carrying seeds of trees into disturbed forest, thus aiding forest regeneration. They produce a quiet chittering call when feeding together in a fruiting tree and when in their roosts. On Biliran and Leyte, they have a karyotype of $2n = 58$, FN = 66.

STATUS: Widespread and currently stable in most parts of their range, but declining in some areas as a result of habitat destruction by logging, mining, and the expansion of agriculture.

REFERENCES: Alviola et al., 2011; Balete et al., 2011; Duya et al., 2007, 2011; Heaney et al., 1989, 1991, 1999, 2006b; Heideman, 1989; Heideman and Heaney, 1989; Ingle, 1993, 2003; Mudar and Allen, 1986; Peterson and Heaney, 1993; Rickart et al., 1989a, 1993, 2013; Roberts, 2006a, 2006b; Utzurrum, 1995.

Harpyionycteris whiteheadi
harpy fruit bat

Thomas, 1896. Ann. Mag. Nat. Hist., ser. 6, 18:244. Type locality: "highlands . . . at 5000 feet," Mindoro Island.

DESCRIPTION: Total length 130–159 mm; tail 0 mm (absent); hind foot 22–24 mm; ear 20–25 mm; forearm 84–91 mm; weight 99–140 g. A moderately large fruit bat, with a long muzzle; anteriorly projecting upper and lower incisors and canines; thick fur on the dorsal surface of the hind feet; and often with mottled pale spots on the wings, resulting from skin punctures that produce scarring. The fur is a rich dark brown over the entire body. No other Philippine bat has anteriorly projecting upper and lower incisors, and none of equal size have heavy fur reaching along the dorsal surface from the body to the hind feet.

DISTRIBUTION: Philippines only; widespread, but excluding the Palawan Faunal Region and the Batanes/Babuyan group of islands. Only a single record is currently available from Luzon, from Mt. Isarog, Camarines Sur Province.

EVOLUTION AND ECOLOGY: Most closely related to *Harpyionycteris celebensis* from Sulawesi, and more distantly to *Boneia*, also from Sulawesi (chapter 11, Family Pteropodidae). *Harpyionycteris whiteheadi* usually are restricted to primary and lightly disturbed forest but occasionally occur in nearby agricultural areas, from sea level to at least 1900 m elevation in the southern Philippines; the one record from Luzon is based on three individuals captured at 475 m elevation that were feeding on the fruit of *Freycinetia*. In the southern Philippines, they are uncommon in lowland forest, moderately common in montane forest, and uncommon in mossy forest. On southern Negros Island, these bats occur at moder-

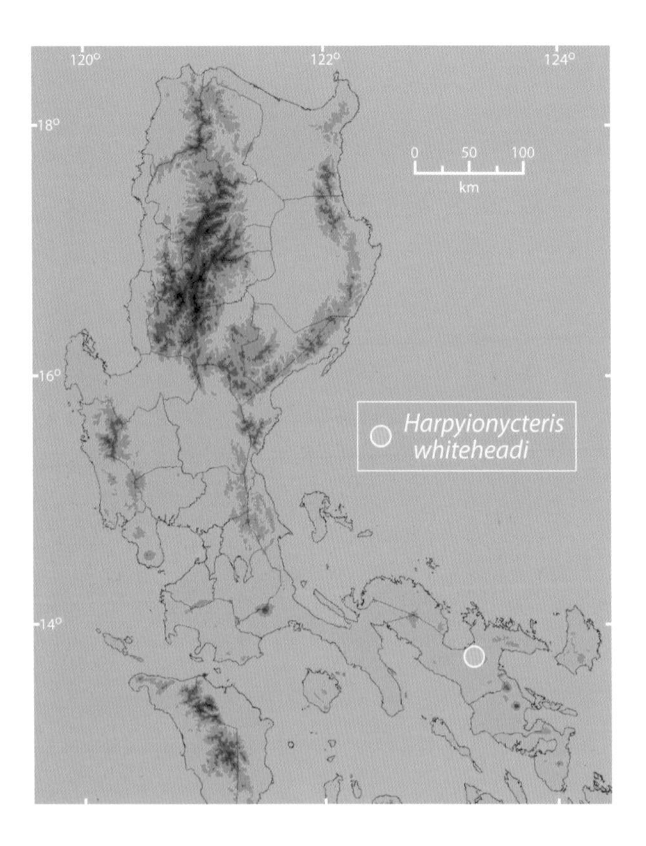

ately low densities, estimated at 0.7/ha at 850–1100 m elevation. They feed on the fruits of figs (*Ficus* spp.) and viny pandans (*Freycinetia*). Females give birth to a single young, probably only once each year; gestation is about 135 days, and lactation lasts for about 120 days. They produce a loud, piercing whistle when flying and a similar, exceptionally loud whistle when captured. They have a karyotype of $2n = 36$, FN = 58.

STATUS: Populations probably are currently stable, because of their primary use of montane forest, which is mostly intact. Locally vulnerable to deforestation.

REFERENCES: Giannini et al., 2006; Heaney, 1991; Heaney et al., 1989, 1999, 2006a, 2006b; Heideman, 1995; Heideman and Heaney, 1989; Rickart et al., 1989a, 1993.

Macroglossus minimus dagger-toothed flower bat

(É. Geoffroy, 1810). Ann. Mus. Hist. Nat. Paris, 15:97. Type locality: Java, Indonesia.

DESCRIPTION: Total length 63–77 mm; tail 0 mm (absent); hind foot 11–18 mm; ear 14–18 mm; forearm 39–43 mm; weight 13–17 g. A small fruit bat, with a long muzzle; a long, narrow tongue; four tiny upper and lower incisors; and especially slender and sharp canines. The fur is short, yellow or golden-brown, and paler ventrally. Glandular tissue on the skin makes a narrow V across the upper chest on most individuals, but this is most conspicuous on adult males. *Eonycteris* are larger (forearm 67 mm or more). *Haplonycteris fischeri* are similar in size (forearm 46–53 mm) but have a short, blunt snout; slightly tubular nostrils; stout canines; and two upper and two lower incisors.

DISTRIBUTION: Found from Thailand to Australia; occur throughout the Philippines, including all parts of Luzon.

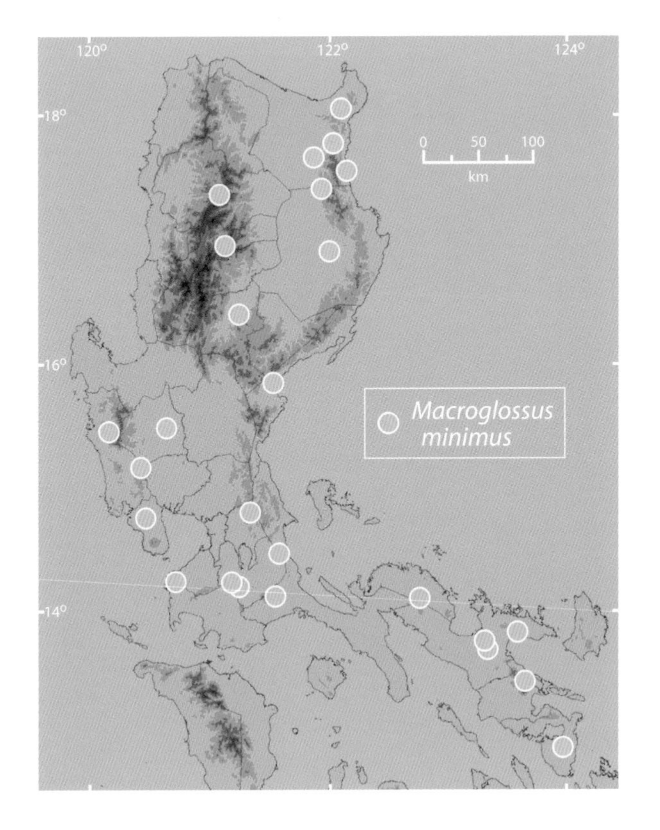

Macroglossus minimus, feeding on sugar-water, photographed on Siquijor Island. Courtesy of J. S. Walsh.

EVOLUTION AND ECOLOGY: The genus *Macroglossus* occurs throughout most of Indo-Australia and is most closely related to *Syconycteris*, a genus that occurs in Australia, New Guinea, Sulawesi, and adjacent islands (chapter 11, Family Pteropodidae). Populations maintain fairly high levels of genetic variation, and genetic similarity is high among the Philippine islands, which implies high rates of gene flow. They occur in virtually every habitat on Luzon: they are abundant in agricultural and other heavily disturbed areas, common in secondary forest, and uncommon in primary forests. On Luzon they are most common from sea level to about 900 m elevation but occur up to 1760 m elevation. Their density usually decreases with increasing elevation. They usually are found in association with wild and domestic bananas (*Musa* spp.) and sometimes roost in the leaves of these plants. They feed primarily on nectar, flowers, and soft, overripe fruits, especially bananas. Home ranges are probably several hundred meters to a kilometer in diameter. They have been found roosting in clusters of dead leaves and in hollows in trees. Females give birth to single young, probably twice per year; gestation is about 120 days. They have a karyotype with $2n = 34$, FN = 62.

STATUS: Abundant and widespread in Indo-Australia; Philippine populations are stable.

REFERENCES: Alviola et al., 2011; Duya et al., 2007; Heaney et al., 1989, 2006a, 2006b; Heaney and Roberts, 2009; Heideman and Heaney, 1989; Heideman and Utzurrum, 2003; Ingle, 1993; Lepiten, 1995; Rickart et al., 1993, 2013; Roberts, 2006a; Sedlock et al., 2008, 2011; Utzurrum, 1992.

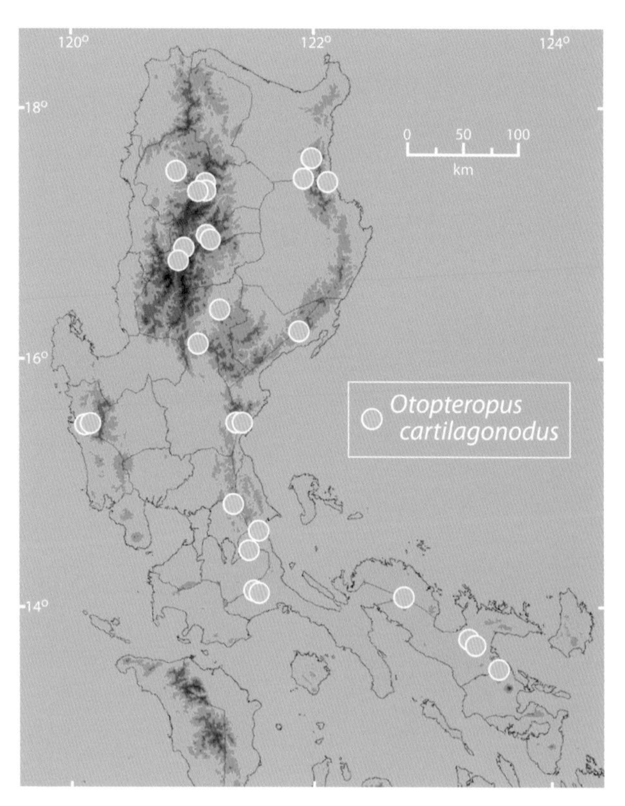

Otopteropus cartilagonodus
Luzon pygmy fruit bat

Kock, 1969. Senckenberg. Biol., 50:333. Type locality: Sitio Pactil, [southern Bauko Municipality], Mountain Province, Luzon Island.

DESCRIPTION: Total length 70–80 mm; tail 0 mm (absent); ear 12–14 mm; forearm 44–50 mm; weight 15–19 g. A small fruit bat, with exceptionally large eyes; a short, blunt snout; short, tubular nostrils; no tail; and only two upper and two lower incisors. The fur is long, soft, and grayish-brown, slightly paler ventrally than dorsally. There are small bits of soft white tissue at the front and back of each ear. Males have an indented circular gland (ca. 1 cm in diameter) on the upper abdomen during breeding season. *Haplonycteris fischeri* are similar in size (forearm 46–53 mm) but lack the bits of white tissue on the ears and have a band of pale fur on the dorsal surface of the forearm. *Cynopterus brachyotis* are larger (forearm 60–70 mm); lack the bits of white tissue on the ears; and have a short tail.

DISTRIBUTION: Known only from Luzon Island, but probably occur in nearly all hilly and mountainous areas that retain primary or secondary forest.

EVOLUTION AND ECOLOGY: *Otopteropus* is most closely related to *Haplonycteris* and *Alionycteris*, both of which also occur only in the oceanic portions of the Philippines. Their closest relatives (*Chironax* and *Penthetor*) occur principally on the Sunda Shelf of Southeast Asia (chapter 11, Family Pteropodidae). Although the species was poorly known prior to the 1980s, they are widespread and common on Luzon. They occur in primary and secondary forest and in mixed forest/cropland in lowland, montane, and mossy forest, from near sea level to 2530 m elevation, but they sometimes are absent below 1000 m elevation. They are generally absent from agriculturally dominated landscapes, especially in the lowlands. They are also absent from some mountains that are rather isolated and of moderate elevation, such as Mt. Natib and Mt. Pinatubo (the latter was devastated by a 1991 eruption). *Haplonycteris fischeri*, which are ecologically similar, occur at lower elevations, and the two species rarely overlap. *Otopteropus cartilagonodus* are usually the most abundant (and sometimes the only) fruit bat above 1000 m elevation on Luzon. Their abundance generally rises with increasing elevations up to 2200–2500 m, and then declines. Often they are especially common in mossy forest. These bats probably feed on small fruits from trees, shrubs, and vines, often including figs (*Ficus*), melostomes (*Melastoma*), and pepper (*Piper*). Males usually occur at higher elevations than females, and often only one sex will be found at a given locality. Females produce a single young each year, with all females in a given area giving birth within a period of a few days, after a period of post-implantation developmental delay. They occasionally produce a soft, chittering call when feeding in a group in a fruiting tree and when they are roosting together. They have a karyotype with $2n = 48$, FN = 62.

STATUS: Widespread on Luzon and apparently stable, because of their primary use of middle- and upper-elevation forest.

REFERENCES: Alviola et al., 2011; Balete et al., 2011, 2013a; Duya et al., 2007; Heaney et al., 1999, 2004; Heideman et al., 1993; Kock, 1969b; Mickleburgh et al., 1992; Mudar and Allen, 1986; Rickart et al., 1999, 2013; Ruedas et al., 1994; Sedlock et al., 2008, 2011; Utzurrum, 1992.

Ptenochirus jagori
greater musky fruit bat

(Peters, 1861). Monatsb. Preuss. Akad. Wiss. Berlin, p. 707. Type locality: Daraga, Albay Province, Luzon Island.

DESCRIPTION: Total length 118–134 mm; tail 4–17 mm; hind foot 16–23 mm; ear 18–23 mm; forearm 72–88 mm; weight 65–87 g. A fairly large fruit bat, with a broad, dark head; short, stout muzzle; and four upper and two lower incisors. Adults have a shoulder ruff of fur, usually with a gland beneath the ruff that produces a yellow oily material that stains the ruff; this is especially apparent on adult males. A distinctive odor of "sweet musky cinnamon" is usually present. Males are slightly larger and darker than females. *Cynopterus brachyotis* are smaller (forearm 60–70 mm) and have four upper and four lower incisors. *Rousettus amplexicaudatus* are similar in size (forearm 83–91 mm) but have a long muzzle; short fur; and a longer tail (17–23 mm). *Eonycteris robusta* and *E. spelaea* also have a long muzzle; short fur; longer tails (20–28 mm and 12–20 mm, respectively); and lack a claw on the second digit.

DISTRIBUTION: Restricted to the Philippines, but absent in the Batanes/Babuyan and Palawan Faunal Regions; occur throughout Luzon.

EVOLUTION AND ECOLOGY: *Ptenochirus* is closely related to *Cynopterus* and *Megaerops* (chapter 11, Family Pteropodidae). There are two species of *Ptenochirus*:

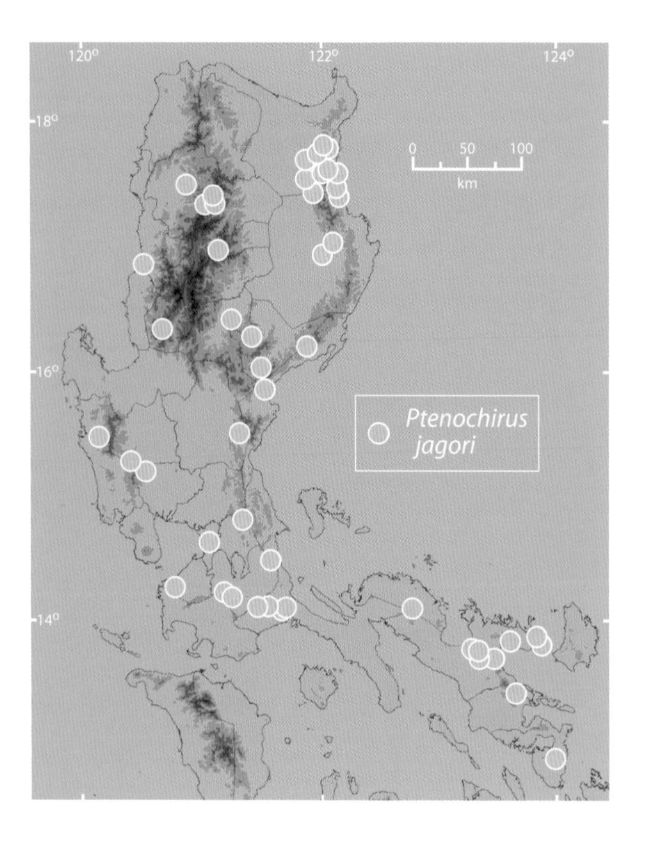

P. jagori occurs through most of the oceanic Philippines, and *P. minor* is present only in the Mindanao Faunal Region. The amount of genetic variation within any given population is generally high, and gene flow levels within current islands is high, but the different Pleistocene island groups each have a genetically distinctive population, which implies somewhat limited gene flow between islands. *Ptenochirus jagori* are one of the most abundant and habitat-tolerant bat species in the Philippines: they are abundant in primary and secondary lowland forest, usually present in rural agricultural areas, and often are present in urban parks and residential areas. They range from sea level to at least 1950 m elevation but are most common in lowland forest, uncommon in montane forest, and usually absent in mossy forest. On Negros, their density has been estimated from 1 to more than 3/ha. Females give birth to a single young twice each year, with most females in a given area giving birth within a period of about a week. Gestation is around 120 days, and lactation takes place for about 90 days. They can live for at least five years, and perhaps eight years, with annual survivorship of ca. 70%. They appear to live in small groups made up of one adult male, several adult females, and several juveniles and/or subadults; young adult males sometimes form small bachelor colonies. Adult males have home ranges about half the size of those of adult females. They feed heavily on figs and other fruits, including bananas (*Musa* spp.). They are among the most important bats for spreading seeds into disturbed habitats. Roosting sites have been found in hollow trees and in shallow caves in limestone and earthen banks. They produce a loud, repeated call when disturbed that is sometimes described as a "rubber duckie" call; it is similar to that of *Cynopterus brachyotis* but is slightly deeper in tone. All individuals have a gland near the junction of the lower neck and shoulder that produces an oily yellow fluid, creating the bats' musky odor; the gland and oily material are most conspicuous in adult males. They have a karyotype with $2n = 44$, FN = 56.

STATUS: Large, generally stable, widespread populations, tolerant of heavily disturbed habitat. They are hunted in some rural areas, with hunters often using fishhooks hung from lines near fruiting trees, but this seems to have little impact on populations of *Ptenochirus jagori*.

REFERENCES: Almeida et al., 2011; Alviola et al., 2011; Balete et al., 2011; Catibog-Sinha, 1987; Duya et al., 2007; Heaney and Roberts, 2009; Heaney et al., 1989, 1999, 2004, 2005, 2006a, 2006b; Heideman, 1995; Heideman and Heaney, 1989; Heideman and Powell, 1998; Ingle, 1992, 1993, 2003; Lepiten, 1995; Mudar and Allen, 1986; Reiter and Curio, 2001; Rickart et al., 1989a, 1993, 2013; Roberts, 2006a; Sedlock et al., 2008, 2011; Utzurrum, 1995.

Pteropus hypomelanus common island flying fox

Temminck, 1853. Esquisses Zool. sur la Côte de Guiné, p. 61. Type locality: Ternate Island, Molucca Islands, Indonesia.

DESCRIPTION: Total length 210–240 mm; tail 0 mm (absent); ear 28–32 mm; forearm 128–149 mm; weight 425–450 g. A fairly large flying fox, with a rusty-yellow or golden mantle covering the back of the head and the shoulders. The pelage posterior to the mantle is dark brown, often with scattered pale hairs. The wing membranes are dark brown, nearly black; the hind foot and posterior third of the hind leg are mostly bare skin, with only a few scattered hairs. Males typically are larger than females. *Acerodon jubatus* are larger (forearm 180–195 mm) and have wings that are dark brown and mottled, with paler areas. *Pteropus vampyrus* are larger (forearm 175–190 mm) but otherwise very similar in appearance. *Pteropus pumilus* are smaller (forearm 103–113 mm) and have paler fur. *Desmalopex leucopterus* are similar in size (forearm 135–150 mm) but have pale brown fur; medium to pale brown wings, with pale blotches; and pale brown skin on the face.

DISTRIBUTION: Thailand to Australia, and throughout the Philippines; only a few widely scattered records from Luzon.

EVOLUTION AND ECOLOGY: The genus *Pteropus* is most closely related to *Acerodon* (chapter 11, Family Pteropodidae). The limited available data suggest that *P. hypomelanus* is most closely related to species of *Pteropus* that occur in Australia and New Guinea and on islands in the tropical Pacific. In the Philippines, *P. hypomelanus* occur in heavily disturbed agricultural and residential areas, from sea level to ca. 900 m elevation, often near the coast; they frequently roost on small islands near the coast of larger islands. There are no confirmed records in the Philippines from primary forest. They feed on both domestic and wild fruits, including

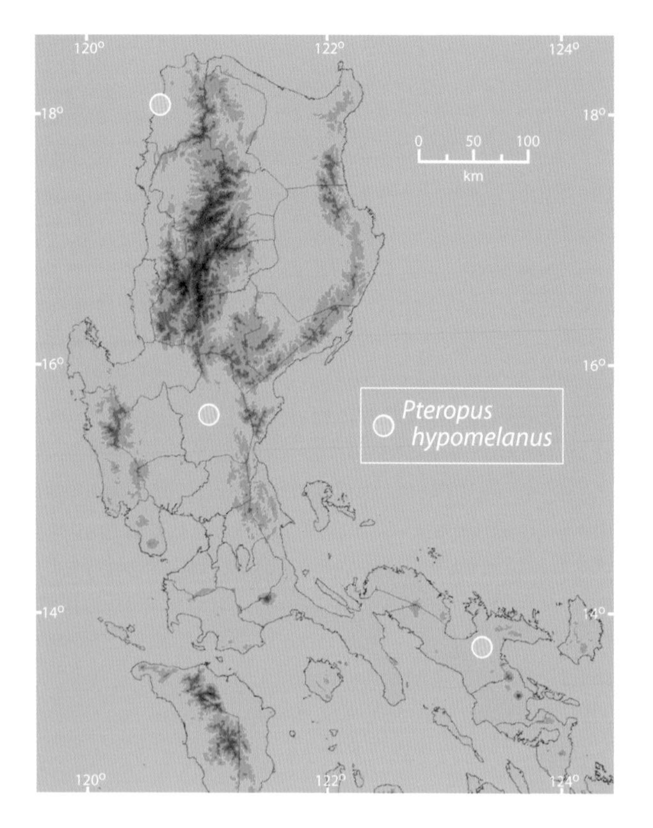

figs (*Ficus* spp.), mangoes (*Mangifera indica*), coconut palms, and chico (*Calocarpum sapota*). They roost in the tops of trees, including coconut palms and broad-leafed trees, in groups of a few to several hundred, often intermixed with *P. vampyrus*. Females give birth to a single young, not more than once per year. The calls are loud and are often heard at night, when a group feeds together in a tree. They have a karyotype with $2n = 38$, FN = 72.

STATUS: Generally widespread and moderately common in the Philippines, but less common than in the past and currently overhunted and declining in some areas. They are hunted with guns, nets, and fishhooks hanging from lines.

REFERENCES: Esselstyn et al., 2008; Heaney et al., 1999; Jones and Kuntz, 2000a; Mickleburgh et al., 1992; Rickart et al., 1993; Utzurrum, 1992.

Pteropus pumilus little golden-mantled flying fox

Miller, 1910. Proc. U. S. Natl. Mus. 38:394. Type locality: Miangas Island [= Palmas Island, between the Talaud Islands and Mindanao], Indonesia.

DESCRIPTION: Total length 155–180 mm; tail 0 mm (absent); hind foot 34–40 mm; ear 23–28 mm; forearm 103–113 mm; weight 145–200 g. Like all flying foxes, the muzzle is long and robust; the eyes are large. A relatively small flying fox, with a soft, dense pelage that varies among individuals from pale gray to honey-brown; old males tend to be the palest. The shoulders and upper back have a fairly long mantle of soft fur; the fur under the chin is medium to pale gray. The wings are medium to dark brown, not mottled. *Desmalopex leucopterus* are larger (forearm 135–150 mm), with paler wings that are mottled with nearly white areas. Other species of *Pteropus* are larger and do not have pale gray or honey-brown fur.

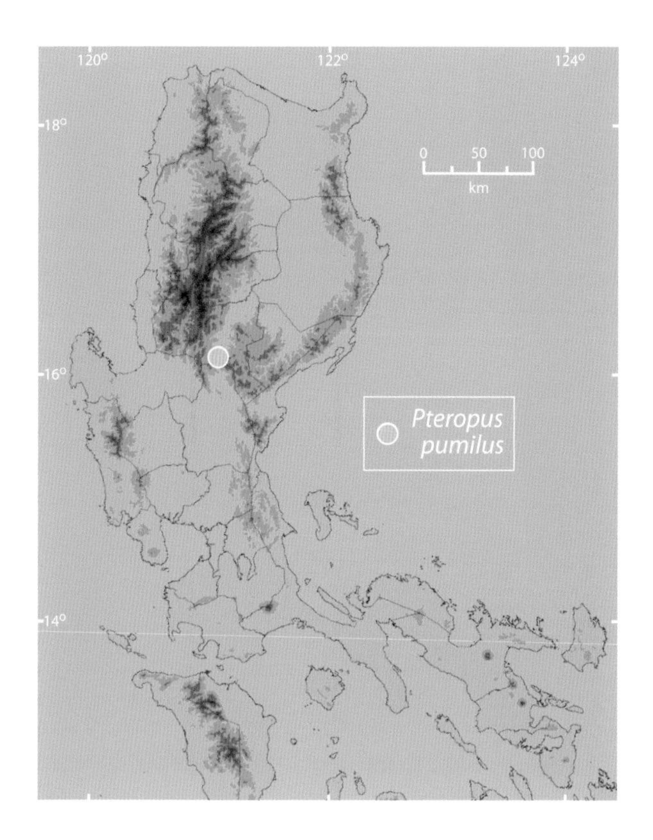

DISTRIBUTION: Occur only in the Philippines, excluding the Batanes/Babuyan and Palawan Faunal Regions but including Miangas Island, Indonesia (adjacent to Mindanao). A single specimen is known from Luzon, from Dalton Pass in Nueva Vizcaya Province.

EVOLUTION AND ECOLOGY: The species of *Pteropus* are most closely related to *Acerodon*, but the relationship of *P. pumilus* to other *Pteropus* is uncertain, and nothing is known of its genetics. Elsewhere in the Philippines, *P. pumilus* are associated with primary and secondary lowland forest, from sea level to about 1100 m elevation (rarely to 1250 m), and are uncommon outside of forest. They are most common on small to medium islands and uncommon to rare on larger islands. One individual in the Visayas was seen roosting in a tree fern. They eat wild figs and other forest fruits. When feeding, they aggregate in small numbers in fruiting trees and make quiet calls. Because they (and other *Pteropus*) fly above the canopy, they are not often caught in mist nets. Dalton Pass, where the single Luzon specimen was captured, is a major migratory pathway where birds and bats move from the central lowlands to the Cagayan River Valley. The rarity of specimens, given their abundance in the central and southern Philippines, implies that they are scarce on Luzon. This species has a karyotype with $2n = 38$, FN = 72.

STATUS: Populations throughout the Philippines have been reduced by some hunting and habitat destruction but are still generally widespread and stable.

REFERENCES: Heaney, 1991; Heaney et al., 1989, 2006a; Heideman and Heaney, 1989; Klingener and Creighton, 1984; Lepiten, 1995; Rickart et al., 1993, 1999; Utzurrum, 1992.

Pteropus vampyrus
giant flying fox

(Linnaeus, 1758). Syst. Nat., 10th ed., 1:131. Type locality: Java, Indonesia.

DESCRIPTION: Total length 245–300 mm; tail 0 mm (absent); hind foot 45–57 mm; ear 33–42 mm; forearm 175–190 mm; weight 630–950 g. A huge fruit bat, probably with the largest wingspan of any bat, though *Acerodon jubatus* is heavier. There are two color morphs. One has an orange-yellow mantle extending from about between the ears to the upper shoulders, with dark russet or golden-brown fur on the chest. The other morph has black or dark brown fur on the entire head, mantle, and chest. In both, the hair of the mantle is longer than that on the posterior two-thirds of the body. The wings are dark brown and evenly colored (without dark or pale blotches). *Acerodon jubatus* are similar in size (forearm 180–195 mm) but have paler wing membranes, with still paler blotches, and a golden cap that begins between the eyes and ends in a narrow V at the nape of the neck or upper shoulders. *Pteropus hypomela-*

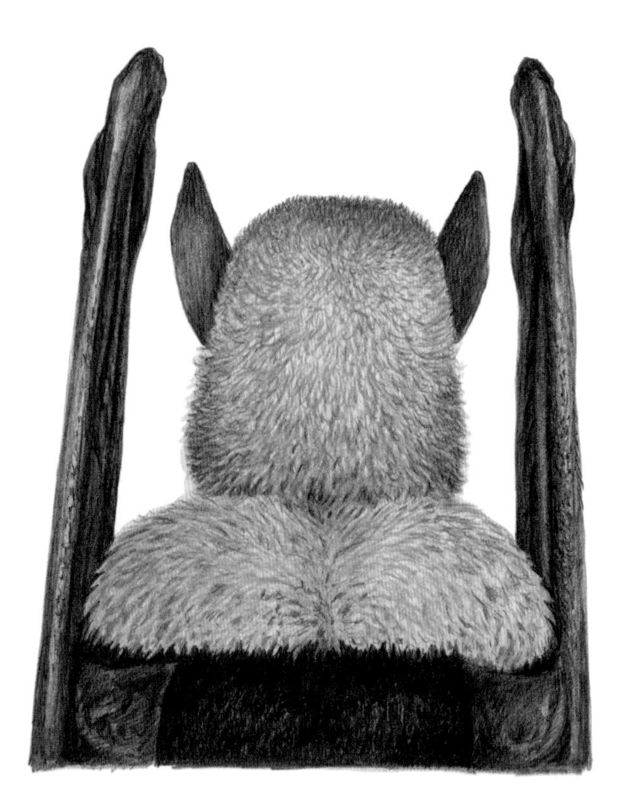

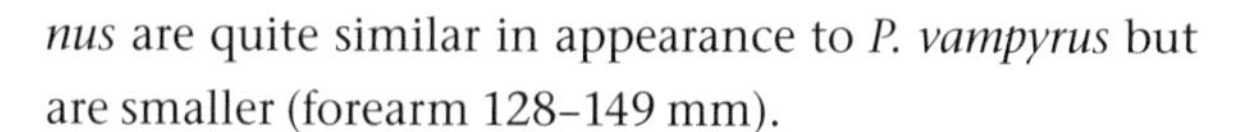

nus are quite similar in appearance to *P. vampyrus* but are smaller (forearm 128–149 mm).

DISTRIBUTION: Indochina to the Lesser Sunda Islands; throughout the Philippines, except in the Batanes/Babuyan Faunal Region. They once occurred throughout Luzon but are now reduced to limited areas.

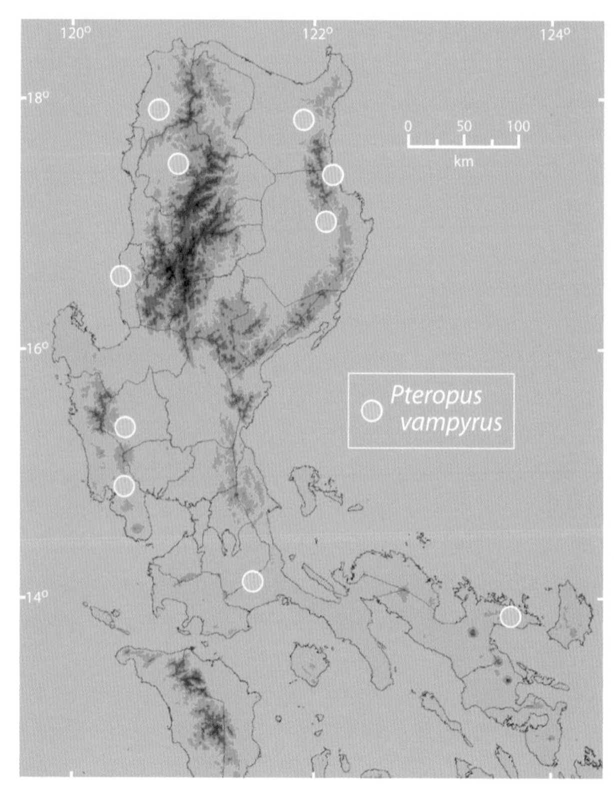

EVOLUTION AND ECOLOGY: The genus *Pteropus* is most closely related to *Acerodon*; *P. vampyrus* is most closely related to *P. giganteus*, which is widespread in South Asia. In the Philippines, *P. vampyrus* are widespread and locally common in primary and secondary lowland forest and nearby agricultural areas, up to at least 1250 m elevation, but they prefer undisturbed and riparian forest. They are recorded as flying 27 km or more per night to feed. They roost in the tops of large trees that they often strip of vegetation, with single colonies including from a few dozen up to ca. 100,000 individuals, often together with *A. jubatus* and/or *P. hypomelanus*. Along the foothills of the Sierra Madre in northeastern Luzon, they often roost quite near the coast, in lowland forest and in areas of ultrabasic soil. (For comments on population sizes and declines, and on possible predators, see the *A. jubatus* species account.) They usually fly above the canopy, and thus are difficult to capture with nets. They eat a wide range of wild and cultivated fruits (especially *Ficus* spp.), flowers, and leaves. Females typically give birth to a single young in April and May. They produce loud calls when feeding together in a fruiting tree.

STATUS: Formerly occurred in many large colonies in the Philippines, including Luzon, but these are now greatly reduced in number and size, due to habitat destruction and overhunting, even within national parks. Currently, hunting is usually done with shotguns. *Pteropus vampyrus* and the other flying foxes are often sold in local markets or within a village.

REFERENCES: Esselstyn et al., 2008; Heideman and Heaney, 1989; Jones and Kunz, 2000b; Lawrence, 1939; Mickleburgh et al., 1992; Mildenstein et al., 2005; Mudar and Allen, 1986; Paguntalan and Jakosalem, 2008; Rabor, 1955, 1977, 1986; Rickart et al., 1993, 2013; Sanborn, 1952; Scheffers et al., 2012; Stier and Mildenstein, 2005; Utzurrum, 1992; van Weerd et al., 2003, 2009, pers. comm.

Rousettus amplexicaudatus
common rousette

(É. Geoffroy, 1810). Ann. Mus. Hist. Nat. Paris 15:96. Type locality: Timor Island, Indonesia.

DESCRIPTION: Total length 137–152 mm; tail 17–23 mm; hind foot 19–21 mm; ear 18–20 mm; forearm 83–91 mm; weight 84–104 g. A medium-sized fruit bat, with adult males substantially larger than females. The dorsal fur is dark brown or gray, rather short and sparse, and often with hairs having a gray tip, which gives them a silvery, frosted appearance. The snout is long and tapering; the ears are long and bluntly pointed. The wing membranes are dark brown, except for being nearly translucent over the white bones, which gives the appearance of white stripes. A claw is present on

the leading edge of the wing (on the second digit). *Eonycteris spelaea* are similar in appearance but lack a claw on the leading edge of the wing; are smaller (forearm 67–80 mm); and have paired kidney-shaped glands lateral to the anus. *Eonycteris robusta* are usually smaller (forearm 67–82 mm) and lack a claw on the leading edge of the wing. *Ptenochirus jagori* are similar in size (forearm 72–88 mm) but have a short, robust muzzle; longer fur; and a shorter tail (4–17 mm).

DISTRIBUTION: Thailand to the Solomon Islands; occur throughout the Philippines, including all of Luzon.

EVOLUTION AND ECOLOGY: *Rousettus* is most closely related to several genera that occur in Africa and, more distantly, to *Eonycteris*. In the Philippines, *R. amplexicaudatus* show high levels of genetic variation within populations. They also exhibit high levels of gene flow overall, but gene flow is higher within populations from a single Pleistocene island group than between island groups, implying that dispersal is somewhat limited. They are abundant and widespread in agricultural

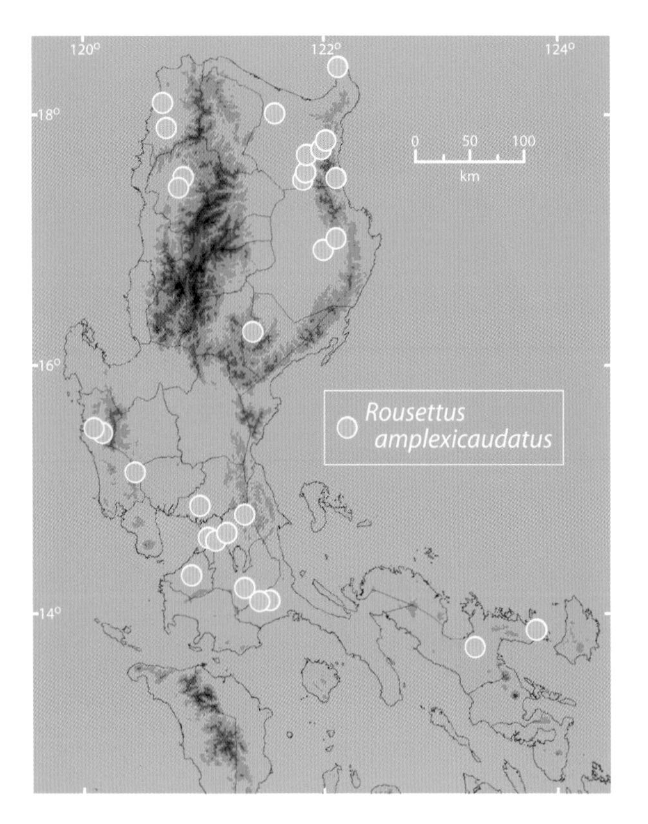

areas up to 700 m elevation, scarce in agricultural areas from 700 m to 1200 m elevation, and absent above 1200 m. They are rarely present in forest, but they often occur in well-vegetated urban areas. All of their known roosting sites are in caves, frequently in association with *E. spelaea*, and they sometimes roost in caves on small islands several kilometers off the coast of a large island. They live in colonies of 100,000 or more per cave when they are not impacted by hunting, but only in colonies in the hundreds or thousands when heavily hunted. In 2007, a colony in a cave on Samal Island (near Davao City) was estimated at 1.8 million individuals. Adult males are less numerous than adult females, and there may be a harem type of mating system. Females probably produce a single young twice each year; gestation is about 150 days and lactation 2.5–3 months, with most females in a given population giving birth at about the same time. They may fly up to 25 km per night to feed. They typically eat overripe fruits, causing less damage in orchards than is often believed. They are docile when handled, rarely biting or producing calls. They have a karyotype with $2n = 36$, FN = 68.

STATUS: Widespread in Indo-Australia; locally abundant in the Philippines, but subjected to intense hunting at some cave roosts, causing massive reductions in numbers.

REFERENCES: Almeida et al., 2011; Balete et al., 2011; Carpenter et al., 2014; Catibog-Sinha, 1987; Duya et al., 2007; Heaney and Roberts, 2009; Heaney et al., 1989, 1991, 1999, 2005, 2006b; Heideman and Heaney, 1989; Heideman and Utzurrum, 2003; Ingle, 1993; Lepiten, 1995; Mould, 2012; Ong et al., 1999; Rickart et al., 1989a; 1993; Sedlock et al., 2008, 2011, 2014b; Utzurrum, 1992.

SECTION 3

Family Emballonuridae: Sheath-Tailed Bats

The 13 genera of this family occur throughout the tropics and much of the subtropics globally; 3 of these genera (*Emballonura*, *Saccolaimus*, and *Taphozous*) are present on Luzon, each with a single species that is widespread in Southeast Asia. As their name implies, all of the species have a tail that has a sheath of skin covering its posterior portion (half or less), with the base of the tail enclosed within the tail membrane and the sheathed portion lying on top of the membrane (chapter 11, "Introduction to Bats"). They are moderately small bats, ranging from about 5 to 35 g, with fairly large eyes and narrow wings. Their roosting posture is distinctive, with the head held out from the side of a building or cave's wall, giving them an alert appearance.

Like other bats that emit their echolocation calls through their mouth (rather than their nostrils), all of these bats have a simple nose, unadorned by complex

Taphozous melanopogon.

folds of skin. All three species eat insects (especially moths and beetles), and two of them (*Saccolaimus saccolaimus* and *Taphozous melanopogon*) fly fast and high above the ground and any forest canopy. They tend to roost in small groups in caves, large old buildings, and hollow trees; they all typically roost in dimly lit but not entirely dark places, so they are often visible in weak daylight. Members of this family are known to be highly social, and some forage as a group in male-defended territories.

No genetic studies have been conducted on any species of sheath-tailed bat in the Philippines, and their ecology is generally poorly known.

REFERENCES: Bradbury and Vehrencamp, 1977; Ruedi et al., 2012; Simmons, 2005.

Emballonura alecto
Philippine sheath-tailed bat

(Eydoux and Gervais, 1836). Mag. Zool., Paris, 6:7.
Type locality: Manila, Luzon.

DESCRIPTION: Total length 57–68 mm; tail 9–14 mm; hind foot 6–9 mm; ear 13–15 mm; forearm 44–49 mm; weight 5–6.5 g (pregnant females up to 8 g). A small, dark brown bat, with the basal half of the tail enclosed within the tail membrane but the posterior half of the tail perforating the upper surface and lying on top of the tail membrane. The nasal area is simple; the tragus is rather short and club-like; and the eyes are slightly larger than in other small brown insectivorous bats. Some vespertilionids are similar in size and in the color of their fur, but no other small brown bat on Luzon has a tail that perforates the tail membrane.

DISTRIBUTION: Widespread on Borneo, the Philippines, and Sulawesi and associated smaller islands; probably occurs throughout the Philippines except the Batanes/Babuyan Faunal Region, but may be absent from northern Luzon, where no records are currently available.

EVOLUTION AND ECOLOGY: The genus *Emballonura* includes eight species that occur from Burma [= Myanmar] to New Guinea and Samoa; *E. alecto* is most closely related to *E. monticola*, which is found from Burma to Sumatra and Borneo. Philippine sheath-tailed bats have been recorded only in lowland areas of up to 450 m elevation on Luzon, and rarely up to 1000 m elevation elsewhere, in mature and secondary forest and (less commonly) in agricultural areas with scattered remnant forest. Most records are from individuals captured in

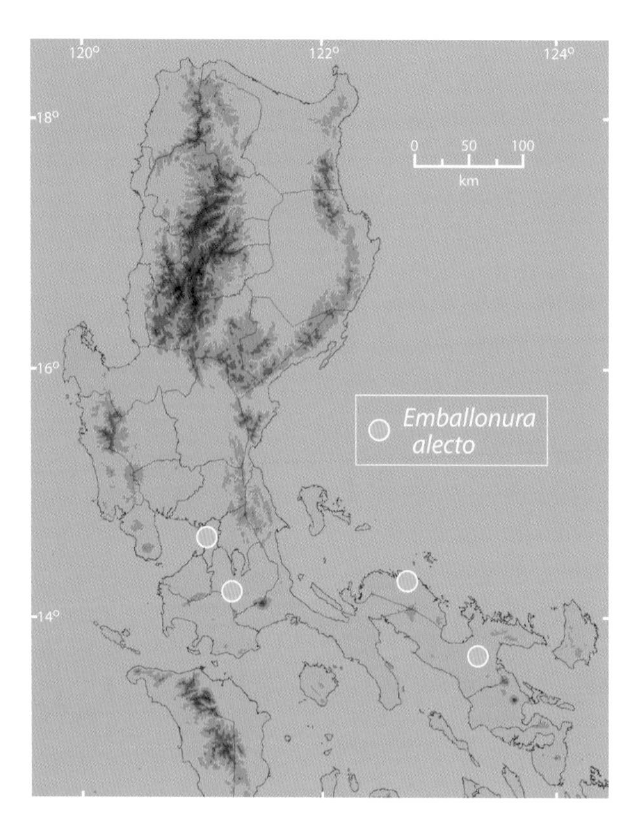

or near caves, under large boulders, in rock crevices, or in man-made tunnels, usually in dimly lit areas rather than in total darkness. Group size is usually small, often 10–25. Females give birth to a single young. The echolocation call is a series of simple, narrow-bandwidth notes, with most energy at ca. 47 kHz.

STATUS: Moderately common in lowland areas with caves, in or near mature or secondary forest and/or mixed agroforestry. They are negatively impacted by roost disturbances caused by the extraction of cave resources, such as guano, edible birds' nests, and the larger bats used by humans as food.

REFERENCES: Heaney et al., 1991, 1999, 2006a; Ingle, 1992; Mould, 2012; Rabor, 1986; Rickart et al., 1993; Ruedi et al., 2012; Sedlock, 2001; Sedlock et al., 2008, 2011, 2014b; Taylor, 1934.

Saccolaimus saccolaimus
pouched bat

(Temminck, 1838). Tijdschr. Nat. Gesch. Physiol., 5:14. Type locality: Java, Indonesia.

DESCRIPTION: Total length 108–117 mm; tail 20–28 mm; hind foot 15–19 mm; ear 16–20 mm; forearm 70–73 mm; weight 28–36 g. A beautiful bat, with a simple nasal region; large eyes; a low, blunt tragus; and a tail that is encased for its basal half within the tail membrane, but pierces the upper surface to lie on top of the membrane for the posterior half. The dorsal fur is black or very dark brown, with scattered small white spots; the ventral fur may be solid black, dark brown, or gray. There is a small pouch on the interior side of the wrist, similar to that of *Taphozous melanopogon* but smaller. There is also a glandular pouch under the chin that is most apparent in adult males; the legs and feet are virtually hairless. *Taphozous melanopogon* are similar in overall appearance but are smaller (forearm 61–65 mm); are medium brown to pale cream in color; and lack a pouch under the chin.

DISTRIBUTION: Very widespread, from India to Timor, New Guinea, and northern Australia; scattered records from the southern and central Philippines. There is only a single confirmed record from Luzon—a specimen captured, photographed, and released in the

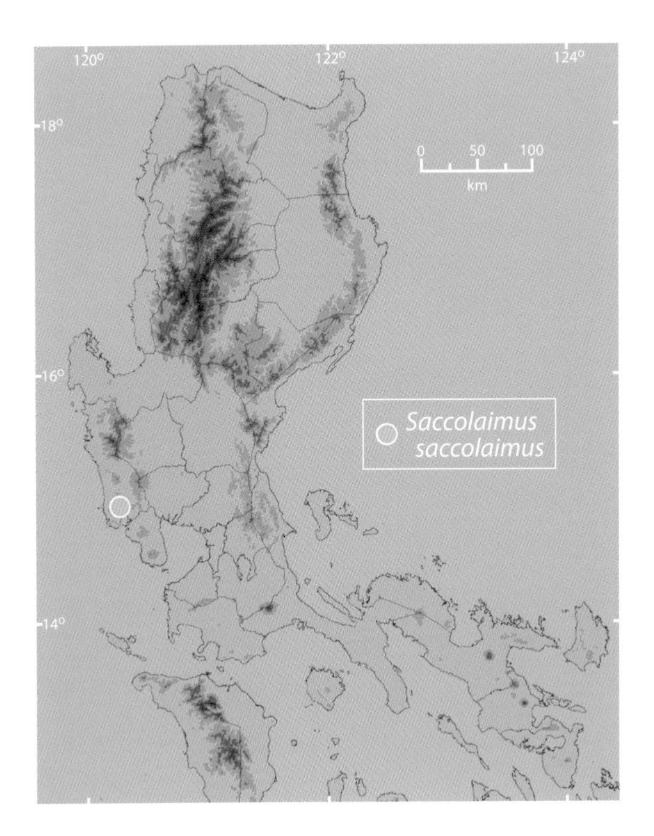

Subic Bay Special Economic Zone, Zambales Province, in December 2010—but there is also a specimen from Pandan, Catanduanes Island, just east of the Caramoan Peninsula.

EVOLUTION AND ECOLOGY: The genus *Saccolaimus* is most closely related to *Taphozous*. Four species are currently recognized: two in Australia and New Guinea, one in tropical Africa, and *S. saccolaimus*. Although *S. saccolaimus* is widespread, this species is poorly known in the Philippines, probably at least in part because they fly high above the ground when foraging and thus are difficult to capture. In the Philippines, they have been recorded from sea level to 800 m elevation. Nearly all of the records come from agricultural and urban residential areas, where they may be common. In Malaysia, colonies ranging from a few dozen to several hundred roost in the eaves of houses, hollow trees, and crevices between boulders. Females give birth to a single young.

STATUS: Poorly known in the Philippines, but may be common.

REFERENCES: Corbet and Hill, 1992; Francis, 2008; Heaney et al., 1991; Medway, 1969.

Taphozous melanopogon black-bearded tomb bat

Temminck, 1841. Monogr. Mamm., 2:287. Type locality: Bantam, West Java, [Indonesia].

DESCRIPTION: Total length 103–111 mm; tail 20–23 mm; hind foot 13–15 mm; ear 20–23 mm; forearm 61–65 mm; weight 23–28 g. An attractive bat, with medium to pale brown fur; large eyes; a simple nasal region; a low, blunt tragus; and a tail that is encased within the tail membrane for the basal half but pierces the membrane and lies on top of the membrane for the posterior half. Adult males are paler than females and have a patch of dark brown hair (a "beard") under the chin; young animals are gray. This species has a distinctive small flap of skin on the interior side of the wrist (at the base of the wing's finger bones). *Saccolaimus saccolaimus* are generally similar, with a similar but smaller wrist pouch, but they are larger (forearm 70–73 mm)

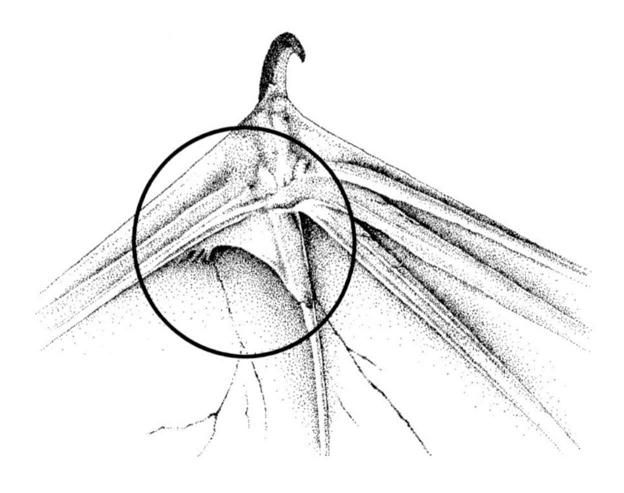

and are very dark brown or black, with scattered small white spots dorsally.

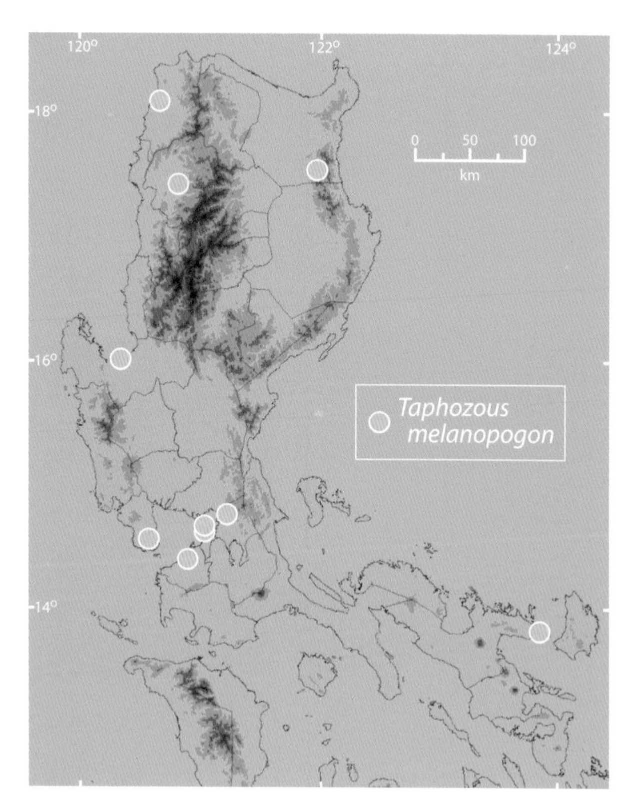

DISTRIBUTION: Widespread from Sri Lanka to Southeast Asia and the Lesser Sunda Islands; occur throughout the Philippines, including Luzon, in appropriate habitat.

EVOLUTION AND ECOLOGY: The genus *Taphozous* is most closely related to *Saccolaimus*. Fourteen species of *Taphozous* are currently recognized, which extend from over much of Africa to Southeast Asia and Australia. *Taphozous melanopogon* are common in urban areas, in areas with limestone caves, and in sea caves, from sea level to about 150 m elevation. There are no records of these bats occurring in mature forest in the Philippines; rather, they appear to prefer agricultural and residential areas. They often roost in shallow portions of caves and in large attics in churches, schools, libraries, and other buildings, where they are easily visible in dim daylight. They roost in colonies of up to ca. 400 individuals, but they cluster in small groups composed of a single adult male with several females and their young; young adult males often form small bachelor clusters. These clusters usually include at least 15 cm between individuals. Females give birth to a single young, which they carry with them for about the first four weeks. They have a karyotype with $2n = 42$, FN = 64.

STATUS: Widespread in Asia; abundant in the Philippines, where they often lives in buildings. They are sometimes hunted in caves.

REFERENCES: Corbet and Hill, 1992; Lawrence, 1939; Medway, 1969; Rickart et al., 1993, 1999; Sanborn, 1952; Taylor, 1934; Wei et al., 2008; Zubaid 1990.

SECTION 4

Family Megadermatidae: Ghost Bats and False Vampire Bats

Only a single species of this family is present within the Philippines. Ghost bats are unmistakable, with their distinctive, enormous ears; prominent, sturdy noseleafs; and large, sharp canines. Unlike most other Philippine insectivorous bats, they are "sit-and-wait" predators: they fly to perches where they hang, wait for their prey to make noise (either by calling or by rustling leaves), and then pounce on them. Like many bats, the females carry their young for several weeks after giving birth, even when they are foraging—but with a twist. Females in this family have both true nipples (in the axillary region) that produce milk and false nipples above the anus; their young spend much of their time holding firmly to the latter, so they are in a head-down position, especially when the mother is flying.

Megaderma spasma.

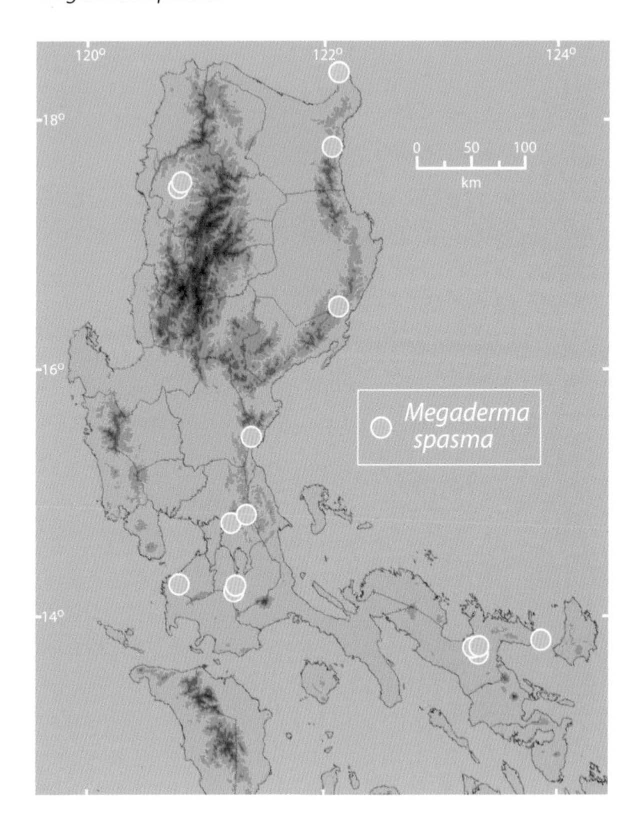

Megaderma spasma common Asian ghost bat

(Linnaeus, 1758). Syst. Nat., 10th ed., 1:32. Type locality: Ternate, Molucca Islands, [Indonesia].

DESCRIPTION: Total length 70–80 mm; tail 0 mm (absent); hind foot 18–22 mm; ear 36–43 mm; forearm 57–63 mm; weight 21–27 g. Females tend to be slightly larger than males. A spectacular bat, with enormous ears that are joined over the forehead; a long, slender tragus; soft gray or gray-brown fur; and no tail. The eyes are moderately large (the largest of any Philippine insectivorous bat); a prominent noseleaf is present. Other large-eared Philippine bats do not have ears that are joined over the forehead and all lack a long, slender tragus.

DISTRIBUTION: From India to the Molucca Islands; occur throughout the Philippines, including all of Luzon, except the Batanes/Babuyan Faunal Region.

EVOLUTION AND ECOLOGY: Only two species of *Megaderma* are known; the other species, *M. lyra*, occurs widely in continental South Asia and Southeast Asia. No genetic studies have been conducted on Philippine populations. *Megaderma spasma* live in lowland primary and disturbed forest, from sea level to 750 m elevation (rarely to 1100 m). They are common in bamboo thickets, mixed agroforest, secondary forest, and primary forest. They usually roost in small groups in caves, tree hollows, and hollow logs, and infrequently in vacant houses. They rarely forage more than 3 m above the ground. A study on Mt. Makiling documented 10 orders of insects eaten by them as prey, with Coleoptera (beetles), Hemiptera (cicadas), and Orthoptera (katydids and crickets) representing over 90% of their prey. All of these are large insects (9–78 mm in length) that are generally cryptic and well hidden but emit loud sounds, suggesting that acoustic cues produced by the prey are important to *M. spasma* in hunting, rather than echolocation calls emanating from the bats. Moths, roaches, and mantises (all of which are nearly silent) are also eaten, but less commonly. Small frogs (*Platymantis*), lizards (flying lizards, gekkos, and skinks), and birds (*Zosterops*) made up no more than 1% of their food items. These bats sometimes take their prey back to their roost to consume them, dropping inedible parts (wings, shells, legs, etc.) below them. Their abundance is usually low, perhaps because of their specialized feeding on large prey. This species has a karyotype with $2n = 46$, FN = 64.

STATUS: Widespread; locally common to uncommon in primary forest and secondary forest.

REFERENCES: Balete, 2010; Balete et al., 2011, 2013a; Esselstyn et al., 2004; Heaney et al., 1991, 1999; Ingle, 1992, 1993; Lawrence, 1939; Lepiten, 1995; Mould, 2012; Rabor, 1986; Rickart et al., 1993, 1999; Sanborn, 1952; Sedlock, 2001; Sedlock et al., 2008; Taylor, 1934.

SECTION 5

Family Hipposideridae: Roundleaf Bats

With 9 genera and over 80 species, the Hipposideridae is one of the largest and most diverse families of bats. This diversity is especially remarkable, given that they are absent from the Americas—they live only in Africa, Eurasia, Indo-Australia, and associated islands. With 2 genera and at least 10 species, the Philippines ranks high as a center of diversity for the family. At least 7 species live on Luzon, including 4 Philippine endemic species, and up to 5 or 6 may occur at a single forested locality, especially when caves are present. Hipposiderids are associated principally with lowland forest, generally up to about 800 m elevation, though some live higher; many of them often or usually roost in caves. This association with caves has made them vulnerable to population declines, because of rampant disturbances of caves for resource extraction (e.g., guano, birds' nests, cave formations, and limestone), tourism, and direct hunting.

Hipposideros obscurus.

Like other insectivorous bats that emit their sonar pulses through their nostrils, these bats have complex noseleafs that allow them to direct the sonar pulses. Careful inspection shows that the noseleaf for each species is unlike the others, which allows these bats to use different resources and coincidentally permits for identification of the various species. Their wings tend to be relatively short and broad, associated with their slow, agile flight.

Because roundleaf bats are typically difficult to capture with mist nets, their taxonomy, ecology, distribution, and conservation status have been poorly known, and it is likely that some of the "species" treated here actually represent species-groups. Increasing use of harp traps has increased the rate at which we learn about them, but much remains unknown about this remarkable group of bats.

REFERENCES: Esselstyn et al., 2012b; Ingle and Heaney, 1992; Sedlock, 2001; Sedlock et al., 2008, 2011, 2014a, 2014b; Simmons, 2005.

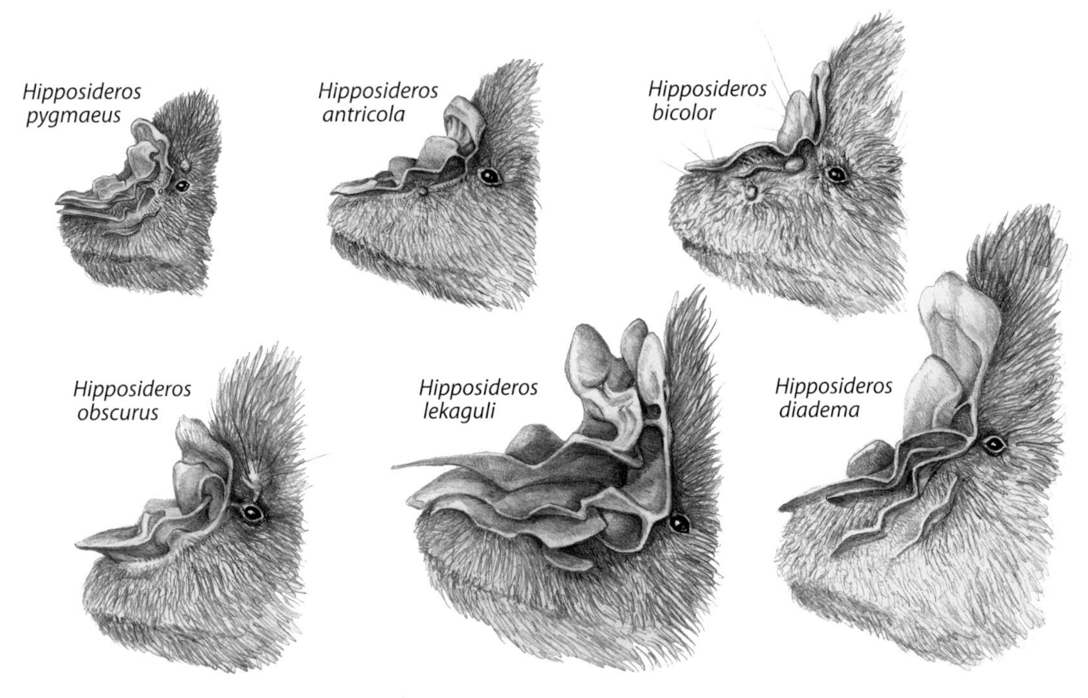

Luzon *Hipposideros*, approximately to the same scale.

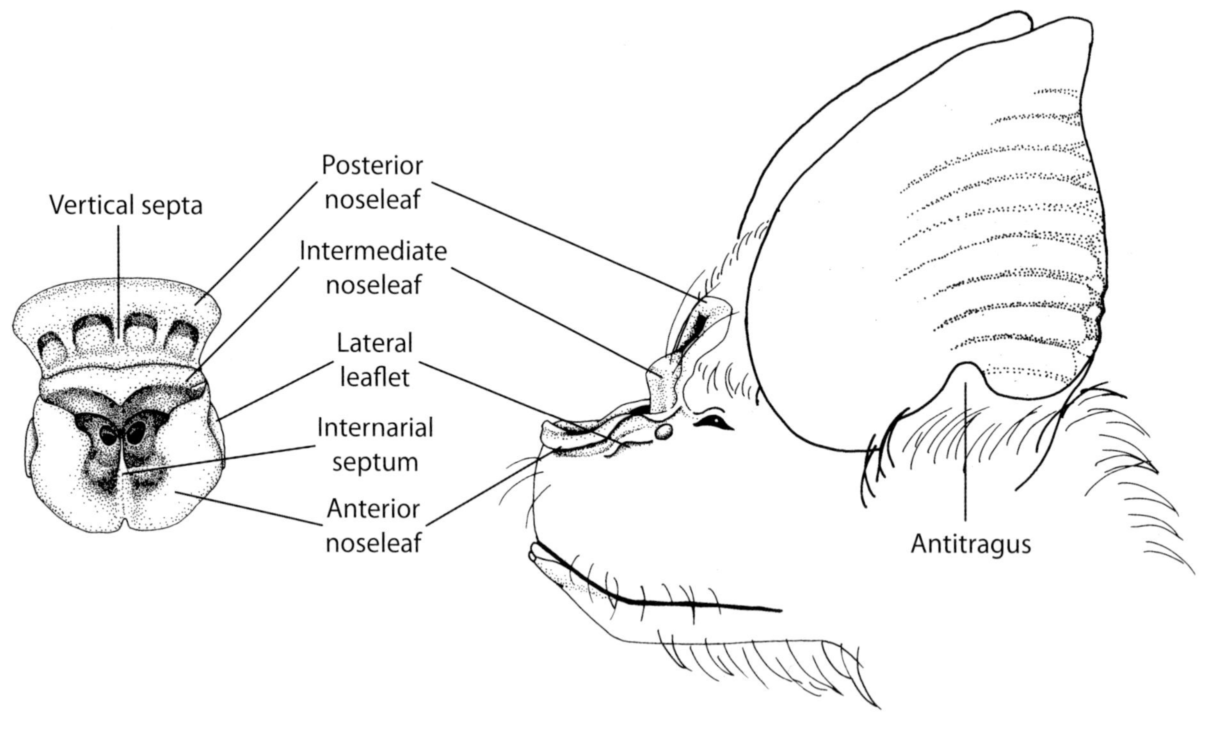

Parts of a *Hipposideros* noseleaf. Redrawn from Ingle and Heaney (1992).

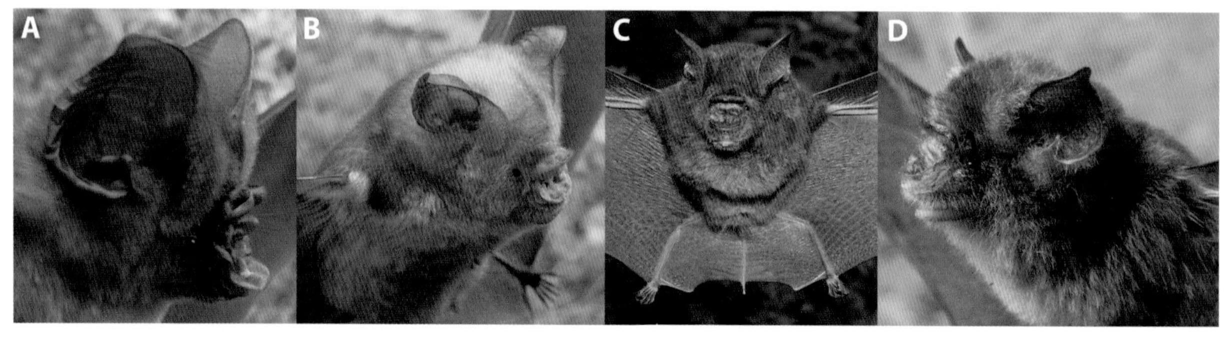

Four *Hipposideros*: A: *H. lekaguli*. B: *H. diadema*. C: *H. obscurus*. D: *H. pygmaeus*.

Coelops hirsutus
Philippine tailless roundleaf bat

(Miller, 1911). Proc. U. S. Natl. Mus., 38:395. Type locality: "Alag River, opposite mouth of Egbert River, Mindoro."

DESCRIPTION: Total length 35–42 mm; tail 0 mm (absent); hind foot 6–8 mm; ear 13–16.5 mm; forearm 33–36.5 mm; weight 2.4–3.6 g. A very small bat, with woolly gray fur; large rounded ears that lack a tragus; very small eyes; and no tail. The noseleaf is complex, with several tiny lappets. Other equally small bats on Luzon either lack a noseleaf (Emballonuridae and Vespertilionidae) or have a tall spear above the nostrils (*Rhinolophus*; chapter 11, "Introduction to Bats").

DISTRIBUTION: Specimens have been recorded from Cebu, Mindanao, Mindoro, and throughout much of Luzon.

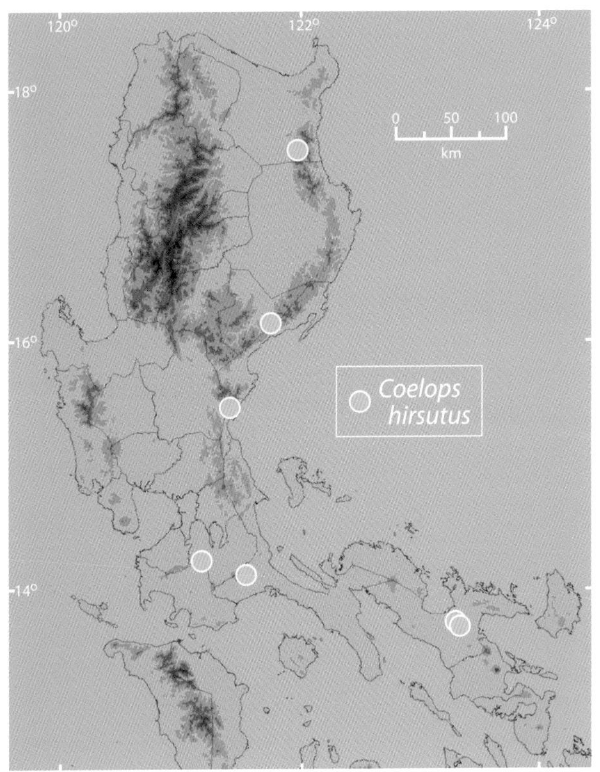

EVOLUTION AND ECOLOGY: Little is known about the evolution of *Coelops*, but it is likely that *C. hirsutus* are most closely related to *C. robinsoni*, from the Malay Peninsula and Borneo. The ecology of *C. hirsutus* is also poorly known, in part because they are rarely captured except with harp traps. They forage beneath the forest canopy, feeding on small invertebrates. They have been recorded in secondary and primary lowland and montane forest, from 100 m to 1500 m elevation; the limited data from studies using harp traps suggest that they are more common in lowland forest (below 800 m) than at higher elevations. Roosting sites include shallow caves along rivers and hollow trees. Females give birth to a single young.

STATUS: Poorly known; apparently widespread in lowland and montane forest.

REFERENCES: Balete et al., 2011; Corbet and Hill, 1992; Duya et al., 2007; Heaney et al., 2006b; Sedlock et al., 2008, 2011; Simmons, 2005.

Hipposideros antricola
Philippine dusky roundleaf bat

Peters, 1861. Monatsb. K. Preuss. Akad. Wiss. Berlin, p. 709. Type locality: Paracali, Luzon Island.

DESCRIPTION: Total length 71–81 mm; tail 27–36 mm (usually 28–32 mm); hind foot 7–9 mm; ear 17–20 mm; forearm 38–43 mm; weight 5–7 g. A small bat, with large ears and a prominent noseleaf. The noseleaf is small; not as wide as the muzzle; and without lateral leaflets. The internarial septum is swollen at the base. The pelage color varies greatly within a population, from dark brown to orange. The dorsal pelage is distinctly bicolored, dark brown (or orange) at the tips but pale tan or nearly white on the basal two-thirds. The ventral pelage is similar, but the tips are not as dark and the base is darker tan. *Hipposideros pygmaeus* are smaller (total length 61–65 mm; ear 11–13 mm) and have two pairs of leaflets lateral to the anterior noseleaf. *Hipposideros bicolor* are slightly larger (forearm 42–46 mm), with a single pair of lateral leaflets. *Hipposideros coronatus* (which occur only in the Mindanao Faunal Region) are larger (forearm 47–51 mm); have an elongated rostrum; and lack pigment on the lower leg and sole of the foot. *Hipposideros obscurus* are larger (forearm 42–47 mm); the tail is shorter (19–24 mm); and they have

two pairs of lateral leaflets. All species of *Rhinolophus* have a noseleaf with a prominent dorsal spear.

DISTRIBUTION: Occur only in the Philippines; widespread (Greater Luzon, Greater Mindanao, Greater Negros-Panay, Mindoro, and Siquijor), with the exception of the Palawan Faunal Region.

EVOLUTION AND ECOLOGY: Recent molecular genetic studies, combined with morphology and echolocation calls, have shown that *Hipposideros ater*, as this taxon was defined until recently, is actually composed of several distinct species. As now defined, *H. antricola*, previously considered a subspecies of *H. ater*, lives only in the oceanic portions of the Philippines and is most closely related to a series of species from Indochina and the Sunda Shelf. *Hipposideros antricola* have been recorded from sea level to 1280 m elevation in lowland and montane forest, with most records below 500 m elevation. They are most readily captured in "tunnel traps" (made from mist nets) or harp traps; those captures suggest that these bats vary from uncommon to fairly abundant. They roost in caves in primary and secondary forest as well as in agricultural areas, and in man-made tunnels in secondary forest. A single young is born each year. The karyotype has $2n = 32$, FN = 60.

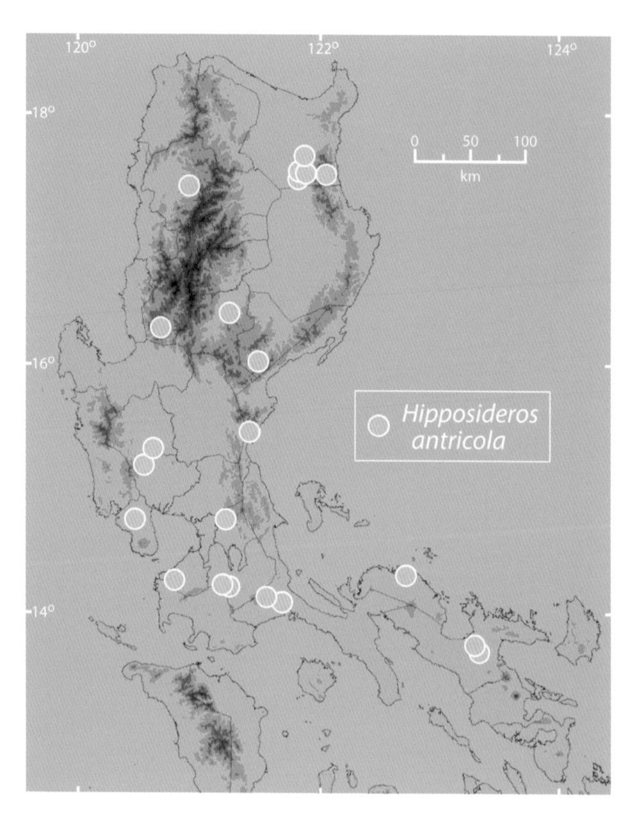

STATUS: Geographically widespread and fairly common in disturbed habitat, although they may have declined as a result of destruction of lowland forest and disturbance of caves.

REFERENCES: Alviola et al., 2011; Balete et al., 2011; Duya et al., 2007; Esselstyn et al., 2004, 2012b; Heaney et al., 1991; Lawrence, 1939; Mould, 2012; Rickart et al., 1993, 1999, 2013; Sanborn, 1952; Sedlock, 2001; Sedlock et al., 2008, 2011, 2014b; Taylor, 1934.

Hipposideros bicolor
bicolored roundleaf bat

(Temminck, 1834). Tijdschr. Nat. Gesch. Physiol., 1:19. Type locality: Anjer coast, Java, [Indonesia].

DESCRIPTION: Total length 78–86 mm; tail 31–33 mm; hind foot 7–9 mm; ear 18–20 mm; forearm 42–46 mm;

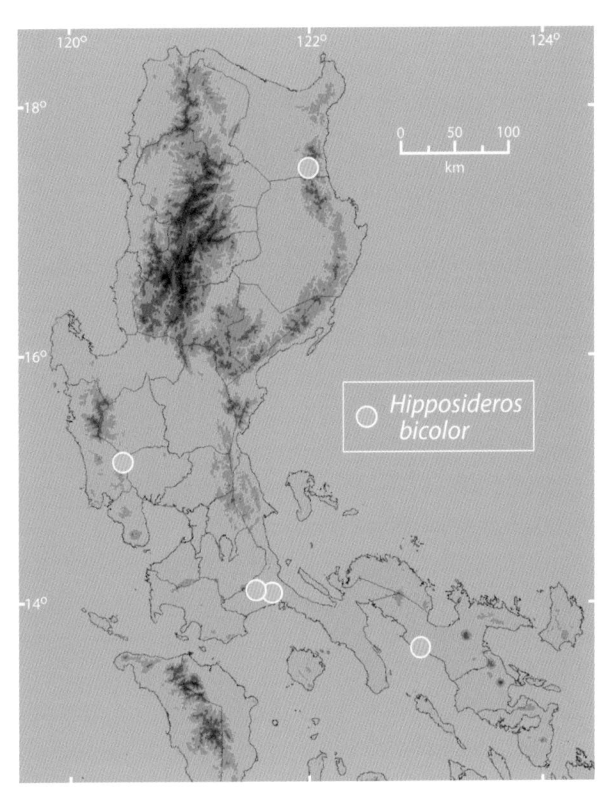

weight 6–10 g. The dorsal pelage is medium to dark brown at the tip, often with reddish tints; the hairs are pale at the base, nearly white to pale cream, giving these bats a bicolored appearance. The ventral pelage is pale off-white to pale cream at the base, with a slight darkening at the tips of the hairs. The noseleaf is small; is not as wide as the muzzle; and has a single lateral leaflet (sometimes small and inconspicuous) on each side. The internarial septum is not swollen at the base. *Hipposideros pygmaeus* are smaller (total length 61–65 mm; forearm 39–45 mm; ear 11–13 mm) and have two pairs of leaflets lateral to the anterior noseleaf. *Hipposideros antricola* are slightly smaller (forearm 38–43 mm) and lack lateral leaflets. *Hipposideros obscurus* are similar in the length of the forearm (42–47 mm), but the tail is shorter (19–24 mm) and they have two pairs of lateral leaflets. All species of *Rhinolophus* have a noseleaf with a prominent dorsal spear.

DISTRIBUTION: As currently defined, occur from India to Timor; appear to be widespread in the Philippines, with scattered records from all parts of Luzon.

EVOLUTION AND ECOLOGY: As currently defined, *Hipposideros bicolor* is clearly a species-group, but the limited studies that have been conducted do not allow precise definition of actual species. Philippine populations have sometimes been considered to represent an endemic subspecies, *H. bicolor erigens*; further study is needed. The ecology of this species is poorly known in the Philippines, in part because they are not readily captured in mist nets. They usually occur in well-developed secondary or primary forest and feed on insects beneath the forest canopy. They have been taken from near sea level to 750 m elevation, often (but not always) in the vicinity of caves or roosting within caves. Females give birth to a single young.

STATUS: Appear to be widespread in the Philippines, but usually uncommon; sometimes present in heavily disturbed forest. They may be impacted by disturbances to caves and by lowland habitat destruction.

REFERENCES: Esselstyn et al., 2012b; Lawrence, 1939; Sedlock et al., 2008, 2011, 2014b.

Hipposideros diadema
diadem roundleaf bat

(É. Geoffroy, 1813). Ann. Mus. Hist. Nat. Paris, 20:263. Type locality: Timor Island, [Indonesia].

DESCRIPTION: Total length 128–151 mm; tail 39–51 mm; hind foot 16–21 mm; ears 28–32 mm; forearm 77–88 mm; weight 33–54 g. Males are slightly larger than females. This is the largest insectivorous bat species in the Philippines. The pelage is long, soft, and dark brown dorsally, though paler at the base. There is a small patch of pale brown or cream fur anterior to the point where the leading edge of the wing attaches to the body, and a similarly colored strip along the side of the body from about the midpoint of the wing to the base of the hind legs. The noseleaf is large and prominent, with three or four pairs of lateral leaflets. *Hipposideros lekaguli* are smaller (forearm 64–72 mm) and have a posterior noseleaf that has three inflated pockets. All other species of *Hipposideros* are even smaller (forearm less than 52 mm) and have no more than two pairs of lateral leaflets. All *Rhinolophus* have a noseleaf with a prominent dorsal spear. *Megaderma spasma* also have a prominent dorsal spear and ears that connect over the forehead.

DISTRIBUTION: As currently defined, occur from Burma [= Myanmar] to the Solomon Islands; present throughout the Philippines, including Luzon, except in the Batanes/Babuyan Faunal Region.

EVOLUTION AND ECOLOGY: A member of a species-group that occurs from India and Sri Lanka to Australia and the Solomon Islands. These bats forage in primary and disturbed lowland forest and agricultural areas, from sea level to 1000 m elevation (rarely to 2200 m). They vary from uncommon to moderately common, with their greatest abundance tending to be at low elevations (up to ca. 500 m). They feed on insects, especially on fairly large beetles, moths, crickets, and katydids; in Australia they are reported to occasionally consume small birds. They roost in hollow trees, caves, and man-made tunnels; they are often present in caves that are frequently disturbed, indicating a high level of tolerance. In Malaysia, their home ranges are roughly 1 km^2. Females give birth to a single young. They have a karyotype with $2n = 32$, FN = 60.

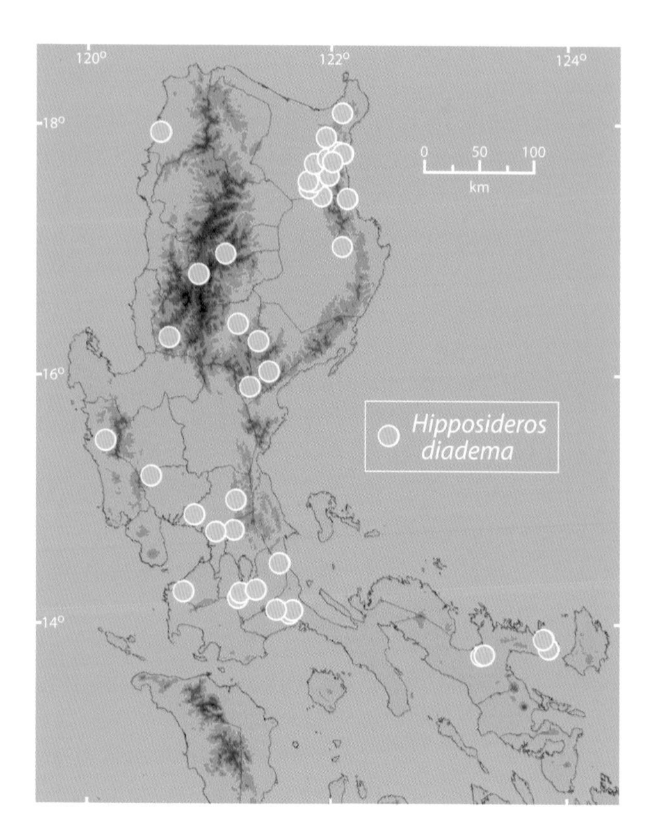

STATUS: Widespread and common.

REFERENCES: Alviola et al., 2011; Duya et al., 2007; Esselstyn et al., 2004; Heaney et al., 1999, 2006b; Ingle, 1992; Lepiten, 1995; Mould, 2012; Pavey and Burwell, 1977; Rickart et al., 1993, 1999; Sanborn, 1952; Sedlock, 2001; Sedlock and Weyandt, 2009; Sedlock et al., 2008, 2011, 2014b.

Hipposideros lekaguli
large Asian roundleaf bat

Thonglongya and Hill, 1974. Mammalia, 38:286. Type locality: Phu Nam Tok Tak Kwang, Kaeng Khoi, Saraburi, Thailand.

DESCRIPTION: Total length 115–124 mm; tail 34–47 mm; hind foot 11–13 mm; ear 26–32 mm; forearm 64–72 mm; weight 17–22 g. One of the most distinctive bats in the Philippines, with a large noseleaf, the posterior portion of which is distinctly trilobed and inflated into pockets. There are two large, complex, lateral leaflets. *Hipposideros diadema* are larger (forearm 77–88 mm) and have a simple posterior noseleaf; all other species of *Hipposideros* are substantially smaller. All species of *Rhinolophus* have a dorsal spear arising from the posterior noseleaf.

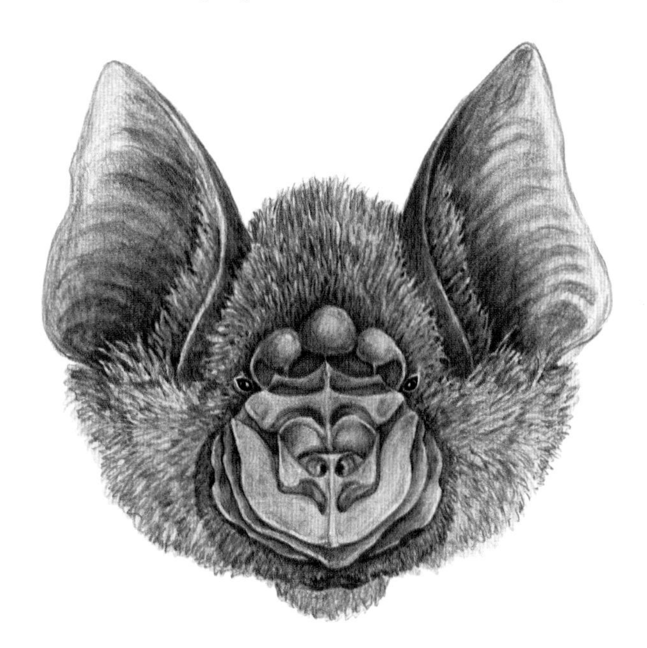

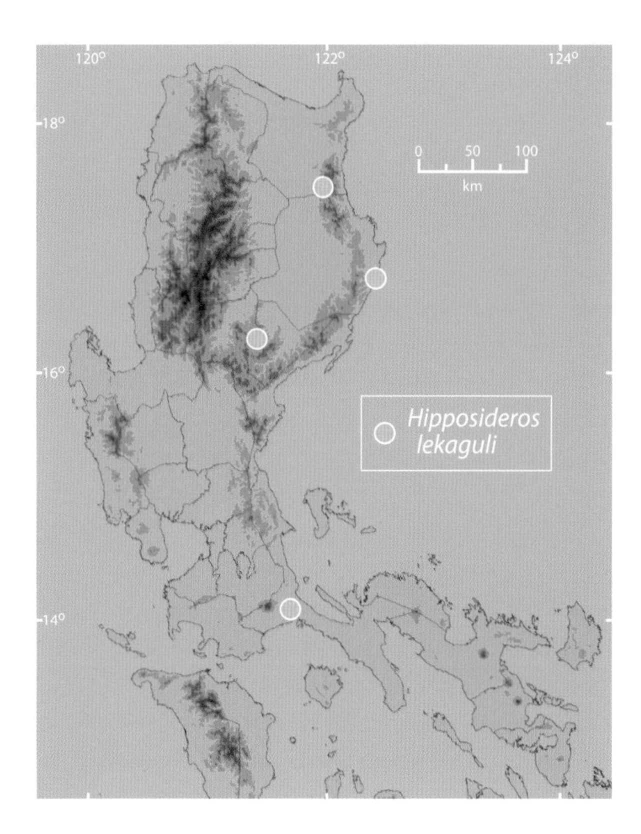

DISTRIBUTION: Currently known only from Thailand, peninsular Malaysia, Mindoro, and Luzon, where it is widespread.

EVOLUTION AND ECOLOGY: Most closely related to *Hipposideros diadema* and other species widespread in Indo-Australia. *Hipposideros lekaguli* are poorly known overall. Philippine records range from sea level to 325 m elevation. They occur in various habitats, including ultrabasic forest, agricultural areas near a river, and mixed secondary forest and coffee plantations over limestone near caves at low elevation. In two areas on Luzon with extensive limestone caves, they were fairly common. Lactating females have been taken in June.

STATUS: Uncertain; geographically widespread but seemingly dependent on lowland caves, which are often heavily disturbed. Additional field studies are needed.

REFERENCES: Balete et al., 1995; Lekagul and McNeely, 1977.

Hipposideros obscurus
Philippine forest roundleaf bat

(Peters, 1861). Monatsb. K. Preuss. Akad. Wiss. Berlin, p. 707. Type locality: Paracale, [Camarines Province], Luzon.

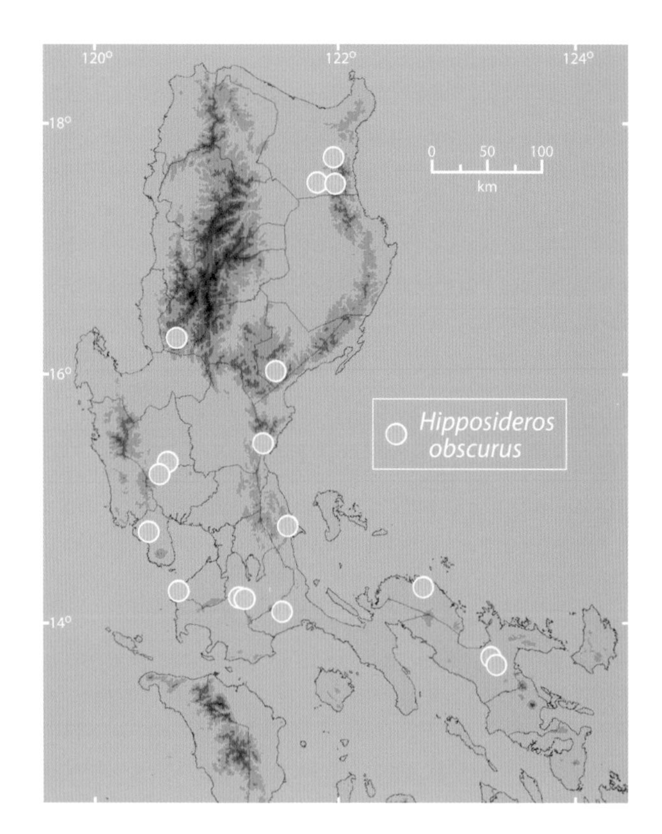

DESCRIPTION: Total length 72–82 mm; tail 19–24 mm; hind foot 10–12 mm; ear 18–20 mm; forearm 42–47 mm; weight 7–12 g. Adult males are slightly larger than adult females. The dorsal pelage is dense and soft, dark brown at the tips and base, with a paler band in the middle of each hair shaft. Some individuals have a patch of creamy-brown fur from the base of the ear to the shoulder and above the upper edge of the wing membrane, in a pattern reminiscent of that on *Hipposideros diadema*. The ventral pelage is darker at the base of the hairs than at the tips. The fur on the head often has a reddish hue. The muzzle is broad and sturdy. The noseleaf has one or two pairs of lateral leaflets. *Hipposideros antricola* are usually smaller (forearm 38–43 mm); and both *H. antricola* and *H. bicolor* have longer tails (27–36 mm and 31–33 mm, respectively). The former have no lateral leaflets, and the latter have a single pair. *Hipposideros diadema* and *H. lekaguli* are both much larger (forearm 77–88 mm and 64–72 mm, respectively).

DISTRIBUTION: Occur only in the Philippines, where they are widespread, except for the Palawan Faunal Region and Babuyan/Batanes islands. Widely distributed on Luzon.

EVOLUTION AND ECOLOGY: Recent DNA-based studies show this species to be distantly related to *Hipposideros diadema* and *H. lekaguli*, as well as *H. coronatus* (which occurs only in the Mindanao Faunal Region) and *H. larvatus* (which is widespread from India to China and Borneo). Populations of *H. obscurus* on Luzon are genetically similar to those from the Negros-Panay Faunal Region, but different from those on Mindoro and Greater Mindanao; further taxonomic studies are needed. On Luzon, they have been found from sea level to 900 m elevation, usually in lowland forest but sometimes in lower montane forest. They are locally common to uncommon in primary and disturbed forest, most often beneath forest canopy. They have been found roosting in small numbers in caves, culverts and mine shafts, dark cavities in a tree buttresses, and hollow trees and logs. Females give birth to a single young. The echolocation call has a frequency of ca. 116–120 kHz. The karyotype has $2n = 24$ (an unusually low diploid number), FN = 44.

STATUS: Widespread and generally common in disturbed and primary lowland forest.

REFERENCES: Balete et al., 2011; Duya et al., 2007; Esselstyn et al., 2012b; Heaney et al., 1991, 2006b; Ingle, 1992; Lepiten, 1995; Rickart et al., 1993, 1999, 2013; Sanborn, 1952; Sedlock 2001; Sedlock et al., 2008, 2011, 2014b; Taylor, 1934.

Hipposideros pygmaeus
Philippine pygmy roundleaf bat

(Waterhouse, 1843). Proc. Zool. Soc. Lond., p. 67. Type locality: "Philippines."

DESCRIPTION: Total length 61–65 mm; tail 20–25 mm; hind foot 7–8 mm; ear 11–13 mm; forearm 39–45 mm; weight 2.5–4 g. The smallest species of *Hipposideros* in the Philippines, and one of Luzon's smallest bats overall. The dorsal pelage is bicolored, whitish or cream at the base and dark brown at the tips. The ventral pelage is medium brown, darkest at the base. There are two sets of lateral leaflets; the inner pair meet above the mouth (at the bottom of the noseleaf) to form a continuous leaflet. *Coelops hirsutus* are even smaller (forearm 33–36.5 mm) and lack a tail. *Hipposideros antricola* have a longer tail (27 mm or more) and lack lateral leaflets. All other *Hipposideros* are larger. All *Rhinolophus* have a conspicuous spear rising dorsally to the posterior noseleaf.

DISTRIBUTION: Restricted to the Philippines; widespread, with the exception of the Palawan Faunal Region and Babuyan/Batanes islands. There are currently no records from northern Luzon.

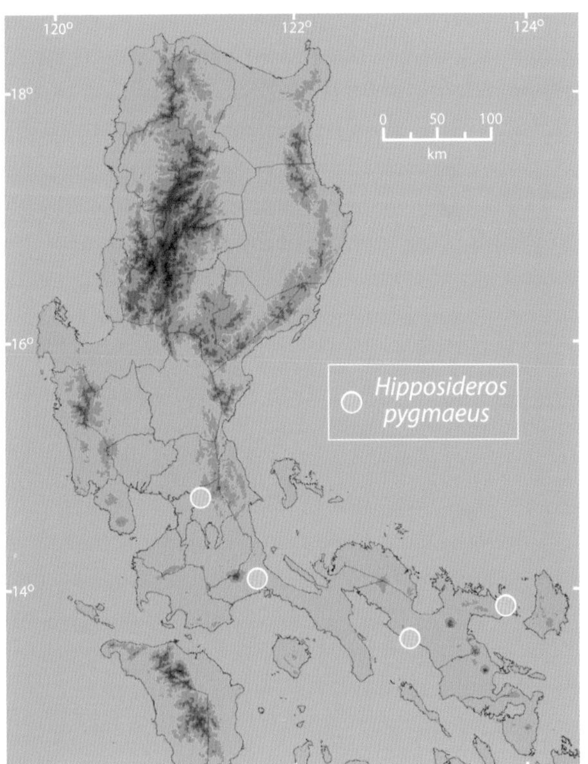

EVOLUTION AND ECOLOGY: This species has traditionally been placed in the *Hipposideros bicolor* species-group, but additional studies are needed. *Hipposideros pygmaeus* have been captured in caves in or near secondary lowland forest and in secondary lowland forest, from sea level to at least 300 m elevation, over karst and ultramafic soils. Where caves and forest (including degraded but regenerating forest) are present, these bats are sometimes common. In the central Philippines, they feed on a wide range of insects (with Lepidoptera and Diptera most strongly represented) and occasionally on spiders. A female pregnant with a single young was recorded in May, and a lactating female was captured in July. They are difficult to capture in mist nets, but harp traps are more successful. More field studies are needed.

STATUS: Widespread, but may be dependent on caves for breeding, and therefore negatively impacted by the degradation of caves and of forest over limestone habitats.

REFERENCES: Sanborn, 1952; Sedlock et al., 2014a, 2014b; Taylor, 1934; unpublished FMNH field notes.

SECTION 6

Family Rhinolophidae: Horseshoe Bats

This family includes only one genus, *Rhinolophus*, an indication that all of its members have a similar body plan and appearance. Over 80 species are currently recognized, however, and genetic and fossil evidence suggests that the family may have first diverged from their closest relatives—the Hipposideridae—more than 30 million years ago. These same genetic data indicate that most of the living species have diversified over the course of the past 10 million years, which is broadly the same period as the major uplift, coalescence, and volcanic growth of the Philippine archipelago. The rhinolophids occur from Western Europe to some of the small islands in the western Pacific Ocean, but they are entirely absent from the Western Hemisphere.

All *Rhinolophus* have elaborate folds of skin around the nostrils; these are used to focus the ultrasonic squeaks that the bats use for echolocation. They differ from the Hipposideridae in having a spear, called the connecting process, that projects high above the nostrils; hipposiderids have a simpler noseleaf that lacks this spear. *Megaderma* also have a large noseleaf, but they are much simpler overall. Adult female rhinolophids and hipposiderids have one pair of nipples on the chest, but they also have a pair of false nipples anterior to the anus; these false nipples are simply lappets of skin to which their babies cling when they are carried by the mother when she flies.

In the Philippines, 10 species of *Rhinolophus* are currently recognized, but many of these show evidence of comprising species-groups, and we suspect that a detailed study of their morphology, genetics, and ecology will show that at least twice that number are actually present. Seven of these species (or species-groups) are present on Luzon, and several of them (*R. arcuatus*, *R. inops*, *R. macrotis*, *R. philippinensis*, and *R. virgo*) appear to contain multiple species within Luzon. This is a poorly known group in need of much further study. The results should provide us with insights into not only the extent of mammalian diversity on Luzon, but also evidence of the processes that have produced that diversity in this group of volant but delicate animals. The presence of the seven species (or species groups) shows that colonization from outside the Philippines has resulted in most of the currently recognized diversity (since many rhinolophids have their closest relatives outside the Philippines), but there are hints of much internal speciation as well. What geographic circumstances have promoted speciation? How much ecological separation must exist to allow species to live together? How long do these processes take, and how

Rhinolophus philippinensis.

genetically different do the lineages become? The rhinolophids appear to be a prime group to investigate these and many other questions.

Based on current taxonomy and geographic and ecological data, we know that up to six species of *Rhinolophus* can occur together at some places on Luzon. Some species are common in forest and probably roost inside hollow trees, but diversity in any given forested area on Luzon tends to be limited to about four species. The greatest diversity and abundance are present in areas with extensive caves. Their dependence on caves (for roosting and raising young) make these bats vulnerable to the high levels of disturbance often present from mining for guano and limestone, from tourism, and from harvesting birds' nests for human consumption. Even though *Rhinolophus* are relatively small bats, they are also actively hunted within caves, with humans often using smoke from fires to asphyxiate all of the bats within a chamber. As a result, although none of these bats are thought to be in imminent danger of extinction, most have had their populations reduced and are in need of protection, especially at their roosting sites. All are in need of much additional study.

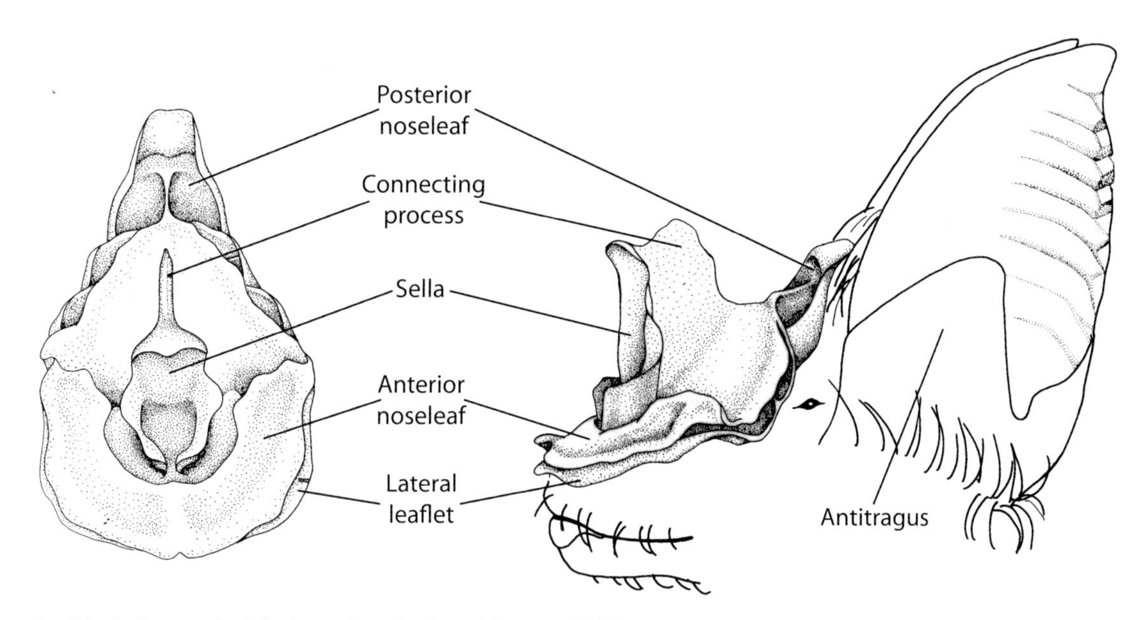

Parts of a *Rhinolophus* noseleaf. Redrawn from Ingle and Heaney (1992).

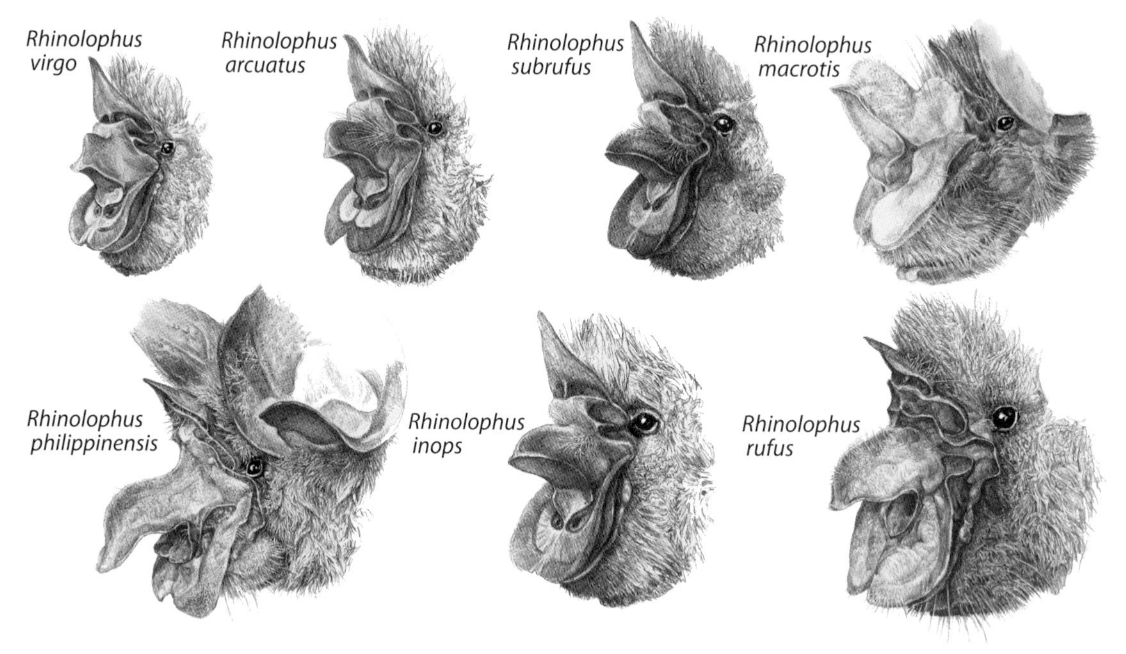

Luzon *Rhinolophus*, approximately to the same scale.

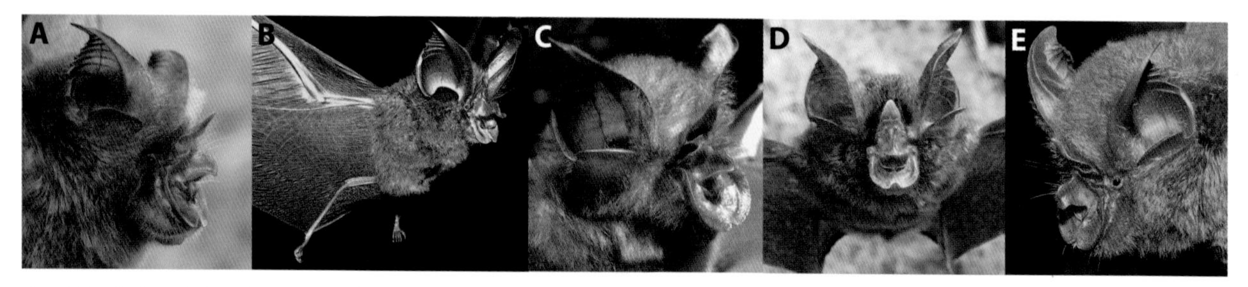

Five *Rhinolophus*. A: *R. rufus*. B: *R. philippinensis*. C: *R. inops*. D: *R. macrotis*. E: *R. virgo*.

REFERENCES: Guillén-Servent et al., 2003; Ingle, 1992; Ingle and Heaney, 1992; Sedlock, 2001; Sedlock et al., 2008, 2009, 2011, 2014a, 2014b; Simmons, 2005; Struebig et al., 2009.

Rhinolophus arcuatus arcuate horseshoe bat

Peters, 1871. Monatsb. K. Preuss. Akad. Wiss. Berlin, p. 305. Type locality: Luzon Island.

DESCRIPTION: Total length 65–83 mm; tail 16–21 mm; hind foot 10–12 mm; ear 18–21 mm; forearm 40–49 mm; weight 6–9 g. A medium-small *Rhinolophus*. The fur is usually medium brown dorsally and slightly paler ventrally. The noseleaf is about as broad as the muzzle, with a single lateral leaflet. *Rhinolophus macrotis* have a forearm of similar length (44–50 mm), but they have a longer tail (27–33 mm) and a much larger noseleaf, sella, and ears (27–29 mm). *Rhinolophus subrufus* are larger (forearm 53–57 mm; weight 12–19 g). *Rhinolophus virgo* are smaller (forearm 37–42 mm; weight 4–7 g) and usually have a yellow tint to the facial skin (rather than brown).

DISTRIBUTION: As currently defined, this species occurs from Sumatra to New Guinea; present throughout Luzon in suitable habitat.

EVOLUTION AND ECOLOGY: Current evidence indicates that this is a complex group of closely related species, rather than a single species. On Luzon, as on most Philippine islands, there appears to be a smaller morph (mean forearm length 45.3 mm) with a narrower noseleaf, and a larger morph (mean forearm length 46.6 mm) with a wider noseleaf; these are genetically distinct. There also appear to be subtle but consistent differences between populations in each Philippine faunal region. Further taxonomic and ecological studies are badly needed. Taken together, the various morphs on Luzon occur from sea level to over 2500 m elevation, which is the widest range for any member of the family, and are present in agricultural lands, second growth, and primary forest. They are often abundant in the lowlands and sometimes common in montane forest; they are usually uncommon or absent in mossy forest. Roosting sites are in caves, crevices, rocky overhangs, and hollow trees. They feed mainly on a diversity of moths but also consume beetles, flies, other small insects, and occasionally spiders. Females give birth to a single young. They are more easily captured in harp traps than in mist nets. Individuals of the small morph have an echolocation call frequency of 69.8 kHz, while the large morph have a frequency of 66.0 kHz, with females calling at a slightly higher frequency than males within a given morph. Specimens of the large morph from Camarines Sur Province have a karyotype of $2n = 58$, FN = 60.

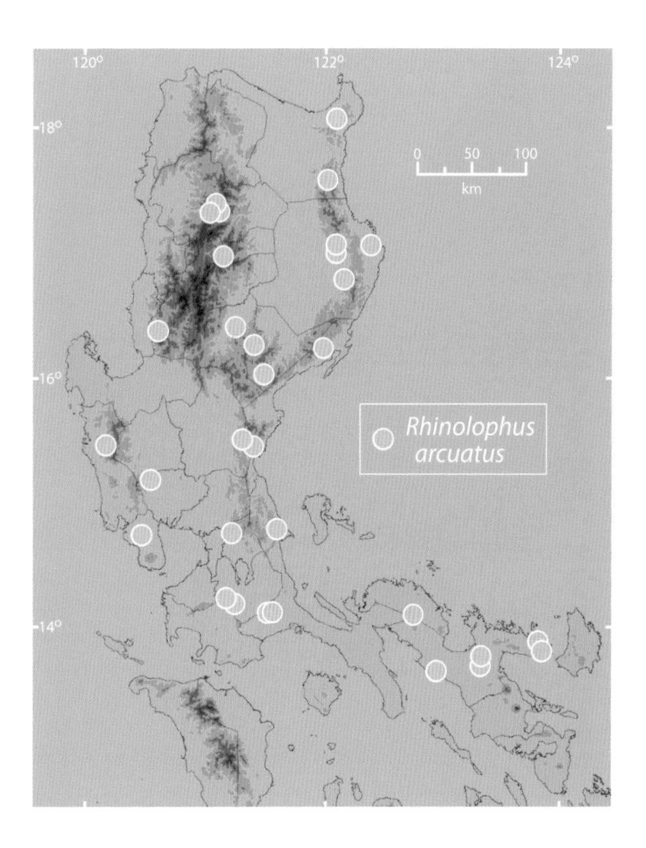

STATUS: Widespread and locally common as currently defined, but uncertain because of taxonomic uncertainty.

REFERENCES: Alviola et al., 2011; Balete et al., 2011, 2013a; Duya et al., 2007; Esselstyn et al., 2004; Heaney et al., 1991, 1998, 2004, 2006a, 2006b; Ingle and Heaney, 1992; Lepiten 1995; Mould, 2012; Patrick et al., 2013; Rickart et al., 1993, 1999, 2013; Sanborn, 1952; Sedlock 2001; Sedlock and Weyandt, 2009; Sedlock et al., 2008, 2011, 2014a, 2014b.

Rhinolophus inops
Philippine forest horseshoe bat

K. Andersen, 1905. Ann. Mag. Nat. Hist., ser. 7, 16:284, 651. Type locality: Todaya, 1325 m elevation, Mt. Apo, Davao Province, Mindanao.

DESCRIPTION: Total length 76–95 mm; tail 17–25 mm; hind foot 11–16 mm; ear 22–28 mm; forearm 50–57 mm; weight 10–18 g. A moderately large *Rhinolophus*. The dorsal fur is medium brown, with rusty-red tints, and paler at the base. The ventral fur is slightly paler. The ears are moderately large; the noseleaf is about the same width as the muzzle, and there are two lateral leaflets. *Rhinolophus arcuatus* are smaller (forearm 49 mm or less; weight 6–9 g) and have only one pair of lateral leaflets. *Rhinolophus macrotis* have a shorter forearm (44–50 mm) and a longer tail (27–33 mm), with proportionately larger ears (27–29 mm). *Rhinolophus subrufus* are usually slightly larger (forearm 53–57 mm; weight 12–19 g), with a longer tail (22–29 mm); the noseleaf has one pair of lateral leaflets and is proportionately wider.

DISTRIBUTION: Occur only in the oceanic Philippines; widespread on Luzon.

EVOLUTION AND ECOLOGY: The limited current evidence suggests that this species is closely related to *Rhinolophus arcuatus* and *R. subrufus*. There is substantial geographic variation in populations referred to this species, and *R. inops* may represent a species complex; further study is needed. They occur in primary and secondary lowland and montane forest, from sea level to 1950 m elevation on Luzon, usually decreasing in abundance with increasing elevation, although they typically are absent from agricultural areas and heavily disturbed vegetation. They feed on a range of insects, but most heavily on Lepidoptera and Diptera. They roost in forest areas (perhaps in hollow trees) and in caves, sometimes forming large colonies along with *R. arcuatus* and *R. virgo*. Females give birth to a single young. They are easily captured in both harp traps and mist nets, though more readily in the former. The

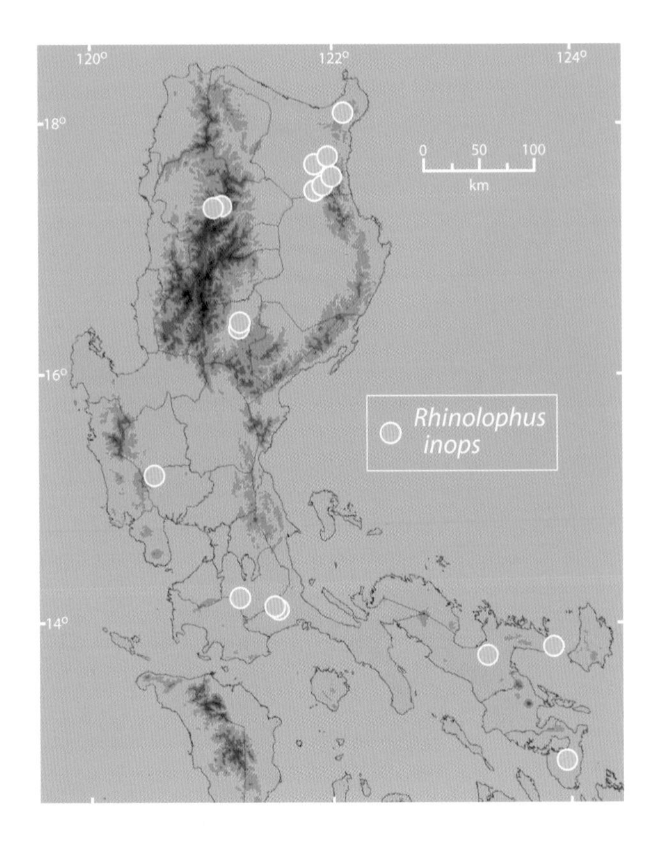

echolocation call of individuals from Laguna Province has a mean frequency of 55.1 kHz. A specimen from Leyte Island has a standard karyotype of $2n = 58$, FN = 60 that is identical to those seen for *R. arcuatus*.

STATUS: Locally common and widespread.

REFERENCES: Alviola et al., 2011; Duya et al., 2007; Heaney et al., 1991, 1999, 2004, 2006a, 2006b; Rickart et al., 1993, 1999; Sedlock, 2001; Sedlock et al., 2008, 2011, 2014a.

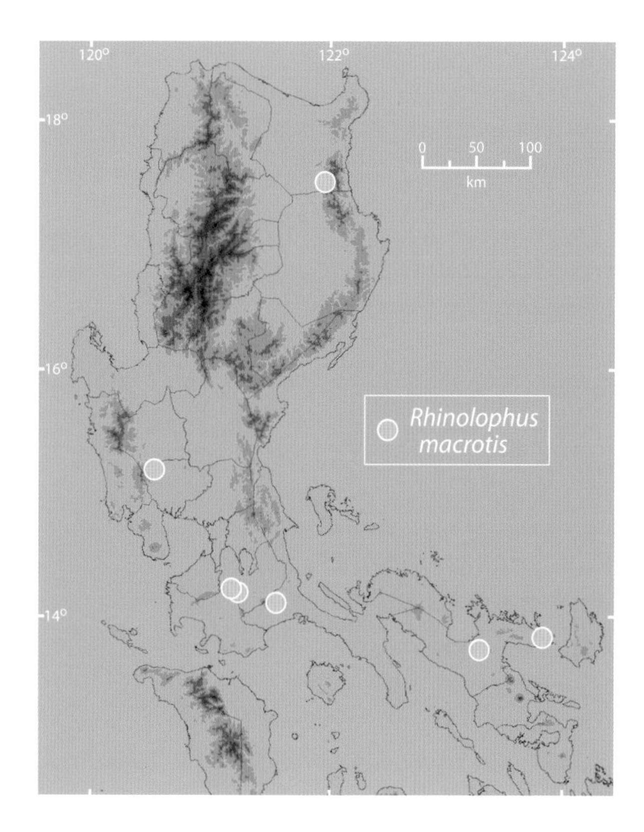

Rhinolophus macrotis
big-eared horseshoe bat

Blyth, 1844. J. Asiat. Soc. Bengal, 13:485. Type locality: Nepal.

DESCRIPTION: Total length 77–85 mm; tail 27–33 mm; hind foot 8–10 mm; ear 27–29 mm; forearm 44–50 mm; weight 6–8 g. A medium-small *Rhinolophus*. The dorsal pelage color varies from dark brown to bright orange and is often grayer ventrally. The ears are very large relative to the body size; the noseleaf and sella are broad, with an especially large connecting process; the body is slender. *Rhinolophus virgo* are smaller (forearm 37–42 mm) and have proportionately smaller ears (15–21 mm). Some *R. arcuatus* have a similar forearm length, but smaller ears (21 mm or less) and shorter tails (21 mm or less). *Rhinolophus philippinensis* are larger (forearm 53–58 mm), with even larger ears (30–38 mm) and a larger noseleaf. Other *Rhinolophus* are larger.

DISTRIBUTION: India to Sumatra and the Philippines, probably throughout Luzon in suitable habitat.

EVOLUTION AND ECOLOGY: As currently defined, this is certainly a species-group, and there may be more than one such group on Luzon; much study is needed. Philippine populations are sometimes assigned to the subspecies *Rhinolophus macrotis hirsutus*. They are closely related to *R. philippinensis*. On Luzon, they have been recorded in primary and secondary lowland forest, from sea level to 620 m elevation, usually in close association with the caves or man-made tunnels where they roost; they sometimes roost with *Hipposideros an-*

tricola. Females give birth to a single young. They feed on insects below the forest canopy, but no details are available. Individuals from Laguna Province have an echolocation call with a mean frequency of 55.1 kHz.

STATUS: Uncertain because of taxonomic uncertainty. May be widespread and stable in forest, but usually at low density.

REFERENCES: Esselstyn et al., 2004; Heaney et al., 1998, 2006b; Ingle and Heaney, 1992; Sedlock, 2001; Sedlock et al., 2011, 2014b; Simmons, 2005.

Rhinolophus philippinensis enormous-eared horseshoe bat

Waterhouse, 1843. Proc. Zool. Soc. Lond., p. 68. Type locality: Luzon.

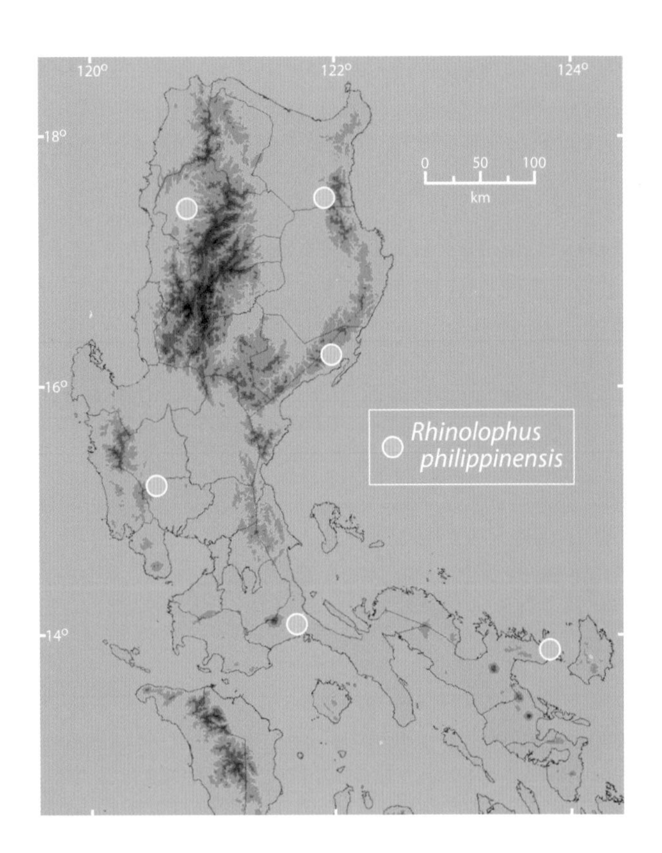

DESCRIPTION: Total length 85–93 mm; tail 29–37 mm; hind foot 10–12 mm; ear 30–38 mm; forearm 53–58 mm; weight 9–12 g. A medium-sized *Rhinolophus*, with huge ears and noseleaf; the connecting process is especially enlarged. The pelage is usually dark brown, and paler ventrally. Only *R. macrotis* have ears of similar relative proportions, and they are much smaller (forearm 44–50 mm). *Rhinolophus subrufus* have a forearm of similar length (53–57 mm), but much smaller ears (24–27 mm) and a smaller noseleaf. *Rhinolophus inops* also have a forearm of similar length (50–57 mm), but have smaller ears (22–27 mm), a smaller noseleaf, and a shorter tail (17–25 mm).

DISTRIBUTION: As currently defined, occur from Borneo and the Philippines to Australia; widespread on Luzon.

EVOLUTION AND ECOLOGY: This taxon represents a species complex, probably including multiple distinct lineages on Luzon and elsewhere within the Philippines; much study is needed. They are closely related to *Rhinolophus macrotis*. On Luzon, *R. philippinensis* have been recorded in primary and secondary lowland and montane forest, from sea level to 1100 m elevation; they appear to occur at low density everywhere, but they also decline in abundance with increasing elevation. They are often associated with large, deep caves, but they also have been taken in some forested areas where caves are not known to be present. Females give birth to single young.

STATUS: Uncertain because of taxonomic uncertainty. As currently defined, widespread in Southeast Asia, but usually at low density and often dependent on caves as roosting sites.

REFERENCES: Duya et al., 2007; Lepiten, 1995; Ruedas et al., 1994; Sedlock et al., 2014b.

Rhinolophus rufus
large rufous horseshoe bat

Eydoux and Gervais, 1836. In Laplace, Voy. autour du monde par les mers de l'Inde . . . corvette . . . *La Favorite* 5 (Zool.), part 2:9. Type locality: Manila, Luzon Island.

DESCRIPTION: Total length 105–121 mm; tail 20–36 mm; hind foot 15–18 mm; ear 32–36 mm; forearm 66–74 mm; weight 29–39 g. The largest *Rhinolophus* species in the Philippines and one of the largest in the genus. The pelage is dark to medium brown, sometimes with a red or orange tint. The ears and noseleaf are small relative to other *Rhinolophus*; the noseleaf is thicker (more fleshy) than in other species; the eyes are larger than in most *Rhinolophus*. No other *Rhinolophus* is as large and generally robust.

DISTRIBUTION: Philippines only; recorded from scattered localities throughout the oceanic Philippines, with records from most parts of Luzon.

EVOLUTION AND ECOLOGY: Generally poorly known. Probably closely related to *Rhinolophus subrufus* and *R. inops*. Recorded in primary and secondary lowland forest, either in or near fairly large, deep caves, from sea level to at least 350 m elevation on Luzon, and up to 600 m elevation elsewhere. They are locally common. Females give birth to single young. A specimen from Catanduanes Island had a karyotype of $2n = 40$, FN = 60.

STATUS: Common in and around large lowland caves with adjacent secondary or primary forest, but at low

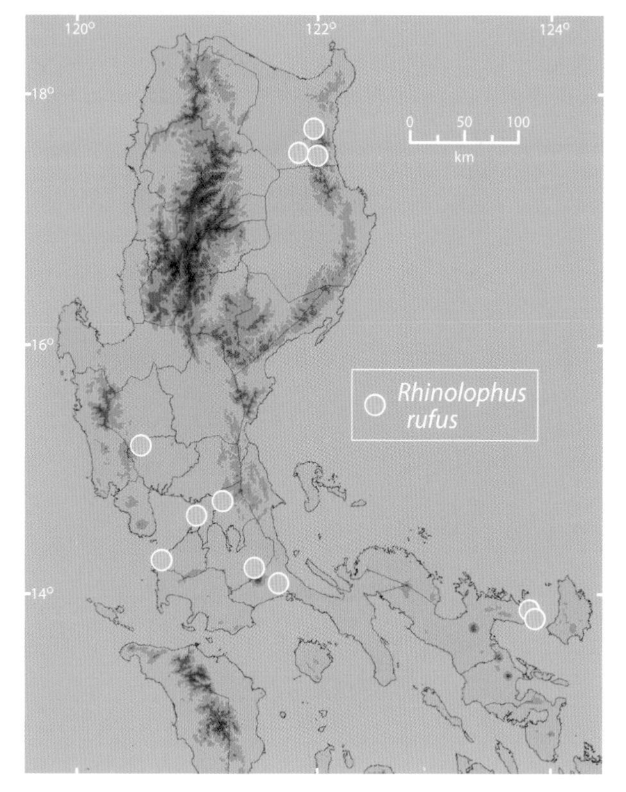

density or absent elsewhere, and probably impacted by the mining of limestone. Further study is needed.

REFERENCES: Csorba et al., 2003; Heaney et al., 1991, 1998; Lawrence, 1939; Rickart et al., 1999; Sanborn, 1952.

Rhinolophus subrufus small rufous horseshoe bat

K. Andersen, 1905. Ann. Mag. Nat. Hist., ser. 7, 16:283. Type locality: Manila, Luzon Island.

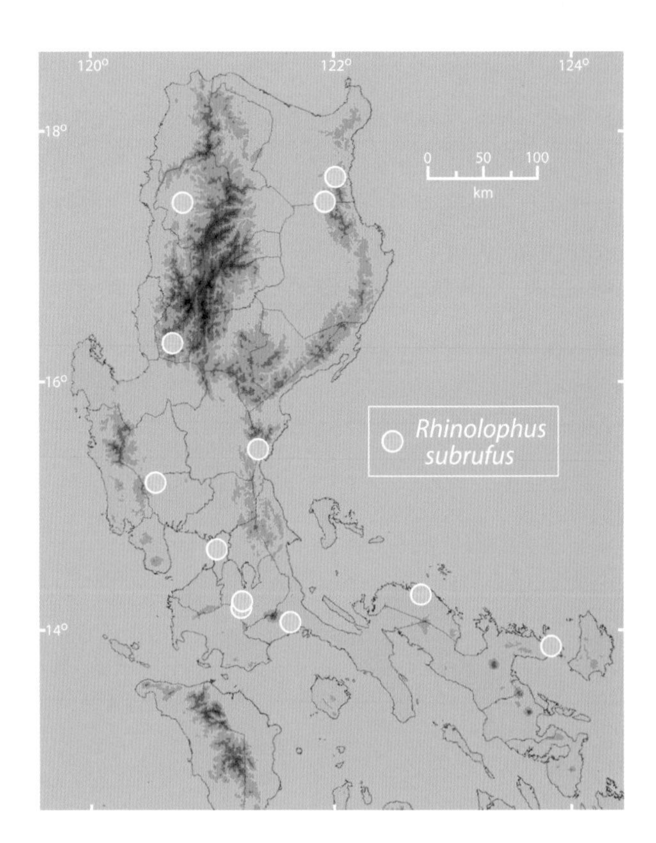

DESCRIPTION: Total length 82–92 mm; tail 22–29 mm; hind foot 12–13 mm; ear 24–27 mm; forearm 53–57 mm; weight 12–19 g. A medium-large *Rhinolophus*. The dorsal fur is reddish-brown and has a slightly woolly/crinkled appearance; the fur is slightly paler ventrally. The noseleaf is slightly wider than the muzzle; one lateral leaflet is present. *Rhinolophus inops* are similar, but they are slightly smaller (forearm usually 50–54 mm, rather than usually 54–56 mm); have a shorter tail (usually 18–23 mm); lack the woolly/crinkled appearance of the dorsal fur; and have a noseleaf equal to or less than the width of the muzzle. *Rhinolophus philippinensis* have a forearm of similar length (53–58 mm), but have much larger ears (30–38 mm) and a larger noseleaf.

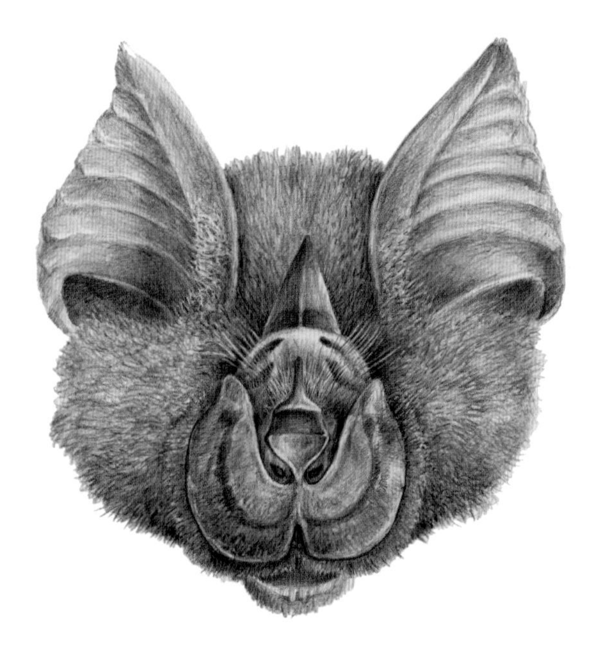

DISTRIBUTION: As currently known, occur only in the oceanic Philippines, with scattered records from throughout Luzon.

EVOLUTION AND ECOLOGY: Poorly known. They are similar and closely related to *Rhinolophus inops*; taxonomic and ecological studies are needed. *Rhinolophus subrufus* are recorded from near sea level to about 735 m elevation, with some records from inside or near caves and others from primary forest distant from caves. They are locally common in areas with extensive caves and are either rare or absent elsewhere. Females give birth to a single young. They often are found together with *R. philippinensis* and *R. rufus*.

STATUS: Unknown because of the combination of taxonomic uncertainty and limited ecological data. They may be dependent on caves.

REFERENCES: Balete et al., 2011, 2013a; Heaney et al., 1998; unpublished data and specimens in FMNH.

Rhinolophus virgo yellow-faced horseshoe bat

K. Andersen, 1905. Proc. Zool. Soc. Lond., 1905:88. Type locality: Pasacao, [Camarines Sur Province], Luzon Island.

DESCRIPTION: Total length 63–72 mm; tail 17–23 mm; hind foot 7–9 mm; ear 15–21 mm; forearm 37–42 mm; weight 4–7 g. A small *Rhinolophus*. The pelage is usually a medium brown, and paler ventrally. Most individuals have a yellow tint to the skin on the face. Some other *Rhinolophus* have forearms of similar size, but they have a broader head and noseleaf; a narrower lateral leaflet; a tuft of hair (usually) on the noseleaf at the base of the connecting process; and brown skin on the face (rather than usually yellow). *Rhinolophus macrotis* have a longer forearm (44–50 mm); have much larger ears (27–29 mm); and have a larger noseleaf and sella. Other *Rhinolophus* are larger.

DISTRIBUTION: Occur only in the Philippines, but widespread within the archipelago, including Luzon.

EVOLUTION AND ECOLOGY: Closely related to *Rhinolophus celebensis*. Geographic variation is present; *R. virgo* may represent a species complex. They often are common from sea level to ca. 800 m elevation in primary and secondary lowland, and uncommon in montane forest from elevations of ca. 800 m to 1050 m. They often roost in caves, culverts, and crevices, and sometimes in darkened cavities in trees. Unlike other Philippine *Rhinolophus*, they seem to roost in small groups, rather than forming large colonies. They are more easily captured in harp traps than in mist nets. They forage below the canopy, feeding on insects (mostly Lepidop-

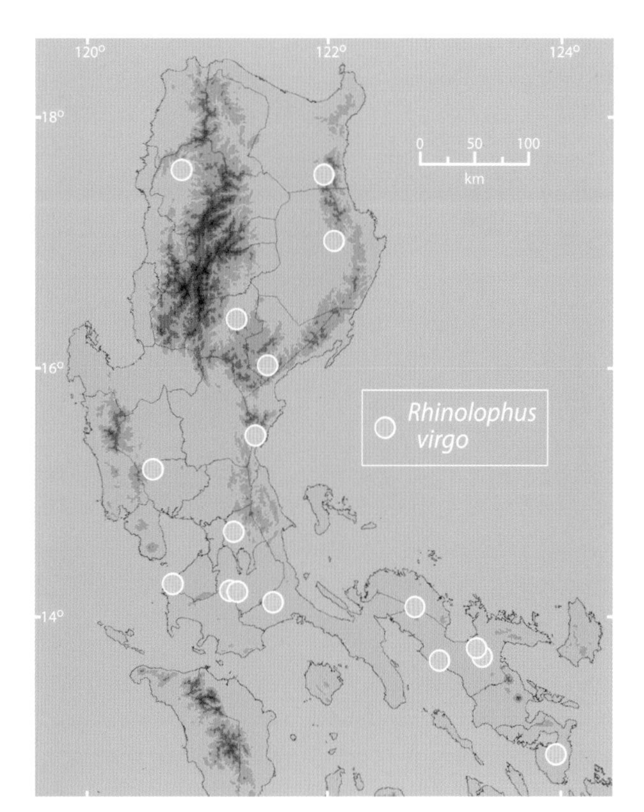

tera and Diptera) and occasionally on spiders. Females give birth to a single young. Individuals from Laguna Province have an echolocation call with a mean frequency of 85.6 kHz.

STATUS: Widespread and moderately common.

REFERENCES: Alviola et al., 2011; Balete et al., 2011; Corbet and Hill, 1992; Duya et al., 2007; Esselstyn et al., 2004; Heaney et al., 1991, 2004, 2006a, 2006b; Mould, 2012; Rickart et al., 1993; Sedlock, 2001; Sedlock et al., 2008, 2011, 2014a; Taylor, 1934.

SECTION 7

Family Vespertilionidae: Evening Bats

The 22 species of this family make this the most diverse family of bats on Luzon, and second only to the family Muridae among Luzon mammals overall. All of the species in the Vespertilionidae eat insects or other arthropods, often catching them in flight using their large tail membrane as a "basket" in which to scoop up the bugs. These bats produce strong ultrasonic squeaks that serve in their echolocation system, allowing them to navigate and catch their prey after dark, as do all other insectivorous bats. Like the emballonurid and mollossid bats, they emit these squeaks through their mouths, not through their nostrils (as do the hipposiderids and rhinolophids), so the vespertilionids have simple noses and faces. All of the vespertilionid species have fairly large ears, small eyes, and a proportionately long tail that lies entirely within the tail membrane (i.e., not extending beyond the membrane or lying in a sheath on top of the membrane).

All of the vespertilionids are small bats, with the largest having a forearm that is about 55 mm long at most and weights of no more than about 20 g. Among these are some of the smallest bats in the Philippines: the tiny bamboo bats of the genus *Tylonycteris*, which weigh only 2.7–3.7 g as adults. Only *Coelops hirsutus*

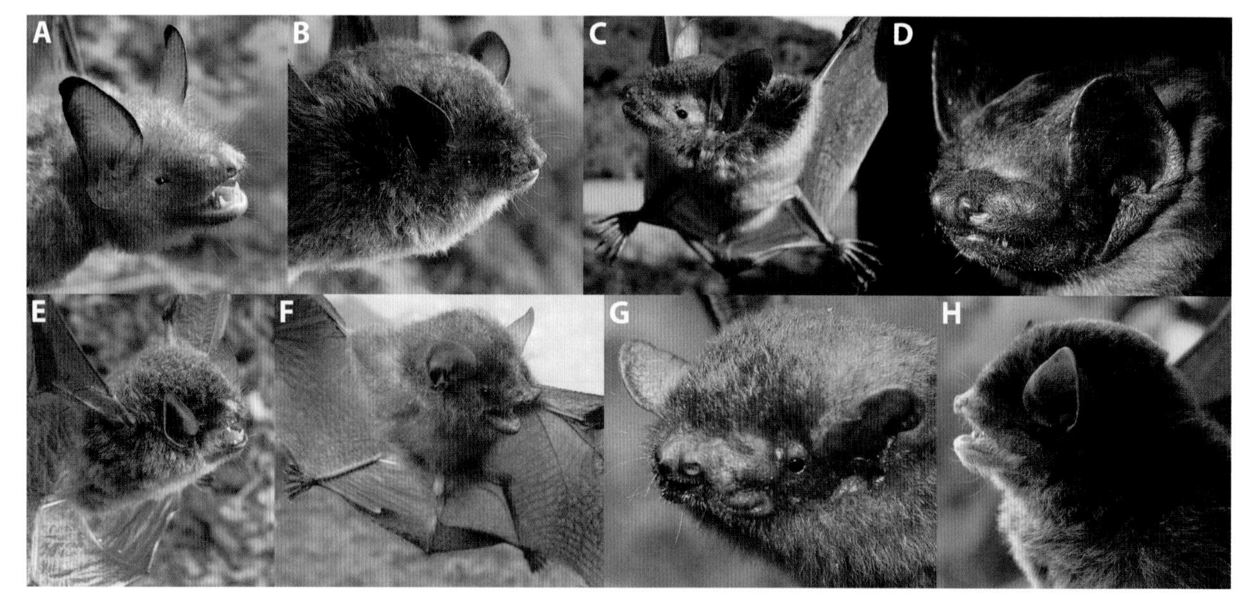

Eight vespertilionids. A: *Myotis rufopictus.* B: *M. muricola.* C: *M. macrotarsus.* D: *Philetor brachypterus.* E: *Kerivoula papillosa.* F: *Murina cyclotis.* G: *Tylonycteris pachypus.* H: *Miniopterus* sp. *Philetor brachypterus* photo courtesy of P. Olsen.

is equally small. The family includes several distinct lineages, often defined as separate subfamilies. The genera *Murina* and *Harpiocephalus* are in the subfamily Murininae: they have stout, broad heads; short, tubular nostrils; and broad, sturdy teeth that they use for chewing beetles and other insects with heavy shells. The Miniopterinae includes only *Miniopterus*: the species in this genus have long, sleek, dark fur and are often the most abundant bats in caves. The Kerivoulinae on Luzon is represented by *Kerivoula*, which are slender bats with funnel-shaped ears that have a long, slender tragus. Over half of the species (about 14) of evening bats are members of the subfamily Vespertilioninae, occurring from sea level to near the tops of the highest mountains and feeding in ecologically diverse habitats on a wide range of insects. Most of the bats that roost in houses on Luzon are members of this subfamily, and others within it roost in banana leaves or in bamboo stems, so, among all of the Luzon bats, they are some of the most familiar to people there. This subfamily also includes the most colorful Philippine bat, the spectacular *Myotis rufopictus*.

As with the other insectivorous bats, the evening bats are difficult to capture with mist nets, because of their sensitive sonar system; thus many are poorly known. Because there are so few specimens, their taxonomy is often uncertain. Although none of these species are currently considered to be endemic just to Luzon, and only one to the Philippines, we suspect that as more studies are conducted, many species-complexes will be identified and separated into their several component species, some of which will be restricted to the Philippines (and perhaps to Luzon alone). The difficulty in capturing evening bats also makes them hard to study ecologically. As with so many Luzon mammals, much remains to be learned!

REFERENCES: Heaney et al., 2012; Simmons, 2005; Taylor, 1934.

Falsistrellus petersi
chocolate pipistrelle

(Meyer, 1899). Abh. Zool. Anthrop.-Ethnology. Mus. Dresden, 7(7):13. Type locality: Minahassa, North Sulawesi, [Indonesia].

DESCRIPTION: Total length 88–92 mm; tail 35–39 mm; hind foot 8–9 mm; ear 14–16 mm; forearm 38–40 mm; weight 6.0–6.6 g. The dorsal fur is long, somewhat woolly, and dark chocolate brown, often with pale tips that give the pelage a frosted appearance. The ventral fur is pale on the distal portion and dark at the base. The tragus is short, broad, and blunt; the nostrils are simple (not tubular). The head is longer and narrower and the ears are longer than in *Pipistrellus* species, which are superficially similar. *Pipistrellus javanicus* have a shorter ear (11–14 mm) and forearm (32–37 mm); a shorter, broader muzzle; and their dorsal pelage lacks a frosted appearance. *Philetor brachypterus* have short, shiny fur; a short, blunt muzzle; a flattened head; narrow wings; and are larger (total length 98–113 mm; weight 11–15 g). *Nyctalus plancyi* have a similar appearance but are larger (total length 123–124 mm; forearm 48–49 mm).

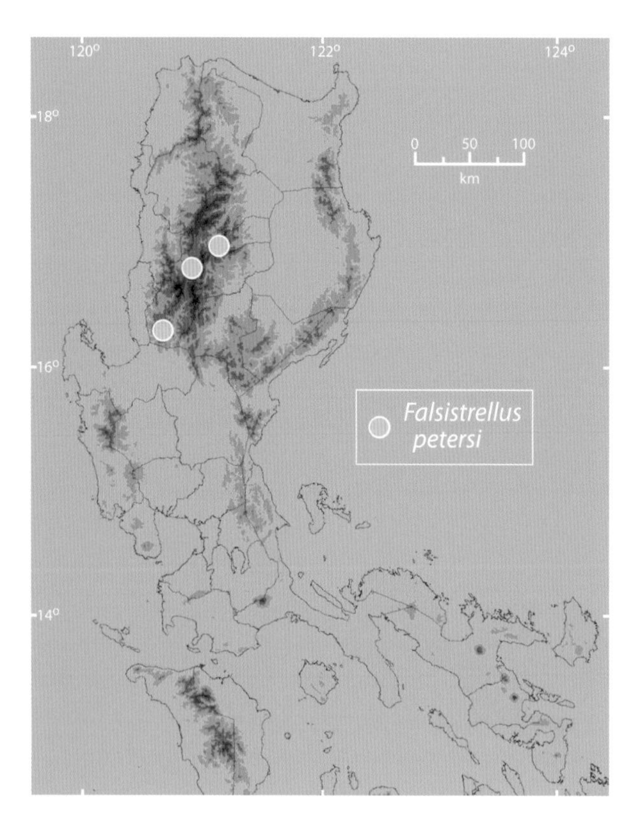

Myotis and *Kerivoula* have high foreheads and a long, slender tragus.

DISTRIBUTION: Sulawesi, Molucca Islands, and the Philippines; records from Luzon and Mindanao.

EVOLUTION AND ECOLOGY: This genus, formerly included in *Pipistrellus*, is most closely related to *Philetor*, *Hypsugo*, and *Tylonycteris*, and only distantly to *Pipistrellus*. Other species in the genus occur from Sri Lanka and Burma [= Myanmar] to Australia. The few specimens from Luzon were taken at elevations from 1250 m to 2530 m in both primary and heavily disturbed mossy forest and pine forest. We saw preserved individuals of this species that had flown into a wooden building at 1500 m elevation (in Barlig, Mountain Province), and we captured two in a wooden building at 2128 m elevation (on Mt. Data), both in open (i.e., unforested) residential areas. We suspect that they commonly roost inside houses at high elevations in the Central Cordillera, and that the small insectivorous bats we often saw flying above towns in the area were this species.

STATUS: Poorly known; may be common at high elevations.

REFERENCES: Corbet and Hill, 1992; Francis, 2008; Francis and Hill, 1986; Heaney et al., 2012; Sedlock et al., 2008.

Harpiocephalus harpia hairy-winged bat

(Temminck, 1840). Monogr. Mamm., 2:219. Type locality: SE side Mt. Gede, Java, [Indonesia].

DESCRIPTION: Total length 103–120 mm; tail 40–49 mm; hind foot 10–13 mm; ear 15–18 mm; forearm 44–49 mm; weight 12–24 g. The dorsal pelage is long, thick, soft, and usually bright orange; it extends onto the forearm and base of the wing membrane and covers most of the dorsal surface of the tail membrane. The ventral pelage is duller and grayer. The head is broad, with heavy canines and molariform teeth. The ears are rounded and generally funnel shaped, with a long, pointed tragus that has a notch near the base

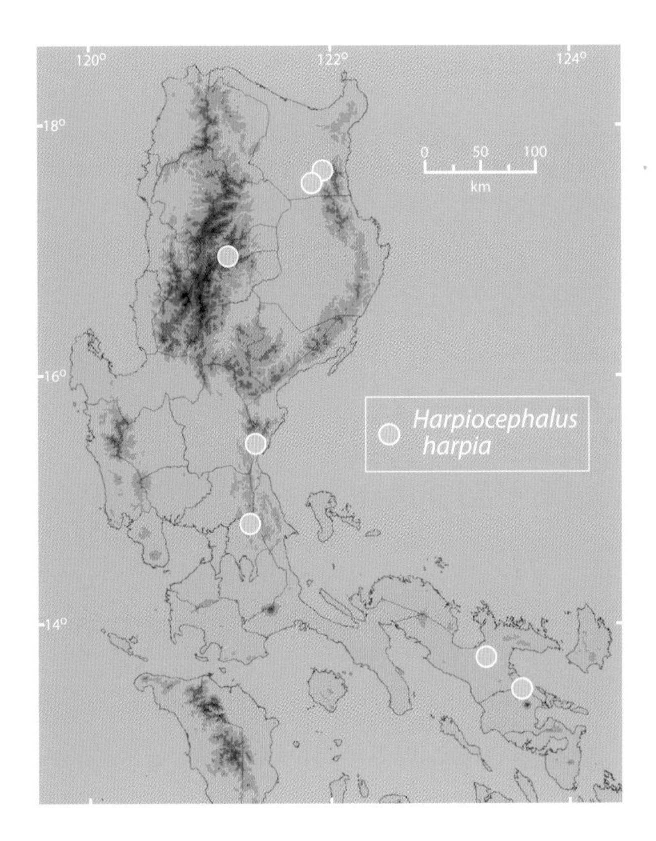

on the posterior edge. The nostrils are elongated into short tubes. *Murina cyclotis* are similar but smaller, with a forearm length of 39 mm or less. *Nyctalus plancyi* and *Scotophilus kuhlii* are both similar in size but have simple nostrils; shorter, sleeker fur; and a tail membrane that has only a small area of hair near the base of the tail on the dorsal surface.

DISTRIBUTION: India to Indochina and Taiwan, Java, the Moluccas, and the Sunda Islands; widespread in the Philippines, probably including all of Luzon.

EVOLUTION AND ECOLOGY: A member of the subfamily Murininae, which consists only of the genera *Harpiocephalus* and *Murina*; these genera occur widely from India to China, Siberia, Borneo, New Guinea and Australia. *Harpiocephalus harpia* are infrequently captured and thus are poorly known. They have been documented on Luzon in primary and disturbed lowland, montane, and mossy forest, from 300 m to 2480 m elevation. They are not associated with caves and probably roost in trees. Males are larger than females. Studies elsewhere suggest that they feed on beetles.

STATUS: Widespread in southern and Southeast Asia, but a substantial amount of variation is present, and taxonomic study is needed.

REFERENCES: Balete et al., 2011, 2013a; Duya et al., 2007; Francis, 2008; Heaney et al., 1999; Ingle and Heaney, 1992; Rickart et al., 1993; Smith and Xie, 2008.

Kerivoula papillosa papillose woolly bat

(Temminck, 1840). Monogr. Mamm., 2:220. Type locality: Bantam, Java, [Indonesia].

DESCRIPTION: Total length 90–109 mm; tail 44–50 mm; hind foot 10–12 mm; ear 15–18 mm; forearm 40–47 mm; weight 9–10 g. A rather delicate bat, with long and woolly fur; dorsally moderately dark brown; the underparts slightly paler; all hair is dark at the base. The ears are delicate and funnel shaped, with a long, slender tragus. The nostrils are simple, not tubular. *Kerivoula whiteheadi* have a shorter tail (34–42 mm) and shorter

forearm (29–33 mm). *Miniopterus schreibersii* are similar in size (forearm 42–46 mm), but have dark brown fur and a short, blunt tragus. *Myotis macrotarsus* are similar in size (forearm 44–50 mm), but have pale ventral fur and huge feet (14–17 mm).

DISTRIBUTION: As currently defined, this species ranges from Indochina and the Sunda Shelf to Sulawesi and the Philippines; poorly known on Luzon.

EVOLUTION AND ECOLOGY: *Kerivoula* is placed (with *Phoniscus*) into the subfamily Kerivoulinae; molecular evidence suggests that it is most closely related to the Murininae. Recent compilations have listed 19 species of *Kerivoula*, but data from recent studies suggest that many more species may be present; more detailed research is needed. The genus occurs from Africa to South Asia, Southeast Asia, and New Guinea. *Kerivoula papillosa* have been recorded on Luzon from 350 m to ca. 500 m elevation. All records are from secondary lowland forest, usually along trails, where these bats are spotty in distribution but locally common. In Malaysia, they have been found roosting in hollow trees and bamboo in groups of two to seven; limited radio-tracking studies suggest that these bats stay within 100–300 m of their primary roost. They are more easily captured in harp traps than in mist nets.

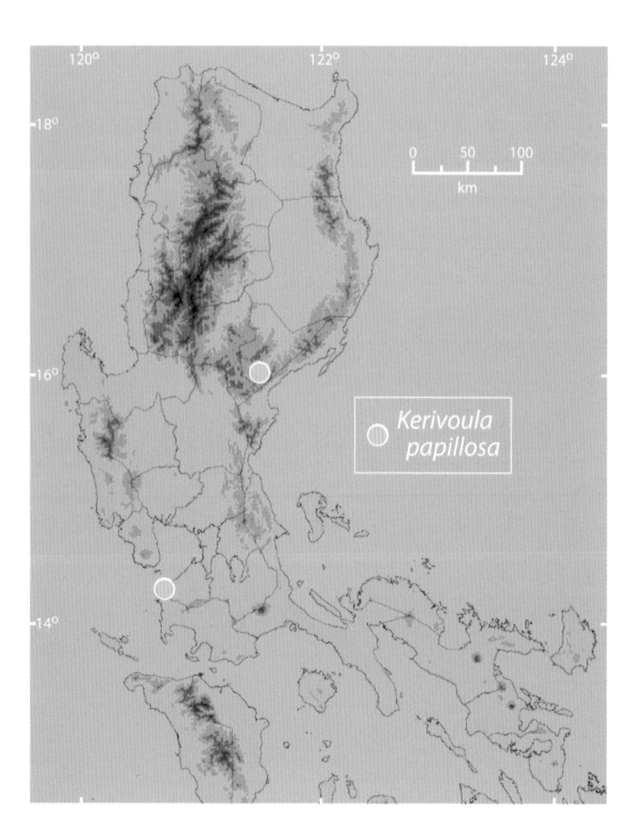

STATUS: Widespread; all Philippine records are from secondary lowland forest.

COMMENT: Recent specimens are similar to *Kerivoula papillosa* from Vietnam (Duya et al., 2007), but referral to this species is tentative. Further study is needed.

REFERENCES: Anwarali Khan et al., 2010; Duya et al., 2007; Francis, 2008; Kingston et al., 2006; Payne et al., 1985; Simmons, 2005; unpublished field notes and specimens in FMNH.

Kerivoula whiteheadi
Whitehead's woolly bat

Thomas, 1894. Ann. Mag. Nat. Hist., ser. 6, 14:460.
Type locality: Molino, Isabela Province, Luzon.

DESCRIPTION: Total length 74–86 mm; tail 34–42 mm; hind foot 7–9 mm; ear 14–18 mm; forearm 29–33 mm; weight 3.5–5.0 g. A small, delicate bat, with large, funnel-shaped ears; a long, slender tragus; and simple nostrils. The dorsal fur varies from reddish brown (typical on Mindanao) to dark brown (typical on Luzon); the ventral fur is somewhat paler; all fur is dark at the base. The wing and tail membranes, and the skin covering the phalanges, is dark (on Luzon). *Kerivoula papillosa* are much larger (forearm 40–47 mm). *Myotis muricola* are similar in size (forearm 29–33 mm) but are dark brown dorsally, and have smaller ears (12–14 mm), with a shorter, less elongate tragus. *Pipistrellus javanicus* are also similar in size (forearm 32–37 mm), but are dark brown dorsally; have small ears (11–14 mm), with a short, blunt tragus; and have a broad muzzle.

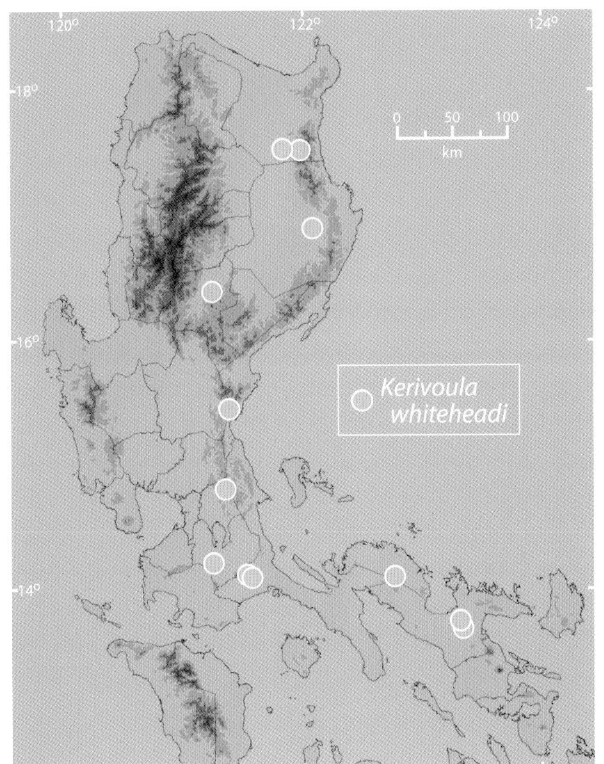

DISTRIBUTION: Southern Thailand to Borneo and the Philippines; occur throughout Luzon.

EVOLUTION AND ECOLOGY: On Luzon, recorded from near sea level to 1465 m elevation in cogon grassland, agricultural areas, abaca plantations, and secondary and primary lowland, montane, and mossy forest, often near creeks. Studies using harp traps have found them to be locally common. A group of 20–30 was found roosting in a cluster of large dead leaves near a river in Borneo. These bats forage low to the ground and fly slowly, so they are difficult to capture in mist nets; most of the recent records are from harp traps. Specimens from Luzon and Mindanao seem to differ consistently in coloration and size; taxonomic study is needed.

STATUS: Widespread in Southeast Asia and fairly common in disturbed habitats and old-growth forest.

REFERENCES: Alviola et al., 2011; Balete et al., 2011, 2013a; Esselstyn et al., 2004; Payne et al., 1985; Sanborn, 1952; Sedlock, 2001; Sedlock et al., 2008, 2011; Simmons, 2005; Taylor, 1934.

Miniopterus australis little bent-winged bat

Tomes, 1858. Proc. Zool. Soc. Lond., p. 125. Type locality: Lifu, New Caledonia.

DESCRIPTION: Total length 75–91 mm; tail 33–42 mm; hind foot 7–10 mm; ear 8–11 mm; forearm 35–37 mm; weight 4–6 g. Like all *Miniopterus*, this species has very dark brown fur that is longer over the shoulders and top of the head than elsewhere on the body. The tragus is short and is blunt and broadest at the tip. The tip of the longest digit on the wing folds back against the adjacent portion of the wing (hence the common name in English). *Miniopterus schreibersii* are very similar but larger (total length 99–111 mm; forearm 42–46 mm). *Miniopterus tristis* are much larger (forearm 51–54 mm). *Myotis muricola* are smaller (total length 65–83 mm; forearm 29–33 mm) and have a long, slender tragus. *Myotis ater* and *My. horsfieldii* are similar in size, but have longer ears (14–15 mm) and the tragus is long and slender. *Pipistrellus javanicus* are similar in size (fore-

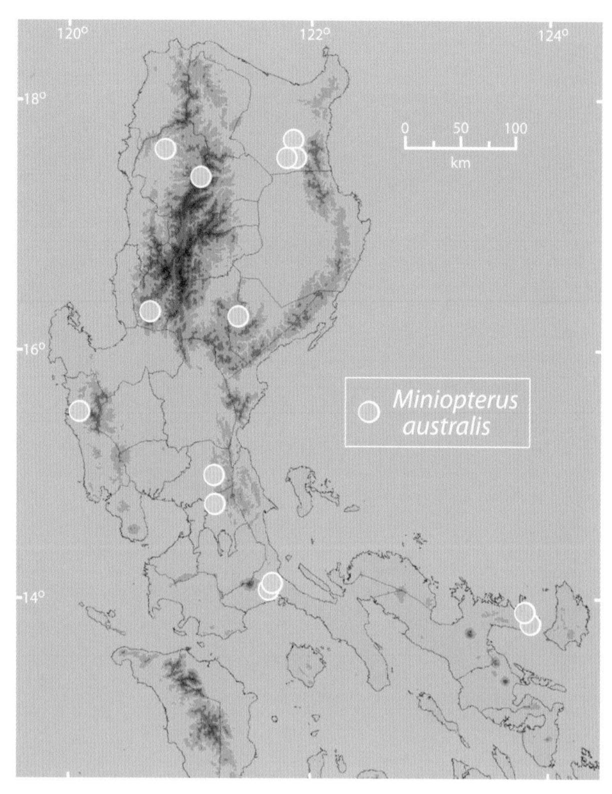

arm 32–37 mm), but have shorter fur; a broad muzzle; and a tragus that is short and blunt but not broadest at the tip.

DISTRIBUTION: India to Australia and throughout the Philippines, including all parts of Luzon.

EVOLUTION AND ECOLOGY: The genus *Miniopterus* is placed in the subfamily Miniopterinae. Only one genus is recognized in this subfamily, but the number of species is highly uncertain. Philippine specimens are currently referred to the subspecies *M. australis paululus*, which is sometimes recognized as a distinct species. These bats roost in caves from sea level to about 1000 m elevation, in agricultural areas and in second growth and primary lowland forest, infrequently including montane forest. They are often the most abundant species in caves in disturbed habitats, with estimates often in the thousands of individuals. They often roost in tight clusters, sometimes intermixed with *M. schreibersii*. The echolocation call of *M. australis* from Laguna Province has a frequency of 61–62 kHz.

STATUS: Geographically widespread and common, but dependent on caves.

REFERENCES: Esselstyn et al., 2004; Heaney et al., 1991; Lepiten, 1995; Rickart et al., 1993; Sanborn, 1952; Sedlock, 2001; Sedlock et al., 2014b; Simmons, 2005; Taylor, 1934.

Miniopterus schreibersii common bent-winged bat

(Kuhl, 1817). Die Deut. Fledermäuse, p. 14. Type locality: Kolumbacs Cave, near Coronini, Banat Mountains, Romania.

DESCRIPTION: Total length 99–111 mm; tail 43–54 mm; hind foot 8–10 mm; ear 12–13 mm; forearm 42–46 mm; weight 9–11 g. Like all *Miniopterus*, this species has dark brown fur that is longer over the shoulders and top of the head than elsewhere on the body. The tragus is short, blunt, and broadest at the tip. As in all *Miniopterus*, the tip of the longest digit on the wing folds back rather tightly against the adjacent portion of the wing. *Miniopterus australis* are smaller (total length 75–91 mm; forearm 35–37 mm). *Miniopterus tristis* are larger (total length 120–129 mm; forearm 51–54 mm). *Myotis ater* and *My. horsfieldii* are smaller (forearm 35–38 mm), with a long, slender tragus. *Myotis macrotarsus* are similar in size (total length 102–111 mm; forearm 44–50 mm), but have a longer ear (16–19 mm), with a long, slender tragus, and much larger hind feet (14–17 mm).

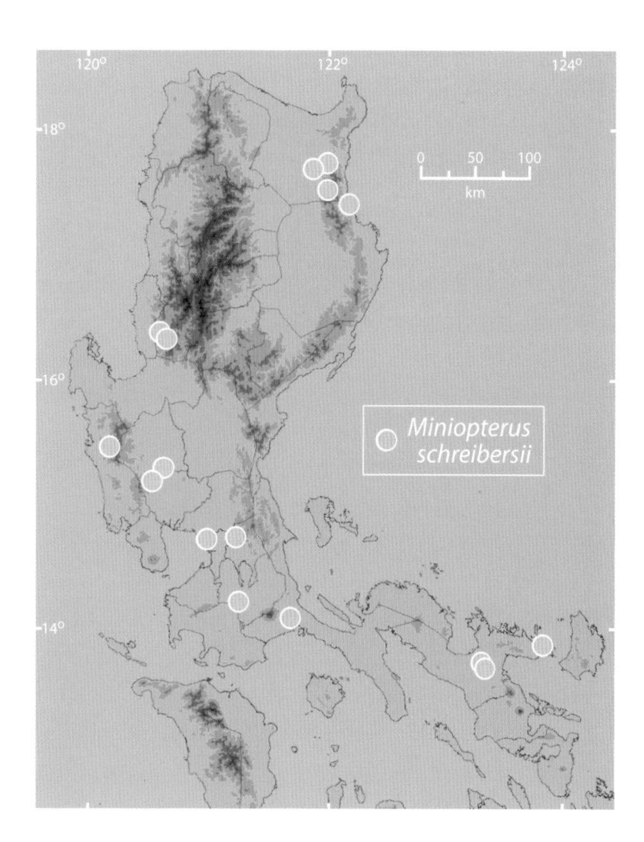

DISTRIBUTION: Europe to the Solomon Islands; throughout the Philippines, including all of Luzon.

EVOLUTION AND ECOLOGY: The taxonomic status of populations referred to this species is highly uncertain; some recent authors refer Philippine specimens to *Miniopterus medius*. On Luzon, this bat usually occurs from sea level to about 1000 m, usually in agricultural areas and in second growth and old-growth lowland forest, but occasionally to about 1700 m in montane forest. They usually roost in caves and do not range far from them, except occasionally roosting in man-made tunnels. They are among the most common bats in many caves, often roosting with *M. australis* in colonies of thousands of individuals. Because of their abundance, they undoubtedly have a major impact on insects in cropland, but no data are available. The echolocation call of *M. schreibersii* from Laguna Province has a mean frequency of 45.6 kHz.

STATUS: Common and widespread, tolerant of disturbance but usually dependent on caves.

REFERENCES: Corbet and Hill, 1992; Esselstyn et al., 2004; Heaney et al., 1991, 1999; 2006b; Lawrence, 1939; Rickart et al., 1993; Sanborn, 1952; Sedlock, 2001; Sedlock et al., 2008, 2014b; Taylor, 1934.

Miniopterus tristis
greater bent-winged bat

(Waterhouse, 1845). Proc. Zool. Soc. Lond., 1845:3.
Type locality: "Philippine islands."

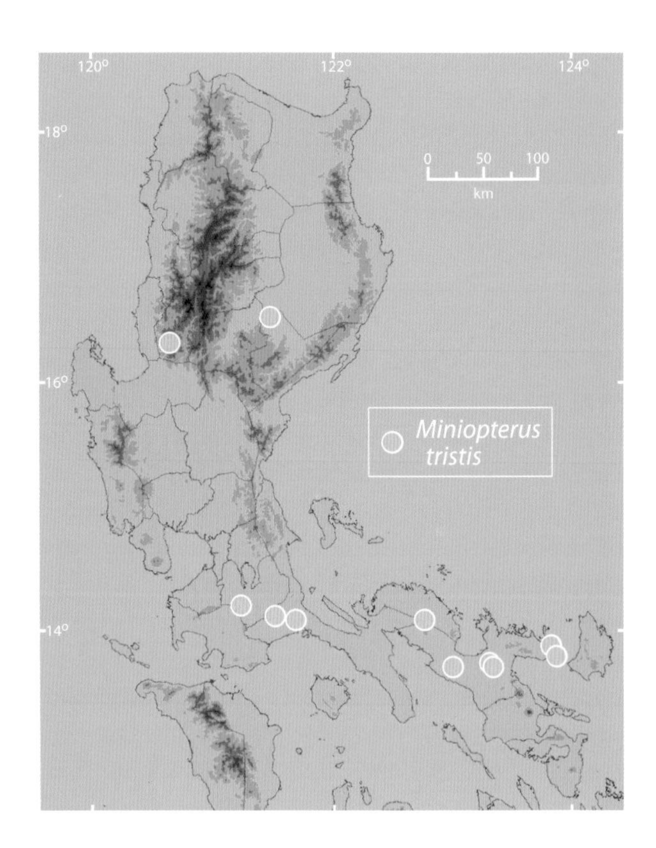

DESCRIPTION: Total length 120–129 mm; tail 53–61 mm; hind foot 11–12 mm; ear 15–16 mm; forearm 51–54 mm; weight 16–21 g. Like all *Miniopterus*, this species has very dark brown fur that is longer over the shoulders and top of the head than elsewhere on the body. The tragus is short, blunt, and broadest at the tip. The tip of the longest digit on the wing folds back rather tightly against the adjacent portion of the wing. *Miniopterus australis* are much smaller (forearm 35–37 mm). *Miniopterus schreibersii* are also smaller (forearm 42–46 mm). *Myotis rufopictus* are similar in size but have a longer forearm (47–55 mm); a longer ear (19–22 mm), with a long, slender tragus; and red or orange fur. *Scotophilus kuhlii* are also similar in size (forearm 49–52 mm) but have short, reddish fur and a shorter tail (42–49 mm).

DISTRIBUTION: Philippines to New Guinea and the Solomon Islands; recorded throughout the Philippines, including Luzon.

EVOLUTION AND ECOLOGY: Usually occur in the lowlands, from near sea level to about 800 m elevation, in secondary and primary lowland forest and agricultural areas. They occasionally occur in montane forest up to 1500 m elevation, however, and appear to be more common in this habitat than the other species of *Miniopterus*. They usually roost in caves, where they often are uncommon but sometimes occur in large numbers (estimated in the thousands). Their echolocation call on Mt. Makiling, Laguna Province, has a mean frequency of 34.6 kHz.

STATUS: Moderately common and widespread, but at least partially dependent on caves.

REFERENCES: Esselstyn et al., 2004; Rickart et al., 1993; Sanborn, 1952; Sedlock, 2001; Sedlock et al., 2008, 2011, 2014b.

Murina cyclotis
round-eared tube-nosed bat

Dobson, 1872. Proc. Asiat. Soc. Bengal, p. 210. Type locality: Darjeeling, India.

DESCRIPTION: Total length 90–101 mm; tail 34–42 mm; hind foot 9–12 mm; ear 14–18 mm; forearm 35–39 mm; weight 8–11 g. The fur is bright brown or rusty-orange brown dorsally and paler ventrally, with a light but conspicuous covering of hair on the dorsal surface of the tail membrane, feet, and forearms; the flight membranes are black. The nostrils are extended laterally into short tubes; the ears are oval and generally funnel shaped; the tragus is long and slender, with a notch at the base. *Murina suilla* are similar but smaller (forearm 33–34 mm; weight 6–7 g). *Harpiocephalus harpia* are similar but larger (forearm 44–49 mm; weight 12–24 g). Among Philippine bats, only *Harpiocephalus* and *Murina* have the combination of a hairy tail membrane; a long, slender tragus; and short, tubular nostrils.

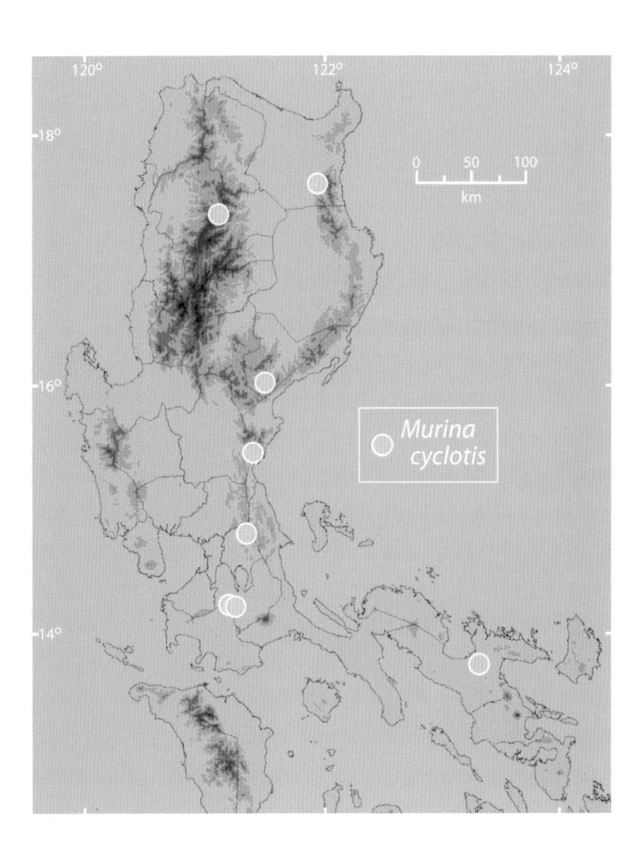

DISTRIBUTION: Sri Lanka to Hainan and Borneo; widespread in the Philippines, including Luzon.

EVOLUTION AND ECOLOGY: More than 18 species in this genus are recognized currently, ranging from India to New Guinea and Australia. Much variation in morphology is present within the Philippines, and as currently defined, *Murina cyclotis* is probably a species-complex rather than a single species. On Luzon, they have been captured in primary and secondary lowland and montane forest, from 100 m to about 1500 m elevation; they have been taken together with *M. suilla* on Mt. Irid and Mt. Mingan. Females often give birth to twins. They are more easily captured in harp traps than in mist nets. They have a karyotype with $2n = 44$, FN = 50.

STATUS: Widespread and moderately common in the Philippines, but much taxonomic uncertainty exists. Detailed taxonomic study is needed.

REFERENCES: Balete et al., 2011, 2013a; Duya et al., 2007; Heaney et al., 1991, 1999, 2006a, 2006b; Kingston et al.,

2006; Lepiten, 1995; Rickart et al., 1993, 1999; Ruedas et al., 1994; Sedlock, 2001; Sedlock et al., 2008; Simmons, 2005.

Murina suilla lesser tube-nosed bat

(Temminck, 1840). Monogr. Mamm., 2:224. Type locality: Tapos, Java, [Indonesia].

DESCRIPTION: Total length 76–78 mm; tail 29–31 mm; hind foot 7–8 mm; ear 14–15 mm; forearm 33–34 mm; weight 6–7 g. The fur is brown or rusty-orange brown dorsally and paler ventrally, with a covering of hair on the tail membrane; the flight membranes are black. The nostrils are extended laterally into short tubes. The ears are of moderate size; have a dark rim; and have a long, slender tragus, with a notch on the posterior edge at the base. *Murina cyclotis* are similar but larger (forearm 35–39 mm; weight 8–11 g). *Harpiocephalus harpia* are much larger (forearm 44–49 mm; weight 12–24 g). In the Philippines, only *Harpiocephalus* and *Murina* have the combination of a hairy tail membrane; a long, slender tragus; and short, tubular nostrils.

DISTRIBUTION: As currently defined, occur in the Malay Peninsula, Borneo, Java, Sumatra, and the Philippines, but much variation is present and multiple species may be present. They have been recorded in most Philippine faunal regions and may be widespread on Luzon.

EVOLUTION AND ECOLOGY: Recorded on Luzon in lightly disturbed lowland forest at 559 m elevation and in transitional lowland-montane forest at 920 m elevation; elsewhere in the Philippines, they have been captured from near sea level to 1900 m elevation in secondary lowland forest/grassland mosaic, secondary forest on limestone, and old-growth lowland, montane, and mossy forest. They have been captured in both mist nets and harp traps. As with other members of the subfamily, they probably feed on beetles.

STATUS: Rather poorly known, but apparently widespread. The taxonomy is uncertain and requires additional study.

REFERENCES: Balete et al., 2011, 2013a; Esselstyn et al., 2004; Kingston et al., 2006; Sedlock et al., unpublished data; Simmons, 2005.

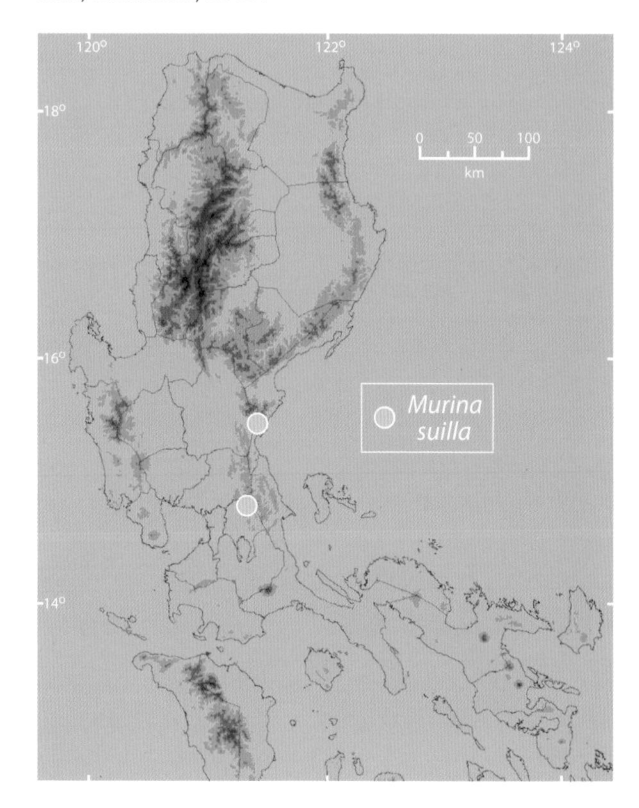

Myotis ater
Peters's myotis

(Peters, 1866). Monatsb. K. Preuss. Akad. Wiss. Berlin, 1866:18. Type locality: Ternate Island, Moluccas, [Indonesia].

DESCRIPTION: Total length 83–92 mm; tail 34–36 mm; hind foot 9–11 mm; ear 14–15 mm; forearm 35–37 mm; weight 6.7–8 g. As with all *Myotis*, the ears are long; the tragus is long and slender, coming to a blunt point; the nostrils are simple (not tubular); the muzzle is not swollen. The dorsal pelage is dark brown, with slightly paler tips; the ventral pelage is similar, but paler overall. *Philetor* and *Pipistrellus* have a short, blunt tragus and a muzzle that appears swollen. *Miniopterus* have the tip of the longest digit on the wing folding back against the adjacent portion of the wing, and have a short, blunt tragus. *Kerivoula* have woolly fur, and funnel-shaped ears with an even longer, more slender tragus. *Myotis muricola* are smaller (forearm 29–33 mm; weight 3–4.5 g). *Myotis horsfieldii* are similar in size (forearm 35–38 mm), but the point of attachment of the posterior edge of the wing on the side of the hind foot differs: in *M. ater*, the wing attaches at the base of the toes, whereas in *M. horsfieldii*, it attaches near the ankle.

DISTRIBUTION: As currently defined, occur from Vietnam and peninsular Malaysia to Sulawesi, the Moluccas, and New Guinea; current Philippine records are only from Culion Island and Luzon (Mt. Isarog, Camarines Sur Province), and possibly from Mt. Kitanglad, Mindanao.

EVOLUTION AND ECOLOGY: *Myotis ater* is closely related to *M. muricola*; they diverged from a common ancestor 3–4 million years ago. There is much uncertainty regarding the number of species in this taxon. We provisionally follow the current listing of Philippine populations as *M. ater nugax*, but we recognize the need for further study. Only two specimens have been captured on Luzon, in mature montane forest at 1125 m elevation on Mt. Isarog. One of these was an adult female with a single embryo.

STATUS: Uncertain because of taxonomic uncertainty and little information on ecology or distribution. Further studies are badly needed.

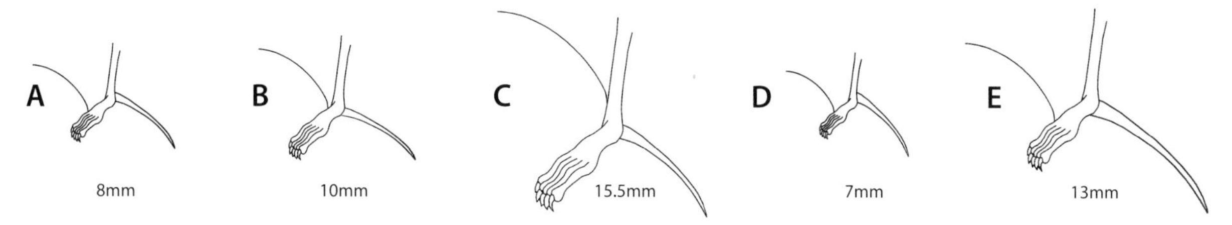

Myotis feet, approximately actual size. A: *M. ater*. B: *M. horsfieldii*. C: *M. macrotarsus*. D: *M. muricola*. E: *M. rufopictus*. Redrawn in part from Ingle and Heaney (1992).

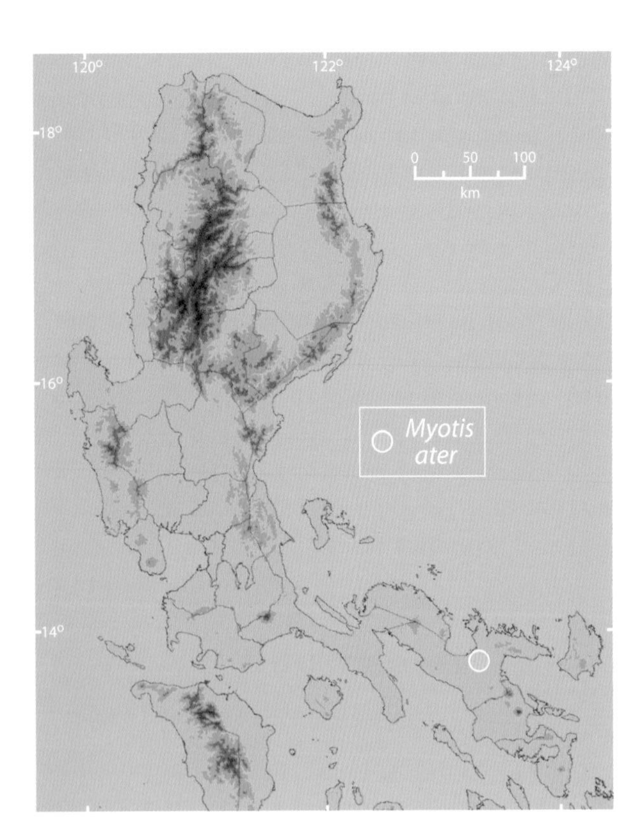

REFERENCES: Corbet and Hill, 1992; Heaney et al., 1998, 1999, 2006b; Sanborn, 1952, Simmons, 2005.

Myotis horsfieldii
common Asiatic myotis, Horsfield's myotis

(Temminck, 1840). Monogr. Mamm., 2:226. Type locality: Mt. Gede, Java, [Indonesia].

DESCRIPTION: Total length 80–88 mm; tail 36–40 mm; hind foot 9–11 mm; ear 14–16 mm; forearm 35–38 mm; weight 5.5–7 g. As with all *Myotis*, the ears are long; the tragus is long and slender, coming to a blunt point; the nostrils are simple (not tubular); the muzzle is not swollen. The dorsal pelage is dark grayish-brown, with slightly paler tips; the ventral pelage is similar, but grayer and paler overall, usually with pale, nearly white fur surrounding the anal region. The ears and feet are proportionately large. The wing membrane attaches to the side of the foot, 1–2 mm below the base of the toes and near the ankle. *Myotis muricola* are smaller (forearm 29–33 mm), and the base of the wing attaches to the side of the foot at the base of the toes. *Myotis ater* are blackish-brown, rather than grayish-brown; the wing membrane attaches at the base of the toes. *Pipistrellus javanicus* have short ears (rarely reaching 14 mm), with a blunt tragus; the muzzle appears swollen. *Miniopterus* have the tip of the longest digit on the wing folding back against the adjacent portion of the wing and have a short, blunt tragus. *Kerivoula* have woolly fur; funnel-shaped ears; and a tragus that is very long, slender, and pointed at the tip.

DISTRIBUTION: Burma [= Myanmar] and southeastern China to the Malay Peninsula, Bali, and Sulawesi; occur throughout the Philippines, including all of Luzon.

EVOLUTION AND ECOLOGY: Related to a group of mostly Asian *Myotis* with large hind feet, including *M. macrotarsus*; they diverged from their common ancestor about 8 million years ago. Philippine populations are sometimes placed in the subspecies *M. horsfieldii jeannei*. On Luzon, they have been recorded in secondary

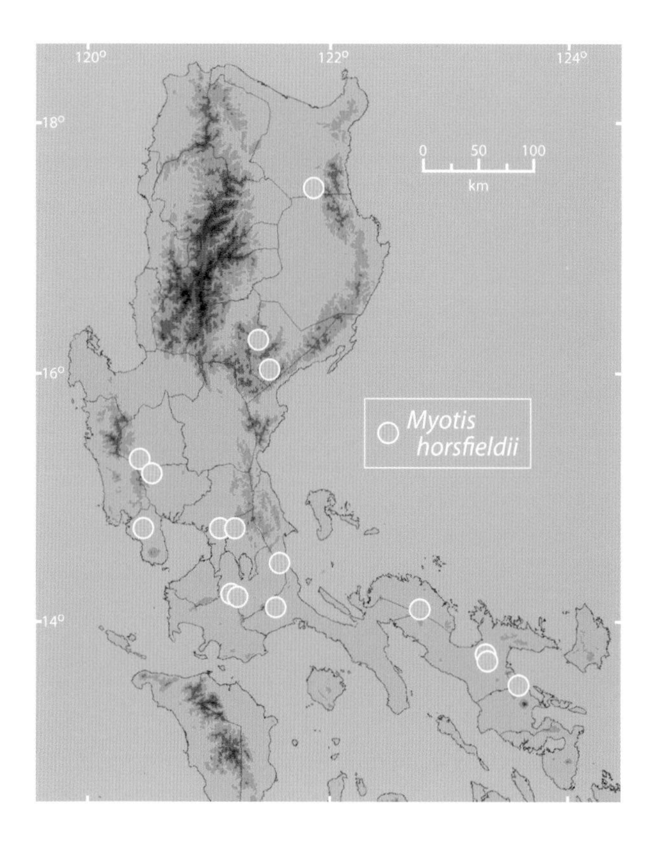

and old-growth lowland, montane, and mossy forest and in agricultural areas, from sea level to about 1400 m elevation. They are often among the most common insectivorous bats in these habitats. They frequently forage on insects that are flying just above the water surface of streams and rivers and are usually absent where streams are lacking. They often roost in caves and tunnels, and they have been found roosting beneath a large rock over a stream. The echolocation call of this species in Laguna Province has a mean call frequency of 47.6 kHz; characteristics of the call imply that it is able to detect prey close to substrates, such as leaves or the water surface. They are easily captured using either mist nets or harp traps.

STATUS: Widespread and moderately common.

REFERENCES: Duya et al., 2007; Esselstyn et al., 2004; Rickart et al., 2013; Ruedi et al., 2013; Sedlock, 2001; Sedlock et al., 2008, 2011; Taylor, 1934.

Myotis macrotarsus
Philippine large-footed myotis

(Waterhouse, 1845). Proc. Zool. Soc. Lond., 1845:5. Type locality: "Philippines."

DESCRIPTION: Total length 102–111 mm; tail 45–54 mm; hind foot 14–17 mm; ear 17–19 mm; forearm 44–50 mm; weight 11–14 g. As with all *Myotis*, the ears are long; the tragus is long and slender, usually coming to a moderately sharp point; the nostrils are simple (not tubular); the muzzle is not swollen. The feet are proportionately large; the base of the wing attaches above the ankle, not on the side of the foot. The dorsal pelage is pale grayish-brown; the ventral pelage is pale yellow or nearly white. *Myotis horsfieldii* are darker, smaller (forearm 35–38 mm); have smaller feet (9–11 mm); and the base of the wing attaches at the base of the foot, near the ankle. *Miniopterus* have the tip of the longest digit on the wing folding back against the adjacent portion of the wing; have a short, blunt tragus; and have long, fluffy pelage. *Nyctalus plancyi* and *Scotophilus kuhlii* have short ears (17 mm or less); a short, blunt tragus; and a shorter hind foot (14 mm or less).

DISTRIBUTION: Borneo and the Philippines; probably throughout the Philippines, including all parts of Luzon.

EVOLUTION AND ECOLOGY: Related to a group of mostly Asian *Myotis* with large feet; *M. macrotarsus* diverged from its closest relatives about 8 million years ago. Their ecology is poorly known. They appear to roost only in caves (from sea level up to about 100 m elevation on Luzon, though up to 250 m elevation elsewhere), usually in small groups. All captures have been made in areas with limestone caves and nearby rivers and large streams, in both mixed agricultural/regenerating areas and primary lowland forest. Closely related, similar species use their large feet to capture insects off the surface of streams and ponds. This species has a karyotype with $2n = 44$, FN = 50.

STATUS: Locally common, but dependent on locations with both caves and rivers. Focused field studies are needed.

REFERENCES: Esselstyn et al., 2004; Francis, 1985; Md. Nor, 1996; Ruedi et al., 2013; Taylor, 1934.

Myotis muricola
whiskered myotis

(Gray, 1846). Cat. Hodgson Coll. Brit. Mus., p. 4. Type locality: Nepal.

DESCRIPTION: Total length 65–83 mm; tail 32–36 mm; hind foot 6–8 mm; ear 12–14 mm; forearm 29–33 mm; weight 3–4.5 g. As with all *Myotis*, the ears are relatively long; the tragus is long and slender, coming to a blunt point; the nostrils are simple (not tubular); the muzzle is not swollen. The dorsal pelage is dark brown, sometimes with slightly paler tips; the ventral pelage is similar, but grayer and paler overall. The wing membrane attaches to the side of the foot, at the base of the toes. *Myotis ater* are larger (forearm 35–37 mm) and usually slightly darker. *Pipistrellus* have a short, blunt tragus and a muzzle that appears swollen. *Miniopterus* have the tip of the longest digit on the wing folding back against the adjacent portion of the wing and have a short, blunt tragus. *Kerivoula* have woolly fur and

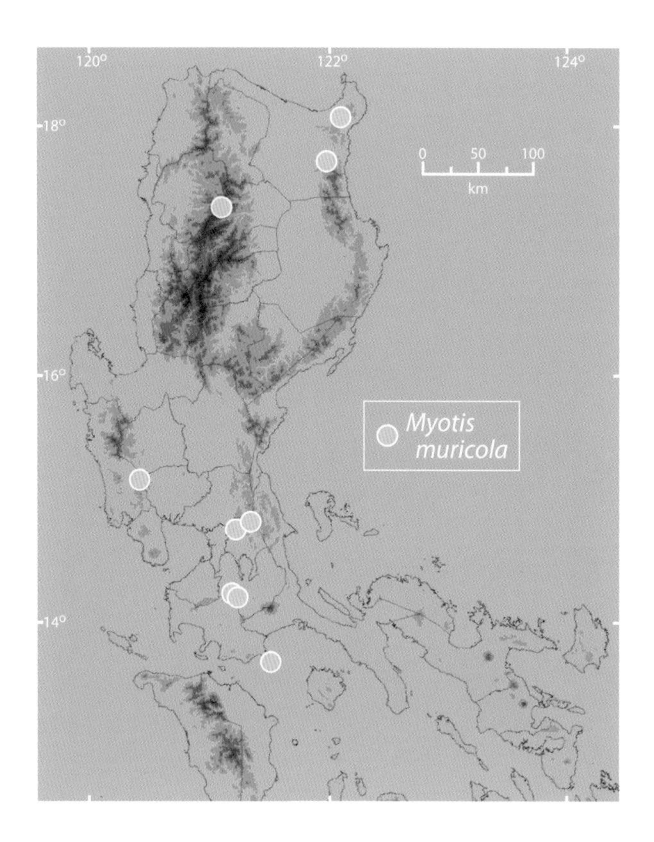

funnel-shaped ears, with a very long, sharply pointed tragus.

DISTRIBUTION: As currently defined, occur from Afghanistan to New Guinea, including all of the Philippines.

EVOLUTION AND ECOLOGY: As currently defined, this is almost certainly a species-group; Philippine populations are sometimes referred to the subspecies *Myotis muricola browni*, but much study is needed. They occur in primary and secondary lowland and montane forest, from near sea level to 1600 m elevation, occasionally including urban areas. Females give birth to a single young. In other countries, *M. muricola* have been found to roost in vegetation, including furled banana leaves, but rarely in caves. The echolocation call of this species in Laguna Province has a mean frequency of 51.8 kHz. They are difficult to capture, perhaps because they are small and are agile fliers.

STATUS: As currently defined, common and widespread in Asia, including Luzon, but the status of populations is uncertain because of taxonomic uncertainty. Detailed studies of phylogenetic relationships and species limits are badly needed.

REFERENCES: Balete et al., 2013a; Duya et al., 2007; Francis, 2008; Heaney et al., 2006b; Ingle, 1992; Ong et al., 1999; Payne et al., 1985; Rickart et al., 1993; Ruedas et al., 1994; Ruedi et al., 2013; Sedlock, 2001; Sedlock et al., 2011.

Myotis rufopictus
Philippine orange-fingered myotis

(Waterhouse, 1845). Proc. Zool. Soc. Lond., 1845:3, 8. Type locality: "Philippines."

DESCRIPTION: Total length 114–120 mm; tail 43–59 mm; hind foot 12–14 mm; ear 19–22 mm; forearm 47–55 mm; weight 9.5–15 g. One of the most strikingly colored bats in Asia. As with all *Myotis*, the ears are long; the tragus is long and slender, coming to a blunt point; the nostrils are simple (not tubular); the muzzle does not appear swollen. The dorsal pelage is usually bright yellow-orange, but sometimes orangish-brown; the ventral pelage is similar, but paler. The wing and tail membranes are black; the skin over the wing bones is orange or yellow-orange. The wing membrane is attached to the side of the foot, at the base of the toes. *Myotis horsfieldii* are dark brown and are smaller (forearm 35–38 mm; weight 5.5–7 g). *Myotis macrotarsus* have pale gray fur and are somewhat smaller (forearm 44–50 mm; weight 11–14 g). *Kerivoula* have woolly fur; have funnel-shaped ears; and are smaller.

DISTRIBUTION: Philippines only, but widespread, probably including all of Luzon.

EVOLUTION AND ECOLOGY: A member of the subgenus *Chrysopteron*, most of which occur in Africa, but the closest relatives of *Myotis rufopictus* occur widely in

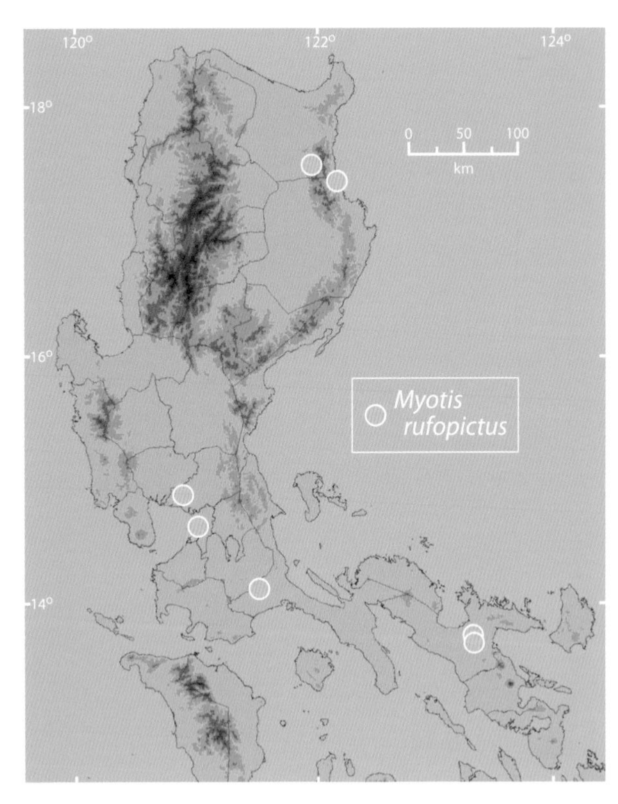

Southeast Asia. *Myotis rufopictus* have been recorded on Luzon in secondary and primary lowland and montane forest, from near sea level to about 1450 m elevation, including forest over limestone and agricultural fields near forest. A specimen dated 1887 was noted as having been captured in Manila. These bats are poorly known and probably difficult to capture using standard sampling techniques. The location of their roosts is unknown, but they have not been taken in caves.

STATUS: Seemingly uncommon, but poorly known. Ecological and phylogenetic studies are badly needed.

REFERENCES: Corbet and Hill, 1992; Csorba et al., 2014; Esselstyn et al., 2004; Heaney et al., 1999; Mudar and Allen, 1986; Ruedi et al., 2013; Sedlock et al., 2008, 2011.

Nyctalus plancyi
Chinese noctule

Gerbé, 1880. Bull. Soc. Zool. France, 5:71. Type locality: Peking, China.

DESCRIPTION: Total length 123–124 mm; tail 42–45 mm; hind foot 12–14 mm; ear 16–17 mm; forearm 48–49 mm; weight 19–20 g. The dorsal fur is rich chocolate brown, dense, and sleek, with a deep reddish tint; the ventral fur is dense but shorter, paler brown at the tips and dark brown at the base. The dorsal pelage extends about 1 cm onto the tail membrane. A conspicuous band of hair extends on the ventral side of the wing membrane to the base of the wrist. The muzzle, face, and ears are pigmented dark brown. The tragus is short and blunt, and curved anteriorly for its upper half. The anteriormost portion of the ear above the eye extends beyond the eye. *Scotophilus kuhlii* are similar in size but have short, reddish-brown dorsal fur; rather thin ventral fur that is pale tan or yellow, and pale at the base; only very sparse hair on the ventral surface of the wing; and an ear that does not extend anterior to the eye. *Harpiocephalus harpia* are similar in size overall but have a shorter forearm (44–49 mm); have nostrils that are somewhat elongated as tubes; and have fur that covers the entire dorsal surface of the tail membrane.

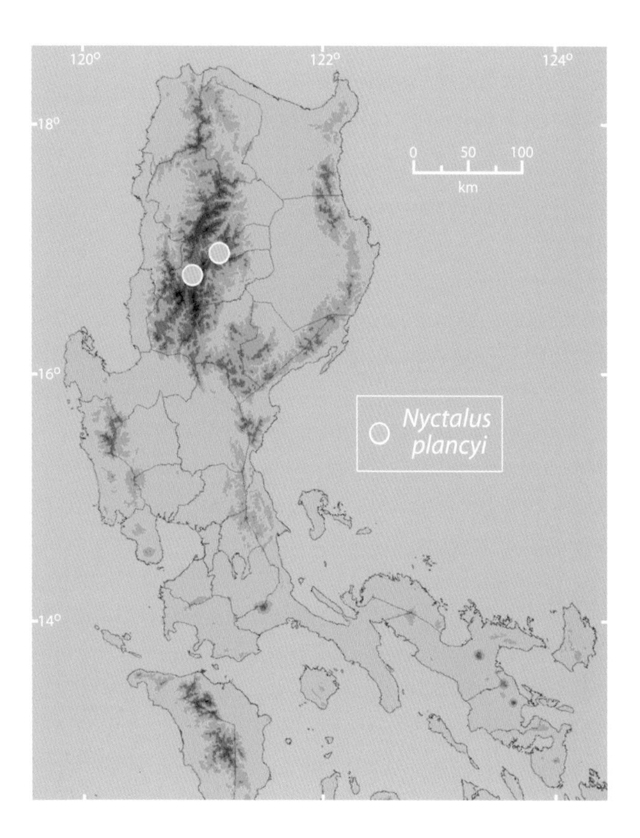

DISTRIBUTION: Eastern and southern China, Taiwan, and the Philippines; records from the Central Cordillera of Luzon.

EVOLUTION AND ECOLOGY: *Nyctalus* is most closely related to *Pipistrellus*, with other species in the genus occurring from Western Europe and the Azore Islands to India, Afghanistan, Burma [= Myanmar], Vietnam, and China. We recently captured three specimens (the first that have come from the Philippines) in the Central Cordillera, from 1730 m to 2310 m elevation in both primary and heavily disturbed mossy forest and pine forest. All were taken in open habitat that had either anthropogenic (e.g., from agricultural fields and human-generated fires) or natural disturbance. *Otopteropus cartilagonodus*, *Ptenochirus jagori*, and *Falsistrellus petersi* were also taken at or near these localities. In China, *N. plancyi* hibernate from November to April, and females typically produce twins after 50–60 days of gestation.

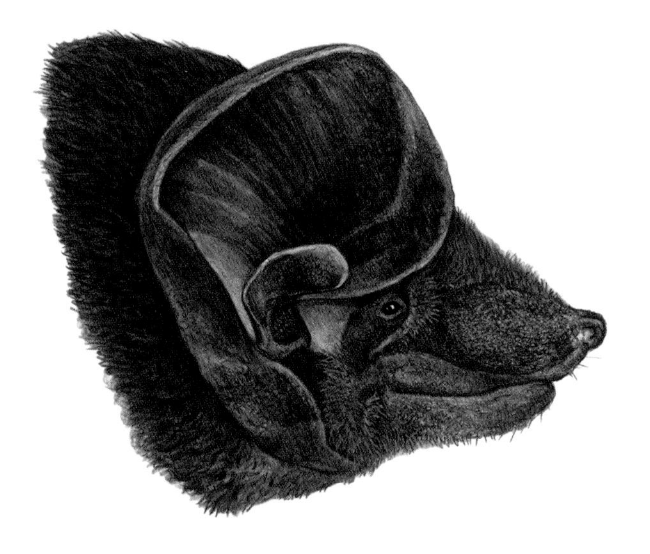

STATUS: Poorly known in the Philippines, but widespread and common elsewhere within their range. They may be common at high elevations in the Central Cordillera. Further research is needed.

REFERENCES: Corbet and Hill, 1992; Heaney et al., 2012; Simmons, 2005; Smith and Xie, 2008.

Philetor brachypterus
narrow-winged pipistrelle

(Temminck, 1840). Monogr. Mamm., 2:215. Type locality: Padang District, Sumatra, [Indonesia].

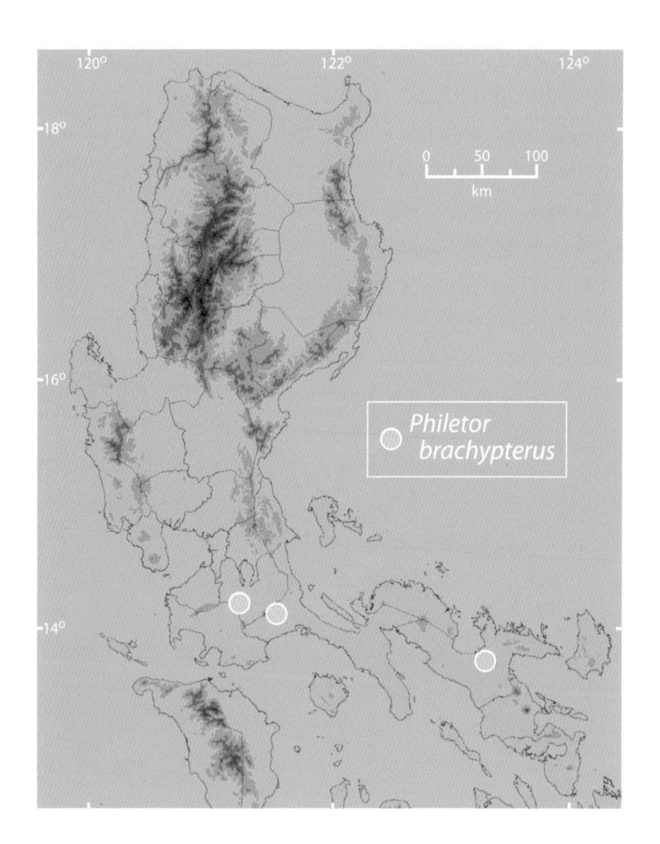

DESCRIPTION: Total length 98–113 mm; tail 27–35 mm; hind foot 8–11 mm; ear 13–15 mm; forearm 33–38 mm; weight 11–15 g. A small, dark brown bat. The fur is short and silky; the muzzle is broad and appears swollen. The ears are broad but low; the tragus is short, blunt, and broadest at the tip. The body is stout, but the wings are noticeably narrow. The head is rather flat, with no noticeable forehead. Both *Tylonycteris* and *Pipistrellus* also have flat heads and a tragus that is rather short and blunt, but *Tylonycteris* are smaller (forearm 27 mm or less), and have a fleshy pad at the base of the thumb. *Pipistrellus javanicus* have a similar forearm length (32–37 mm) and tail length (28–38 mm), but their total length is less (71–91 mm); the ear is shorter (9–14 mm); and they weigh less (4.1–8 g). *Falsistrellus petersi* have a forearm with a similar length (38–40 mm); are shorter in total length (88–92 mm); and have long, dark fur, with pale tips. All *Myotis* have a longer, more slender tragus that comes to a bluntly pointed tip; a distinct forehead; and a relatively narrow (not swollen) muzzle.

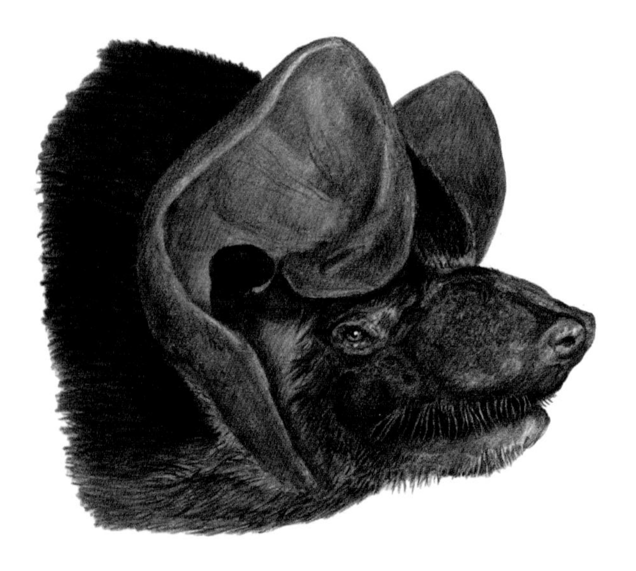

DISTRIBUTION: Documented from Nepal to New Guinea and the Bismarck Archipelago; records from throughout most of the Philippines, but recorded only from the southern portions of Luzon.

EVOLUTION AND ECOLOGY: The genus *Philetor* is closely related to *Tylonycteris*. Only a single species of *Philetor* is currently recognized, but the genus is poorly known and requires further study. On Luzon, *P. brachypterus* have been documented in mixed cultivated and second growth areas and in primary and disturbed lowland forest, from sea level to about 620 m elevation, and up to 1450 m elevation elsewhere in the Philippines. They appear to forage primarily in relatively open areas and typically fly at high speed. Females give birth to a single young. Most have been captured in tunnel traps, but they are occasionally captured in mist nets. The echolocation call of this species from Mt. Makiling, Laguna Province, had a mean frequency of 30.6 kHz.

STATUS: Apparently widespread and moderately common, but poorly known.

REFERENCES: Heaney et al., 1999, 2006b, 2012; Kock, 1981; Rickart et al., 1993; Sedlock, 2001; Sedlock et al., 2008, 2011.

Pipistrellus javanicus
Javan pipistrelle

(Gray, 1838). Mag. Zool. Bot., 2:498. Type locality: Java, [Indonesia].

DESCRIPTION: Two species are currently lumped under this name, with a slowly emerging picture of how to distinguish them. The larger (referred to in some of our publications as "A"): total length 83–91 mm; tail 32–38 mm; hind foot 8–10 mm; ear 11–14 mm; forearm 32–37 mm; weight 6.3–8.0 g. The smaller (referred to in some of our publications as "B"): total length 71–81 mm; tail 28–34 mm; hind foot 6–8 mm; ear 9–12 mm; forearm 32–34 mm; weight 4.1–6.4 g. Both morphs are small brown bats with a flat head (with no detectable forehead); a broad muzzle that appears swollen; and short ears, with a short tragus that is only slightly curved and is rounded at the tip. The fur is dark brown, dark at the base and slightly paler ventrally. Other small brown bats on Luzon have a flat head and short tragus, but some of their other characteristics are different. *Philetor brachypterus* have similar forearm and tail lengths, but the total length is greater (98–113 mm); the ear is longer (13–15 mm); and they are heavier (11–15 g). *Falsistrellus petersi* are similar in size but are somewhat longer (total length 88–92 mm); the forearm is longer (38–40 mm); and the dorsal fur has pale tips that give it a frosted appearance. *Pipistrellus tenuis* are smaller (total length 71–77 mm; forearm 30–32 mm; weight 3.1–3.8 g). *Tylonycteris* also are smaller (forearm 27 mm or less) and have a fleshy pad at the base of the thumb. *Myotis* have proportionately longer ears with a long tragus, a distinct forehead, and a muzzle that does not appear swollen.

DISTRIBUTION: Korea to Java and the Philippines; apparently occur throughout the Philippines, but uncertain because of taxonomic uncertainty. Recorded throughout Luzon.

EVOLUTION AND ECOLOGY: *Pipistrellus* contains over 30 species that occur over very broad portions of the Earth. *Pipistrellus* are superficially similar to *Falsistrellus* and *Philetor*, but they are not closely related; instead, *Pipistrellus* is more closely related to *Nyctalus* and *Glischropus*. Two cryptic species are present in the Philippines, but they have been lumped as *P. javanicus*; detailed study is needed. They are common in urban areas (including parks and other green space within Metro Manila) and in agricultural areas. They are also common in secondary and primary lowland and montane forest, occasionally extending into mossy forest, with a total elevational range from sea level to about 1750 m. They have been found roosting in hollow trees. Females give birth to a single young. The echolocation calls of the two morphs in Laguna Province differ slightly, with mean call frequencies of 43.7 kHz and

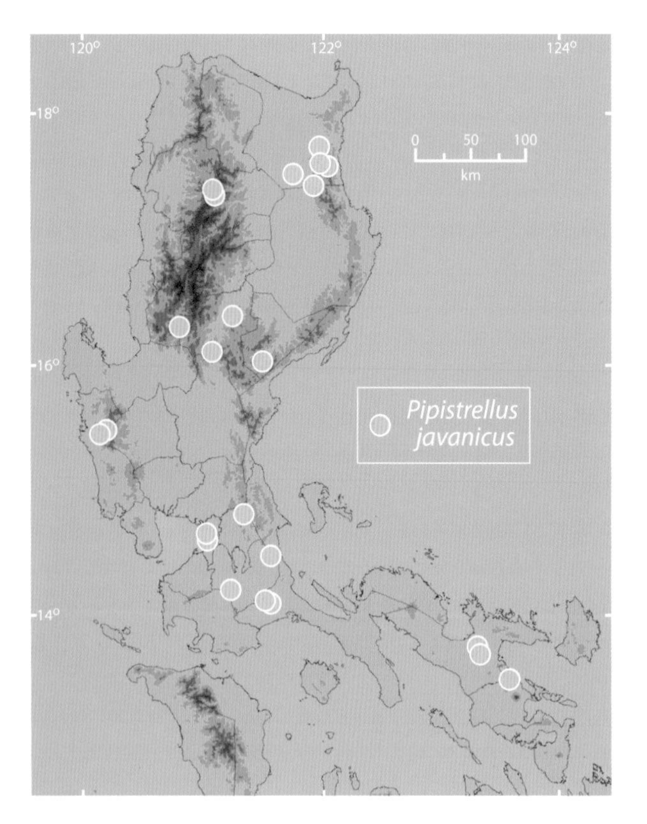

41.9 kHz, respectively. Specimens from Mt. Isarog have karyotypes with $2n = 38$, FN = 48.

STATUS: Common and widespread in eastern Asia, including heavily disturbed habitats.

REFERENCES: Alviola et al., 2011; Balete et al., 2013a; Duya et al., 2007; Esselstyn et al., 2004; Heaney et al., 1999, 2006a, 2006b; Rickart et al., 1999; Sanborn, 1952; Sedlock, 2001; Sedlock et al., 2008, 2011; Taylor, 1934.

Pipistrellus tenuis
least pipistrelle

(Temminck, 1840). Monogr. Mamm., 2:229. Type locality: Sumatra, [Indonesia].

DESCRIPTION: Total length 71–77 mm; tail 26–33 mm; hind foot 5–7 mm; ear 10–12 mm; forearm 30–32 mm; weight 3.1–3.8 g. A small, dark brown bat, slightly paler ventrally. As in all *Pipistrellus*, the muzzle is broad and appears somewhat swollen; the ear and tragus are short; and the skull is rather flat, without a forehead. *Falsistrellus*, *Philetor*, and *Pipistrellus javanicus* are all larger; *Tylonycteris* are smaller. *Myotis muricola* is similar in size, but has a longer ear (12–14 mm), with a longer, tapering tragus; has a pronounced forehead; and has a muzzle that is relatively narrow and does not appear swollen.

DISTRIBUTION: Thailand to Australia; scattered records from the Philippines, with all Luzon records from the Central Cordillera.

EVOLUTION AND ECOLOGY: As currently defined, *Pipistrellus tenuis* may be a species-group rather than a single species; detailed study is needed. On Luzon, this species has been recorded from ca. 1450 m to 2650 m elevation in the Central Cordillera; elsewhere in the Philippines, there are records from as low as 1250 m. They occur in primary and secondary montane and mossy forest, and often are the most common bats in

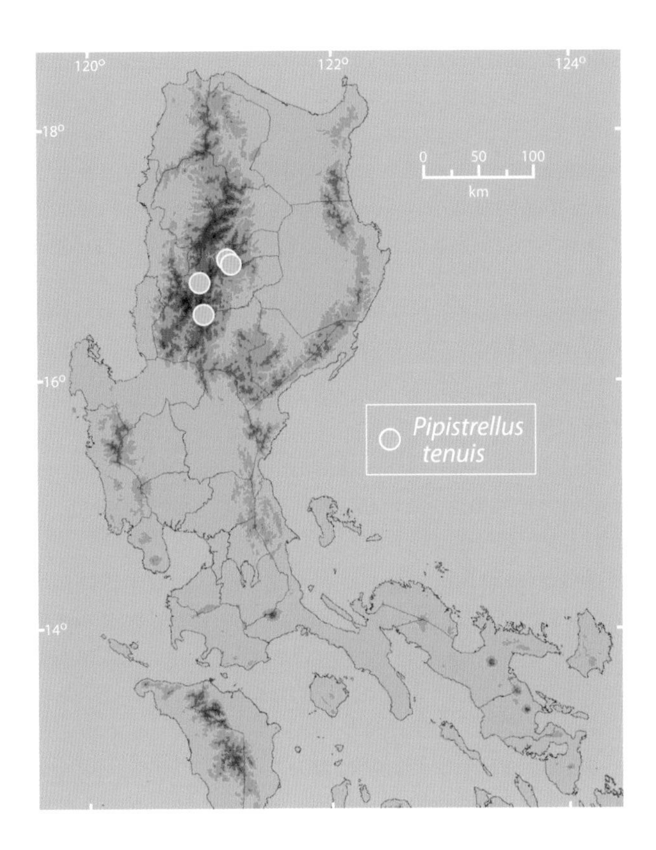

forest at these elevations. Beginning at early dusk, they are often seen foraging below and within the canopy and above shrubby areas and along trails. Females give birth to a single young.

STATUS: Widespread and moderately common.

REFERENCES: Sedlock et al., 2008; Taylor, 1934; unpublished field notes and specimens in FMNH.

Scotophilus kuhlii
lesser Asian house bat

Leach, 1822. Trans. Linn. Soc. Lond., 13:71. Type locality: "India."

DESCRIPTION: Total length 115–127 mm; tail 42–49 mm; hind foot 9–12 mm; ear 14–16 mm; forearm 49–52 mm; weight 17–22 g. A robustly built bat. The dorsal fur is fairly short, glossy, and reddish-brown; the tail membrane is nearly hairless, as the dorsal fur ends abruptly at the base of the tail; there is only sparse hair on the ventral surface of the wing extending from the body to the wrist. The ventral fur is pale tan or yellowish, with hairs that are pale at the base. The tragus is relatively long and slender, curving forward. *Nyctalus plancyi* are similar in size, but the dorsal fur is a rich chocolate brown; the ventral fur is dark at the base; and a conspicuous band of hair extends from the ventral side of the wing membrane to the base of the wrist. Most other insectivorous bats are smaller; none that are similarly large have short, reddish-brown fur and a long, slender, curving tragus. *Taphozous melanopogon* also commonly roost in urban and suburban buildings, but those roosting individuals are easily distinguished, even at a distance (see the species account for that species).

DISTRIBUTION: Pakistan to Taiwan and Bali; found throughout the Philippines, including all of Luzon.

EVOLUTION AND ECOLOGY: The genus *Scotophilus* occurs from Africa to Southeast Asia; related genera are found from Central and South America to Africa and Australia. *Scotophilus kuhlii* are abundant, common

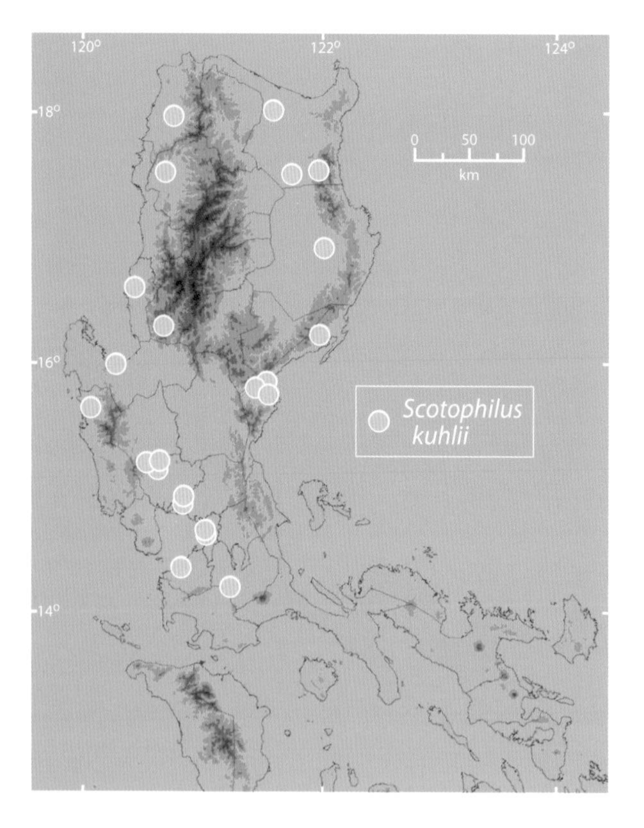

bats in the Philippines, foraging in urban and agricultural areas and in secondary forest near human habitations, from sea level to about 600 m elevation. In the lowlands, they are the most common bats roosting in buildings (especially the attics of schools, churches, and government buildings), often in tightly packed colonies numbering in the thousands. They begin flying at dusk and are often seen flying high and fast as they swoop to catch insects. They occasionally roost in "tents" formed from modified palm fronds, which may originally have been created by *Cynopterus brachyotis*. They undoubtedly have a substantial impact on insect populations in urban and agricultural areas, though few data are available. Females usually give birth to twins once each year. On Negros, most births occur in April. The echolocation call of this species in Laguna Province had a mean call frequency of 20.6 kHz. Specimens from Leyte and Negros had karyotypes with $2n = 36$ and $FN = 48$.

STATUS: Widespread and abundant over much of their range.

REFERENCES: Alcala, 1976; Catibog-Sinha, 1987; Esselstyn et al., 2004; Hoogstraal, 1951; Rickart et al., 1989b, 1993, 1999; Sedlock, 2001; Taylor, 1934.

Tylonycteris pachypus lesser flat-headed bat, lesser bamboo bat

(Temminck, 1840). Monogr. Mamm., 2:217. Type locality: Batam, West Java, [Indonesia].

DESCRIPTION: Total length 61–63 mm; tail 25–27 mm; hind foot 5–6 mm; ear 8–9 mm; forearm 24–26 mm; weight 2.7–3.7 g. The smallest bat species in the Philippines, and one of the smallest in the world. They have glossy reddish-brown fur; a broad and flat head; and a swollen pad at the base of the thumb. *Tylonycteris robustula* are larger (forearm 25–27 mm; weight 5.5–6.0 g) and have darker brown fur. *Pipistrellus tenuis* have dark brown fur; a longer ear (10–12 mm); and a longer forearm (30–32 mm).

DISTRIBUTION: As currently defined, occur from India to the Philippines and the Lesser Sunda Islands; probably present throughout the Philippines, including all

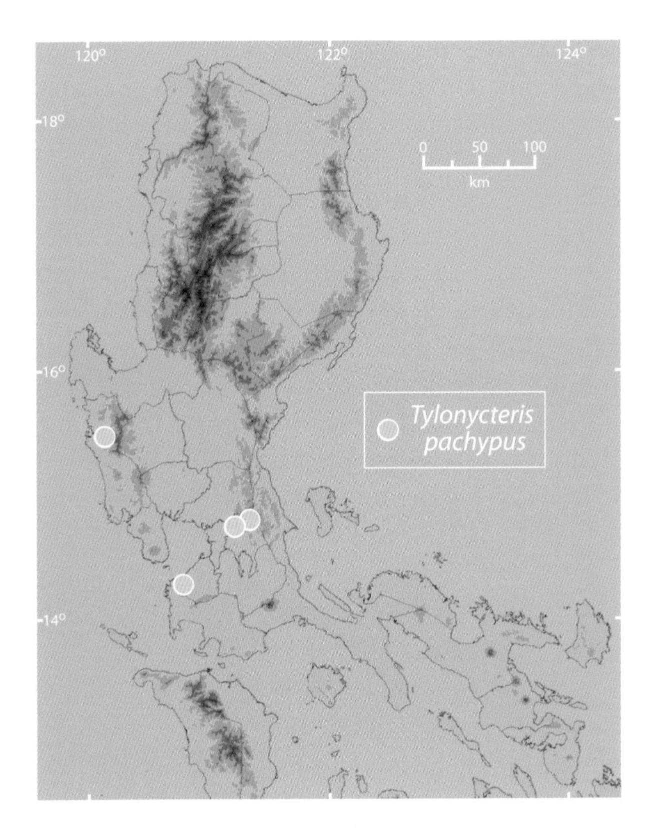

of lowland Luzon, though current records are limited to central Luzon.

EVOLUTION AND ECOLOGY: *Tylonycteris* are closely related to *Philetor*. Only two species of *Tylonycteris* have been recognized traditionally, but recent studies indicate that many more species are present; further study is needed. Philippine specimens are the smallest in this species, and the name *T. pachypus meyeri* is often used for them. These tiny bats are probably common in lowland agricultural areas, from sea level to about 500 m elevation. They roost in small groups inside hollow stems of bamboo, usually in areas with extensive bamboo thickets. Because they are tiny and have slow but agile flight, they are difficult to capture in mist nets and harp traps and thus are poorly known. They often become active at early dusk and are easily visible as they flutter about, but they often are mistaken for moths. In Malaysia, they are said to favor small, live bamboo stems, which they enter through slits as thin as 3.6 mm that have been created by beetles; *T. robustula* are thought to favor larger, dead stems. *Tylonycteris pachypus* roost in groups of 1–16 individuals, and occasionally up to 40. Females usually give birth to twins once each year. Termites are a favored food.

STATUS: Widespread, probably moderately common.

REFERENCES: Balete et al., 2013a; Esselstyn et al., 2004; Francis, 2008; Heaney and Alcala, 1986; Heaney and Balete, pers. obs.; Medway, 1969; Payne et al., 1985.

Tylonycteris robustula
greater bamboo bat

Thomas, 1915. Ann. Mag. Nat. Hist., ser. 8, 15:227.
Type locality: Upper Sarawak, Borneo, [Malaysia].

DESCRIPTION: Total length 68–74 mm; tail 28–31 mm; hind foot 6–7 mm; ear 10–11 mm; forearm 25–27 mm; weight 5.0–6.6 g. One of the smallest bat species in the Philippines. They have glossy dark brown fur; a broad and flat head; and a dark pad at the base of the thumb. *Tylonycteris pachypus* are smaller (forearm 24–26 mm; weight 2.7–3.7 g) and have darker brown fur. *Pipistrellus tenuis* are larger (forearm 30–32 mm). *Myotis muricola* also are larger (forearm 29–33 mm) and have longer ears (12–14 mm); a long, slender tragus; and a high forehead.

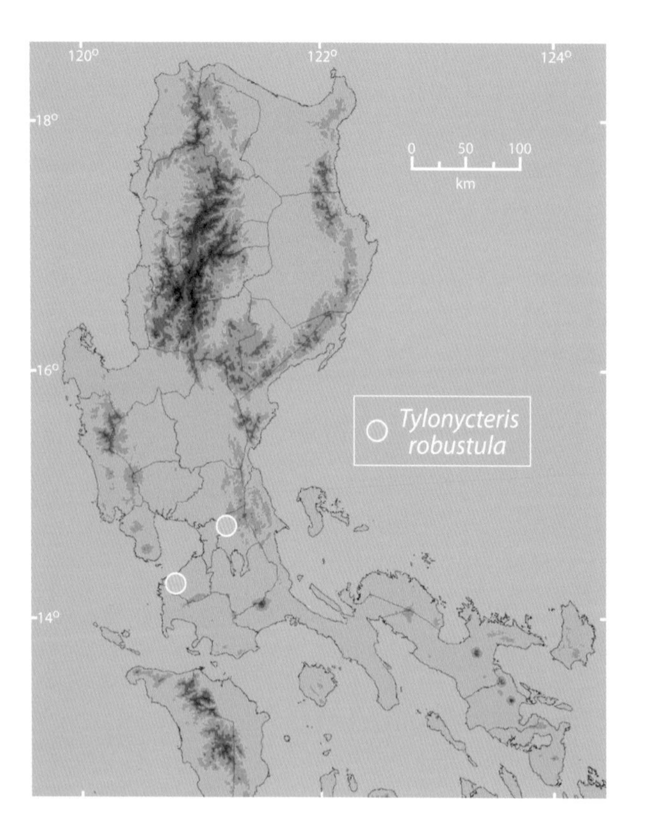

DISTRIBUTION: Southern China to the Lesser Sunda Islands and the Philippines; recorded on Luzon only from a few localities, but probably more widespread.

EVOLUTION AND ECOLOGY: Philippine specimens are smaller than those referred to this species from other areas; further study is needed. These tiny bats occur in disturbed lowland agricultural areas, from sea level to about 500 m elevation. As with *Tylonycteris pachypus*, they live inside standing bamboo, and most specimens have been captured in or near bamboo thickets. In Malaysia, they have been found in groups of one to seven individuals. Females usually give birth to twins once each year.

STATUS: Widespread and probably fairly common, but taxonomic and targeted field studies are needed.

REFERENCES: Francis, 2008; Heaney and Alcala, 1986; Medway, 1969.

SECTION 8

Family Molossidae: Free-Tailed Bats

This is one of the most diverse families of bats, with 16 genera and over 100 species currently recognized. These species occur over most portions of the tropics and subtropics of the world. Only 2 species of this family occur on Luzon, however, and only 2 additional species occur elsewhere in the Philippines. *Chaerophon plicatus* is widespread in South Asia and Southeast Asia; the other, *Otomops* sp., may be a Philippine endemic, though basic taxonomic and phylogenetic studies are needed to determine its status.

Molossids are morphologically distinctive bats in two ways. First, their tail is always moderately long and extends well past the end of the tail membrane, hence the name "free-tailed." Second, the ears meet and are joined to each other over the top of the head. Most also have rather wrinkled lips and complex ears that lack a tragus; no other bats in the Philippines have this unique set of features. Their nostrils are rather simple, with no more than a hint of extended tubes; no noseleaf is present. Their wings are proportionately long and narrow and are associated with their habit of flying fast, usually high above the vegetation. The two species on Luzon are medium-sized bats, with forearm length from about 40 mm to 60 mm, and weights of about 13 g to 23 g.

The two species of molossids on Luzon have greatly differing behaviors and ecology. *Chaerophon plicatus* typically live in vast colonies in lowland caves, in numbers often reaching into the millions. They stream out of the caves at dusk, often clearly visible from a distance as a dark strand reaching from the ground into the sky, shifting fluidly as the bats adjust to winds. They often live near agricultural areas, and their impact on crop pests must be great, though little is known

Chaerophon plicatus.

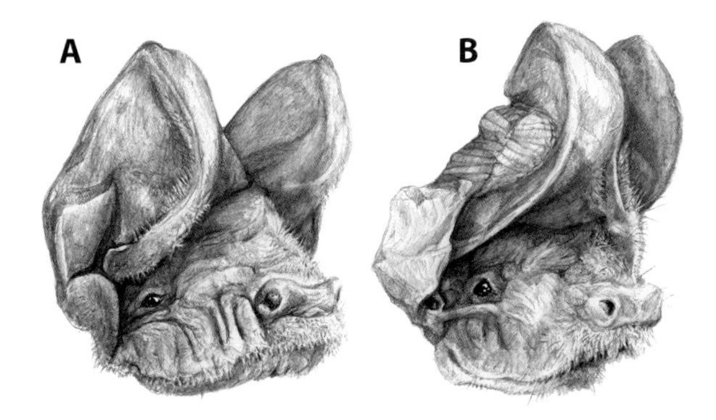

A: *Chaerophon plicatus*. B: *Otomops* sp.

with certainty. The genus *Chaerophon* is diverse and widespread, and because they roost in caves (and can be captured easily), their phylogeny and taxonomy have been well studied.

The other Luzon free-tailed bat, *Otomops* sp., was discovered so recently that phylogenetic studies have not yet been undertaken. Although they appear different from other members of the genus, it is not yet clear if they should be recognized as a distinct species. They probably roost in trees (perhaps in hollow spaces) and are difficult to capture, so they are poorly known—only two specimens have ever been captured on Luzon, to the best of our knowledge.

REFERENCES: Simmons, 2005; Taylor, 1934.

Chaerephon plicatus
wrinkle-lipped bat

(Buchanan, 1800). Trans. Linn. Soc. Lond., 5:261. Type locality: Puttahaut, Bengal, India.

DESCRIPTION: Total length 92–99 mm; tail 33–38 mm; hind foot 8–10 mm; ear 18–20 mm; forearm 41–43 mm; weight 13–16.5 g. Like all molossids, the tail of this species projects beyond the posterior edge of the interfemoral membrane for over half its length; the muzzle lacks a noseleaf and projects well beyond the lower jaw. The ears are very prominent, joining over the top of the head. The upper lip is heavily wrinkled. The fur is short, dense, and dark brown dorsally, and paler ventrally; short, stiff hairs are present on the first toe. *Otomops* sp. is the only other bat species on Luzon with similar ears, but they are larger (forearm 60–61 mm) and have dark pelage, with silver highlights over the shoulders and upper back. Bats in other families do not have a tail that project far beyond the posterior edge of the tail membrane, and only *Megaderma spasma* has ears that are joined over the top of the head.

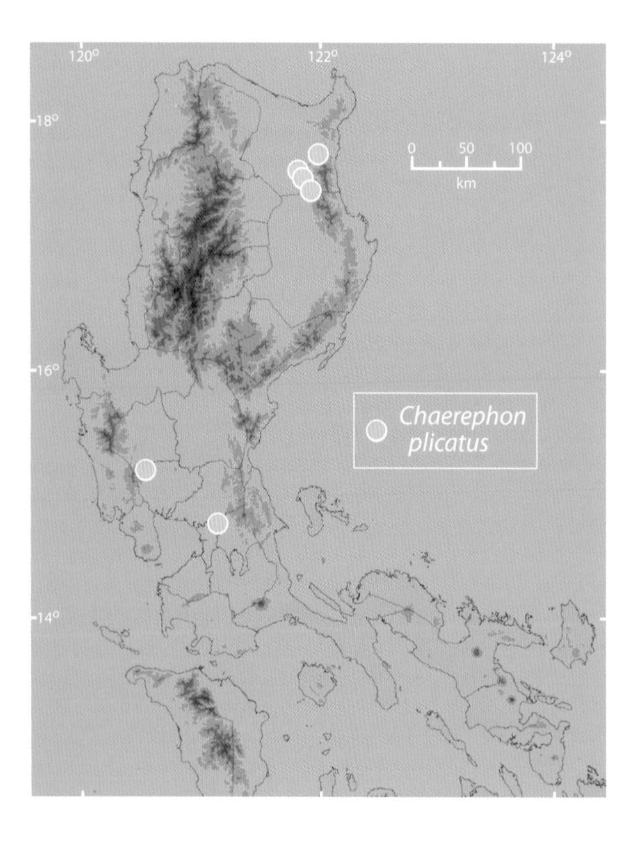

DISTRIBUTION: Occur from India to Bali, Hainan, and Luzon; Luzon records are from areas of extensive limestone in central and northern Luzon.

EVOLUTION AND ECOLOGY: Recorded on Luzon only in or near caves, from sea level to about 400 m elevation, in secondary and primary lowland forest. Their colonies in caves often number in the tens of thousands to millions of individuals; they leave the caves early in the evening in conspicuous, streaming flocks. They fly rapidly and high above any vegetation, often ascending to high altitudes. These bats probably have a large, beneficial impact on crop pests, but little information is available. They were formerly present in a vast colony in the Montalban Caves, Rizal Province, but appear to have been extirpated some time after 1956, when they were still abundant. A single specimen was taken on Clark Air Base in 1956, which implies a colony in that vicinity, but no further information is available. The only population known to remain on Luzon currently is in the Callao Caves, in the vicinity of Peñablanca, Cagayan Province. A video of this colony taken in November 2014 showed hundreds of thousands, and perhaps more than a million, wrinkle-lipped bats departing a cave at sunset, in a long stream that, from a distance, had the appearance of smoke.

STATUS: Widespread in Southeast Asia; in the Philippines, probably vulnerable and declining, due to overhunting and the mining of limestone. A population formerly in a large cave on Leyte was destroyed prior to 1984. Since 1980, reported in the Philippines only in Cagayan Province, Luzon, and in Argao and Carmen Municipalities, Cebu.

REFERENCES: Danielsen et al., 1994; Duya et al., 2007; Lawrence, 1939; Paguntalan, pers. comm.; Rickart et al., 1993; Taylor, 1934.

Otomops sp.
Philippine mastiff-bat

DESCRIPTION: Total length 132–135 mm; tail 40–46 mm; hind foot 11–13 mm; ear 30–34 mm; forearm 60–61 mm; weight 19–23 g. The fur over the head and neck is dense, soft, and chocolate brown; at the shoulders,

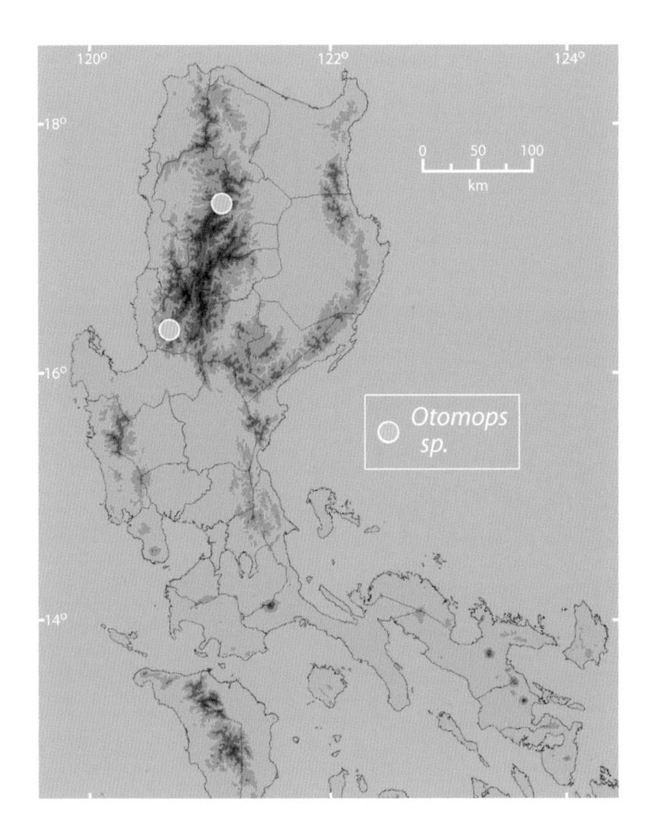

there is a band of pale cream hairs, with the fur posterior to the band a paler brown, which then grades posteriorly to rich chocolate brown. The band of pale fur extends around the upper chest, separating the darker fur of the throat from the slightly paler fur of the chest and abdomen. A band of pale fur also extends along the margin of the wings where they meet the furred torso. The ears are large and complex; they meet over the top of the head. The tail is long, with about half extending beyond the end of the tail membrane. *Chaerophon plicatus* are smaller (forearm 41–43 mm). The only other bat on Luzon with ears that are joined over the head is *Megaderma spasma*.

DISTRIBUTION: Related to *Otomops formosus*, which occurs on Java; Philippine records are from Luzon (in the Central Cordillera) and Mindanao.

EVOLUTION AND ECOLOGY: The genus *Otomops* occurs from Africa and Madagascar to South Asia and New Guinea. Our specimens were obtained recently and are undergoing study; they appear to represent an unnamed species. The Luzon specimens were taken from 1450 m to 1700 m elevation: one in a small grove of large pine trees on a ridgetop surrounded by mature montane forest, and one in open pine forest in a semi-urban area. Their long, relatively narrow wings imply that they fly fast, and many molossids fly high above the ground and vegetation, which may make them difficult to capture.

STATUS: Poorly known and difficult to capture, but apparently widespread within the Philippines.

REFERENCES: Heaney et al., 2003, 2006b; unpublished data and specimens in FMNH.

Glossary

ADAPTIVE RADIATION: The evolutionary process by which an ancestral species gives rise to an array of descendant species exhibiting great ecological, morphological, and/or behavioral diversity.

ALLELOPATHY: The chemical inhibition of one plant (or other organism) by another, through the release of toxic chemicals that inhibit germination or growth.

ALLOPATRIC: Occurring in separate, non-overlapping geographic areas.

ALLOPATRIC SPECIATION: The differentiation of geographically isolated populations into distinct species.

ASH (VOLCANIC): Tiny particles of solidified magma, usually with a glassy texture, ejected from volcanoes during violent eruptions and formed as escaping gases blast the magma into a fine particulate spray that solidifies on contact with the atmosphere (see **TUFF**).

ASSEMBLAGE: The set of plant or animal species found in a specific geographical unit or area.

BIODIVERSITY: The variability of life from all sources, including within and among species and ecosystems.

BIODIVERSITY HOTSPOT: May mean a geographic area with high levels of species richness and/or endemism. Also used to refer to areas that combine high biodiversity with high threat to that biodiversity (e.g., due to habitat loss).

BIOTA: All species of plants, animals, and microbes inhabiting a specified region.

CALDERA: A large, circular depression or crater, often miles across, formed by a volcanic eruption followed by collapse of the ground surface.

CLADE: Any evolutionary branch in phylogeny, based on genealogical relationships.

CLIMAX COMMUNITY: The final stage of ecological succession, where the community has reached equilibrium and the species composition is relatively stable unless disturbed.

COLONIZATION: The arrival of a species on an island (or other defined area) where that species was not already present, especially when the establishment of a breeding population subsequently occurs.

COMPETITION: Negative, detrimental interaction between organisms, caused by their need for a common resource. Competition may occur between individuals of the same species (intraspecific competition) or of different species (interspecific competition).

CONSERVATION: Human intervention, with the goal of maintaining valued biodiversity (including genetic variation, species, ecosystems, landscapes, and natural resources).

CONTINENTAL or **LAND-BRIDGE ISLANDS:** Emergent fragments of the continental shelf, separated from the continents by shallow waters. This separation is often recent, as a result of the postglacial rise of sea level, which isolated the species on the island from conspecific mainland populations.

CONTINENTAL SHELF: The area of shallow sea floor (less than 200 m in depth) adjacent to continents, underlain by continental crust and effectively a submerged extension of the continents.

CRUST: The outermost layer of the Earth's mantle, consisting of either continental crust (20–40 miles thick and mostly of granitic composition) or oceanic crust (usually up to 7 miles thick and mostly of basaltic composition).

DISJUNCTION: A distribution pattern exhibited by taxa with populations that are geographically isolated.

DISPERSAL: The movement of organisms away from an existing area of occurrence to a different, previously unoccupied area.

DISTURBANCE: Any relatively discrete event in time that removes organisms and opens up spaces that can be colonized by individuals of the same or different species.

ECOTOURISM: Tourism based on an interest in observing nature, preferably with minimal ecological impact.

EL NIÑO–SOUTHERN OSCILLATION (ENSO): Climatic anomalies yielding significant changes in rainfall, temperature, humidity, and storm patterns over much of the tropics and subtropics, due to movement of warm surface waters from the western part of the equatorial Pacific basin to the eastern part, disrupting the upwelling along the western coast of South America.

EMPTY NICHE: In an island context, refers to the lack of representation of a particular mainland ecological guild, providing opportunities for colonizing taxa to exploit the vacancy.

ENDEMIC: A taxon restricted to a given specified area, whether a mountain top, island, archipelago, continent, or zoogeographic region.

EPIPHYTE: A plant that grows on another plant and independently derives nutrients and moisture, without being parasitic.

EQUILIBRIUM: A condition of balance between opposing forces, such as between rates of colonization and extinction.

EXTINCTION: The loss of all populations of a species from an island or other specified area, or the global loss of a species.

EXTIRPATION: The loss of a local population, not implying global loss of the species.

FAULT: A planar of gently curved fracture in the Earth's crust where the rocks on either side have shifted measurably. The energy released when a fault shifts produces earthquakes.

FOLIVORE: An herbivorous animal that primarily feeds on plant leaves.

FOREARC: The region at a subducting plate boundary between an oceanic trench and its associated volcanic arc.

FRUGIVOROUS: Feeding mainly on fruits.

GENE FLOW: The movement of genes within a population or between populations, caused by the dispersal of gametes or offspring.

GENERALIST SPECIES: A species that can live under a variety of environmental conditions and can utilize a wide range of resources.

GRANIVOROUS: Feeding mainly on seeds.

HABITAT FRAGMENTATION: Process by which a large, continuous area of habitat is diminished in area and divided into numerous isolated habitat fragments by either natural or anthropogenic forces.

ICE AGES: An informal term for the glaciated periods of the Pleistocene epoch (2.6 million–10,000 years ago), when the Earth was periodically cooler than usual and large areas of the continents were periodically covered by ice sheets.

IGNEOUS ROCK: Any rock formed by the solidification of molten rock, either underground (intrusive rock) or on the Earth's surface (extrusive rock).

IMMIGRATION: In island biology, refers to the process of arrival of a propagule on an island not occupied by that species (see **COLONIZATION**).

INSECTIVOROUS: Feeding mainly on insects and other arthropods.

INVASION: The spread of a naturalized exotic species into natural or semi-natural habitats, hence "invasive species."

KAINGIN: A traditional form of subsistence agriculture that involves clearing and burning forest vegetation, utilizing the resulting ash for fertilizer, and shifting to another area when the land loses fertility.

KARST: Landscape underlain by limestone that has been eroded by dissolution, producing ridges, towers, fissures, sinkholes, and other characteristic landforms.

LAHAR: A destructive landslide on the slope of a volcano, consisting of pyroclastic debris and water.

LATE SUCCESSIONAL SPECIES: Species that occur primarily in or are dominant in the late stages of succession (e.g., mature or nearly mature forest).

LITHOSPHERE: The solid portion of the Earth, consisting of the crust and uppermost portion of the mantle.

MAGMA: Molten rock material formed within the Earth; becomes igneous rock upon cooling and solidification. Magma that erupts onto the Earth's surface is called lava.

MANTLE: The 1800-mile-thick region between the Earth's crust and core, forming more than three-quarters of the volume of the Earth.

MIDOCEAN RIDGE: A long, seismically active submarine ridge system situated in the middle of an ocean basin, marking the site of the upwelling of magma associated with seafloor spreading.

MOLECULAR CLOCK: The extent of molecular divergence (attributable to random mutations), based on the idea that proteins and DNA evolve at a more-or-less constant rate, that is used as a metric for the timing of events within lineage development. Greater confidence can be placed in such molecular clocks if they can be independently calibrated, usually using fossils.

MONOPHYLETIC GROUP: A group of taxa that share a common ancestor, including all descendants of that ancestor; also referred to as a **CLADE**.

NATURAL SELECTION: The process of eliminating from a population, through differential survival and reproduction, those individuals with inferior fitness.

NICHE: The total requirements of a population or species for resources and physical conditions.

NON-NATIVE SPECIES: A foreign (exotic, or alien) species moved by humans to a region outside its natural geographical range or environment.

OCEANIC ISLANDS: Islands (mostly with basaltic foundations), originating from submarine volcanic activity, that have never been connected to the continents. They are initially populated by species that have dispersed to the islands from elsewhere and may be subsequently enriched by speciation.

OCEANIC TRENCH: A deep, linear depression on the ocean floor formed by subduction, where an oceanic plate bends down underneath an adjacent plate to plunge into the Earth's interior (see **SUBDUCTION ZONE**).

OLD-GROWTH FOREST: Mature forest at or near the stable climax state of succession, also called primary forest.

OMNIVOROUS: Feeding on a wide variety of both animal and plant matter.

OPHIOLITE: Rock of the deep ocean floor now found on land, usually emplaced during subduction or continental collision.

PHYLOGENETIC TREE: A line diagram, derived from a phylogenetic analysis, showing the hypothesized branching sequence of a monophyletic taxon and using shared derived character states to determine when each branch diverged.

PHYLOGENY: The evolutionary relationships between an ancestor and all its known descendants.

PIONEER SPECIES: Species that occur during the earliest stages of succession following a major habitat disturbance (contrast with **LATE SUCCESSIONAL SPECIES**).

PLATE TECTONICS: The theory, confirmed by abundant evidence, that the Earth's outer rocky shell is broken up into several dozen individual plates, 50–100 miles thick, that move and interact to produce earthquakes, volcanoes, and most of the major geographical features of the planet.

PLEISTOCENE: The first epoch of the Quaternary period, formally taken as beginning about 2.6 Ma, which ended about 11,500 years BP with the transition to the present Holocene epoch.

PLEISTOCENE ICE-AGE ISLANDS: Islands that occurred during glacial periods of lowered sea level, less fragmented and of larger extent than those that exist currently.

PLIOCENE: The final epoch of the Tertiary period, spanning approximately 5.3–2.6 Ma.

QUATERNARY: The most recent period of the geological timescale, following the end of the Pliocene and continuing to the present day, and comprising the Pleistocene and Holocene epochs.

REGENERATION: Recovery of habitat following disturbance (see **SUCCESSION**).

SEAFLOOR SPREADING: The mechanism whereby new ocean floor is created at midocean ridges as two oceanic plates diverge and magma wells up to fill the gap.

SECONDARY FOREST: Young forest that has developed through natural succession in previously disturbed areas. It may also be used to describe heavily disturbed old-growth forest.

SEDIMENTARY ROCK: A rock formed from eroded pieces of preexisting rocks that were transported by wind, water, or ice and then deposited and cemented together. Also, any rock formed from particles of biological skeletons or shells, or chemically precipitated out of water.

SEISMIC BELT: A long, narrow region with a distinctly high frequency of earthquakes.

SISTER-SPECIES: A pair of species that are each other's closest relatives, sharing a more recent common ancestor than they do with other related species.

SPECIALIST SPECIES: A species that occurs within a specific range of environmental conditions and can utilize a limited range of resources. A species may be specialized in some respects and not in others.

SPECIATION: The process in which two or more contemporaneous species evolve from a single ancestral population.

STRATA: Layers of sedimentary rock originally laid down horizontally but may later be faulted, tilted, or folded by tectonic movements.

STRATOVOLCANO: The composite cone built by a volcano that emits both molten and solid materials (especially ash) and resulting in a steep-sided cone. Most Luzon volcanoes are stratovolcanoes.

SUBCLIMAX COMMUNITY: A community that is prevented from reaching the final climax stage of succession, due to repeated disturbances such as fire or severe storms.

SUBDUCTION: The process in plate tectonics by which lithosphere-bearing oceanic crust is destroyed by the movement of one plate beneath another to form a marine trench. The process leads to the heating and subsequent remelting of the lower plate, in turn generating volcanic activity, typically forming an arc of volcanic islands above the area of subduction.

SUBDUCTION ZONE: The region where a moving oceanic plate bends down beneath an adjacent plate to form an oceanic trench. Characterized by frequent earthquakes and volcanism in the adjacent volcanic arc (see **VOLCANIC ARC**).

SUBSIDENCE: The downward movement of an object relative to its surroundings. Subsidence of the lithosphere can be due to increased mass (e.g., increased ice, water, or rock loading) or the movement of the island away from midocean ridges and other areas that can support an anomalous mass.

SUCCESSION: Refers to directional change in plant or animal communities following either significant ecosystem disturbance or the creation of an entirely new land-surface area (see **DISTURBANCE** and **REGENERATION**).

SUSTAINABLE DEVELOPMENT: Development that meets the needs of the present human generation without compromising the ability of future generations to meet their own needs.

SUTURE: A zone of faults and fractures the marks the line along which two terranes have collided and joined (see **TERRANE**).

SYMPATRIC: Occupying the same geographical region.

SYMPATRIC SPECIATION: The differentiation of two reproductively isolated species, arising from one initial population within the same local area, in which much gene flow potentially could occur (see **ALLOPATRIC SPECIATION**).

SYNTOPIC: Occupying the same local habitat.

TERRANE: A block of Earth's crust, bounded by faults, whose geological history is distinct from adjacent crustal blocks, often because it has traveled from far away. Continents grow by the accretion of terranes to their edges (see **OPHIOLITE** and **SUTURE**).

TERTIARY: A period of geological time lasting from ca. 65 Ma to the beginning of the Quaternary period.

TUFF: A rock composed of pyroclastic particles, that is, particulate material blown into the air during a volcanic eruption (see **ASH** and **LAHAR**).

ULTRABASIC (also **ULTRAMAFIC**): Igneous or meta-igneous rocks (and their derived soils) that have a low silica content and are very rich (ultrabasic) or extremely rich (ultramafic) in magnesium and iron.

VAGILITY: The degree of ability to move actively from one place to another.

VERMIVOROUS: Feeding mainly on earthworms and other soft-bodied invertebrates.

VOLCANIC ARC: A line of active volcanoes that parallels an oceanic trench where subduction is taking place.

Literature Cited

Achmadi, A. S., J. A. Esselstyn, K. C. Rowe, I. Maryanto, and M. T. Abdullah. 2013. Phylogeny, diversity, and biogeography of Southeast Asian spiny rats (*Maxomys*). *Journal of Mammalogy* 94:1412–1423.

Alava, M. N. R., M. L. Dolar, S. Leatherwood, and C. J. Wood. 1993. Marine mammals of the Philippines. *Asia Life Sciences* 2:227–234.

Alcala, A. C. 1976. *Philippine Land Vertebrates*. Quezon City: New Day.

Alcala, A. C., and W. C. Brown. 1969. Notes on the food habits of three Philippine wild mammals. *Silliman Journal* 16:91–94.

Almeida, F., N. P. Giannini, R. DeSalle, and N. B. Simmons. 2011. Evolutionary relationships of the Old World fruit bats (Chiroptera, Pteropodidae): Another star phylogeny? *BMC Evolutionary Biology* 11:281.

Aloise, G., and S. Bertolino. 2005. Free-ranging population of the Finlayson's squirrel *Callosciurus finlaysonii* (Horsfield) (Rodentia, Sciuridae) in south Italy. *Hystrix* 16:70–74.

Alviola, P. A., M. R. M. Duya, M. V. Duya, L. R. Heaney, and E. A. Rickart. 2011. Mammalian diversity patterns on Mt. Palali, Caraballo Mountains, Luzon. *Fieldiana: Life and Earth Sciences* 2:61–74.

Ambal, R. G. R., M. V. Duya, M. A. Cruz, O. G. Coroza, S. G. Vergara, N. de Silva, N. Molinyawe, and B. R. Tabaranza Jr. 2012. Key biodiversity areas in the Philippines: Priorities for conservation. *Journal of Threatened Taxa* 4:2788–2796.

Anonymous. 1876. *Memoria-Catálogo de la Colección de Productos Forestales: Presentada por la Inspección General de Montes de Filipinas en la Exposición Universal de Filadelphia*. Manila: J. Loyzaga.

Anwarali Khan, F. A., S. Solari, V. J. Swier, P. A. Larsen, M. T. Abdullah, and R. J. Baker. 2010. Systematics of Malaysian woolly bats (Vespertilionidae: *Kerivoula*) inferred from mitochondrial, nuclear, karyotypic, and morphological data. *Journal of Mammalogy* 91:1058–1072.

Aurelio, M. A., E. Barrier, C. Rangin, and C. Müller. 1991. The Philippine fault in the Late Cenozoic tectonic evolution of the Bondoc-Masbate-N. Leyte area, central Philippines. *Journal of Southeast Asian Earth Sciences* 6:221–238.

Bachman, S. B., S. D. Lewis, and W. J. Schweller. 1983. Evolution of a forearc basin, Luzon Central Valley, Philippines. *American Association of Petroleum Geologists Bulletin* 67:1143–1162.

Balete, D. S. 2010. Food and roosting habits of the lesser false vampire bat, *Megaderma spasma* (Chiroptera: Megadermatidae), in a Philippine lowland forest. *Asia Life Sciences Supplement* 4:111–129.

Balete, D. S., P. A. Alviola, M. R. M. Duya, M. V. Duya, L. R. Heaney, and E. A. Rickart. 2011. The mammals of the Mingan Mountains, Luzon: Evidence for a new center of mammalian endemism. *Fieldiana: Life and Earth Sciences* 2:75–87.

Balete, D. S., and L. R. Heaney. 1997. Density, biomass, and movement estimates for murid rodents in mossy forest on Mt. Isarog, southern Luzon, Philippines. *Ecotropica* 3:91–100.

Balete, D. S., L. R. Heaney, P. A. Alviola, and E. A. Rickart. 2013b. Diversity and distribution of small mammals in the Bicol Volcanic Belt of southern Luzon Island, Philippines. *National Museum of the Philippines: Journal of Natural History* 1:61–86.

Balete, D. S., L. R. Heaney, and R. I. Crombie. 1995. First records of *Hipposideros lekaguli* Thonglongya and Hill, 1974, from the Philippines. *Asia Life Sciences* 4:89–94.

Balete, D. S., L. R. Heaney, and E. A. Rickart. 2013a. The mammals of Mt. Irid, Southern Sierra Madre, Luzon Island. *National Museum of the Philippines: Journal of Natural History* 1:15–29.

Balete, D. S., L. R. Heaney, E. A. Rickart, R. S. Quidlat, and J. C. Ibañez. 2008. A new species of *Batomys* (Mammalia: Muridae) from eastern Mindanao Island, Philippines. *Proceedings of the Biological Society of Washington* 121:411–428.

Balete, D. S., L. R. Heaney, M. J. Veluz, and E. A. Rickart. 2009. The non-volant mammals of Mount Tapulao, Zambales Province, Luzon. *Mammalian Biology* 74:456–466.

Balete, D. S., E. A. Rickart, and L. R. Heaney. 2006. A new species of shrew-mouse, *Archboldomys* (Rodentia: Muridae: Murinae), from the Philippines. *Systematics and Biodiversity* 4:489–501.

Balete, D. S., E. A. Rickart, L. R. Heaney, P. A. Alviola, M. R. M. Duya, M. V. Duya, T. Sosa, and S. A. Jansa. 2012. *Archboldomys* (Muridae: Murinae) reconsidered: A new genus and three new species of shrew mice from Luzon Island, Philippines. *American Museum Novitates* 3754:1–60.

Balete, D. S., E. A. Rickart, L. R. Heaney, and S. A. Jansa. 2015. A new species of *Batomys* from Luzon Island, Philippines. *Proceedings of the Biological Society of Washington* 128:22–39.

Balete, D. S., E. A. Rickart, R. G. B. Rosell-Ambal, S. Jansa, and L. R. Heaney. 2007. Descriptions of two new species of *Rhynchomys* Thomas (Rodentia: Muridae: Murinae) from Luzon Island, Philippines. *Journal of Mammalogy* 88:287–301.

Bankoff, G. 2006. One island too many: Reappraising the extent of deforestation in the Philippines prior to 1946. *Journal of Historical Geography* 33:314–334.

Barbehenn, K., J. P. Sumangil, and J. L. Libay. 1973. Rodents of the Philippine croplands. *Philippine Agriculturalist* 56:217–242.

Bartels, E. 1964. On *Paradoxurus hermaphroditus javanicus* (Horsfield, 1824). *Beaufortia* 10:193–201.

Bellwood, P. 1997. *Prehistory of the Indo-Malaysian Archipelago*. Hawaii: University of Hawaii Press.

Bellwood, P. 2013. *First Migrants*. Oxford: Wiley-Blackwell.

Berglund, H., J. Järemo, and G. Bengtsson. 2009. Endemism predicts intrinsic vulnerability to nonindigenous species on islands. *American Naturalist* 174:94–101.

Bertolino, S., P. J. Mazzoglino, M. Vaiana, and I. Currado. 2004. Activity budget and foraging behavior of introduced *Callosciurus finlaysonii* (Rodentia, Sciuridae) in Italy. *Journal of Mammalogy* 82:254–259.

Beyer, H. O. 1957. New finds of fossil mammals from the Pleistocene strata of the Philippines. *National Research Council of the Philippines, Bulletin* 41:220–239.

Bintanja, R., R. S. W. van de Wal, and J. Oerlemans. 2005. Modelled atmospheric temperatures and global sea levels over the past million years. *Nature* 437:125–128.

Bird, M. I., D. Taylor, and C. Hunt. 2005. Palaeoenvironments of insular Southeast Asia during the Last Glacial Period: A savanna corridor in Sundaland? *Quaternary Science Reviews* 24:2228–2242.

Borges, P. A. V., and J. Hortal. 2009. Time, area, and isolation: Factors driving the diversification of Azorean arthropods. *Journal of Biogeography* 36:178–191.

Bradbury, J. W., and S. L. Vehrencamp. 1977. Social organization and foraging in emballonurid bats. *Behavioural Ecology and Sociobiology* 2:1–17.

Brigham-Grett, J., M. Melles, A. Andreev, P. Tarasov, R. De-Conto, S. Koenig, N. Nowaczak, V. Wennrich, P. Rosen, E. Haltia, T. Cook, C. Gebhardt, C. Meyer-Jacob, J. Snyder, and U. Herzschuh. 2013. Pliocene warmth, polar amplification, and stepped Pleistocene cooling recorded in NE Arctic Russia. *Science* 340:1421–1426.

Broad, R., and J. Cavanagh. 1993. *Plundering Paradise: The Struggle for the Environment in the Philippines*. Berkeley: University of California Press.

Brown, R. M., C. D. Siler, C. H. Oliveros, J. A. Esselstyn, A. C. Diesmos, P. A. Hosner, C. W. Linkem, A. J. Barley, J. R. Oaks, M. B. Sangulia, L. J. Welton, D. C. Blackburn, R. G. Moyle, A. T. Peterson, and A. C. Alcala. 2013. Evolutionary processes of diversification in a model island archipelago. *Annual Review of Ecology, Evolution, and Systematics* 44:411–435.

Bureau of Mines and Geo-Sciences, Ministry of Natural Resources. 1982. *Geology and Mineral Resources of the Philippines*. Vol. 1, *Geology*. Manila.

Bureau of Mines and Geo-Sciences, Ministry of Natural Resources. 1986. *Geology and Mineral Resources of the Philippines*. Vol. 2, *Mineral Resources, Plate Supplement*. Manila.

Cameron, R. A. D., K. A. Triantis, C. E. Parent, F. Guilhaumon, M. R. Alonso, M. Ibañez, A. M. de Frias Martins, R. J. Ladle, and R. J. Whittaker. 2013. Snails on oceanic islands: Testing the General Dynamic Model of oceanic island biogeography using linear mixed effect models. *Journal of Biogeography* 40:117–130.

Campbell, P., A. S. Putnam, C. Bonney, R. Bilgin, J. C. Morales, T. H. Kunz, and L A. Ruedas. 2007. Contrasting patterns of genetic differentiation between endemic and widespread species of fruit bats (Chiroptera: Pteropodidae) in Sulawesi, Indonesia. *Molecular Phylogenetics and Evolution* 44:474–482.

Campbell, P., C. J. Schneider, A. M. Adnan, A. Zubaid, and T. H. Kunz. 2004. Phylogeny and phylogeography of Old World fruit bats in the *Cynopterus brachyotis* complex. *Molecular Phylogenetics and Evolution* 33:764–781.

Cardoso, P., M. Arnedo, K. A. Triantis, and P. A. V. Borges. 2010. Drivers of diversity in Macaronesian spiders and the role of species extinctions. *Journal of Biogeography* 37:1034–1046.

Carpenter, E.-S., R. Gomez, D. L. Waldien, and R. E. Sherwin. 2014. Photographic estimation of roosting density of Geoffroy's rousette fruit bat *Rousettus amplexicaudatus* (Chiroptera: Pteropodidae) at Montfort Bat Cave, Philippines. *Journal of Threatened Taxa* 6:5838–5844.

Catibog-Sinha, C. 1987. The wild vertebrate fauna of Mount Makiling, Laguna. *Philippine Journal of Science* 116:19–29.

Catibog-Sinha, C. C., and L. R. Heaney. 2006. *Philippine Biodiversity: Principles and Practice*. Philippines: Haribon Foundation.

Conservation International Philippines, Department of Environment and Natural Resources–Protected Areas and Wildlife Bureau, and Haribon Foundation. 2006. *Priority Sites for Conservation in the Philippines: Key Biodiversity Areas*. Quezon City: Conservation International Philippines, Department of Environment and Natural Resources–Protected Areas and Wildlife Bureau, and Haribon Foundation.

Corbet, G., and J. E. Hill. 1992. *The Mammals of the Indomalayan Region*. Oxford: Oxford University Press.

Corlett, R. T. 2010. Invasive aliens on tropical East Asian islands. *Biodiversity and Conservation* 19:411–423.

Croft, D. A., L. R. Heaney, J. J. Flynn, and A. P. Bautista. 2006. Fossil remains of a new, diminutive *Bubalus* (Artiodactyla: Bovidae: Bovini) from Cebu Island, Philippines. *Journal of Mammalogy* 87:1037–1051.

Csorba, G., C. H. Chou, M. Ruedi, T. Gorpol, M. Motokawa, S. Wiantoro, V. D. Thong, N. T. Son, L. K. Lin, and N. Furey. 2014. The reds and the yellows: A review of Asian *Chrysopteron* Jentinck, 1910 (Chiroptera: Vespertilionidae: *Myotis*). *Journal of Mammalogy* 95:663–678.

Csorba, G., P. Ujhelyi, and N. Thomas. 2003. *Horseshoe Bats of the World (Chiroptera: Rhinolophidae)*. Shropshire, UK: Alana Books.

Danielsen, E., D. S. Balete, T. D. Christensen, M. Heegaard, O. F. Jakobsen, A. Jensen, T. Lund, and M. K. Poulsen. 1994. *Conservation of Biological Diversity in the Sierra Madre Mountains of Isabela and Southern Cagayan Province, the Philippines*. Manila: BirdLife International.

Department of Environment and Natural Resources. 2012. *2012 Annual Report, New Conservation Areas in the Philippines Project (NewCAPP)*. UNDP-GEF-DENR/PAWB, www.newcapp.org/cgi-bin/news/archive.php?id=75/ [no longer available].

Détroit, F., E. Dizon, C. Falguères, S. Hameau, W. Ronquillo, and F. Sémah. 2004. Upper Pleistocene *Homo sapiens* from the Tabon cave (Palawan, the Philippines): Description and dating of new discoveries. *Human Palaeontology and Prehistory* 3:705–712.

de Vos, J., and A. Bautista. 2001. An update on the vertebrate fossils from the Philippines. *National Museum Papers* 11:58–62.

de Vos, J., and A. Bautista. 2003. Preliminary notes on the vertebrate fossils from the Philippines. *Proceedings of the Society of Philippine Archeologists* 1:42–62.

Diamond, J. 2014. Human melting pots in Southeast Asia. *Nature* 512:262–263.

Dickerson, R. E. 1928. *Distribution of Life in the Philippines*. Monograph, Philippines Bureau of Science 2. Manila: Bureau of Printing.

Dickinson, E. C., R. S. Kennedy, and K. C. Parkes. 1991. *The Birds of the Philippines: An Annotated Check-List*. British Ornithologists' Union Check-List 12. Tring, Herts., UK: British Ornithologists' Union.

Dizon, E. Z. 2003. New direct dating of the human fossils from Tabon Cave, Palawan, Philippines. *Proceedings of the Society of Philippine Archaeologists* 1:63–67.

Dolar, M. L., W. F. Perrin, B. L. Taylor, G. L. Kooyman, and M. N. R. Alava. 2006. Abundance and distributional ecology of cetaceans in the central Philippines. *Journal of Cetacean Research and Management* 8:93–111.

Duplantier, J. M., and J. B. Duchamin. 2003. Introduced small mammals and their ectoparasites: A description of their colonization and its consequences. Pp. 1191–1194 in *The Natural History of Madagascar*, ed. S. M. Goodman and J. P. Benstead. Chicago: University of Chicago Press.

Duya, M. R. M., P. A. Alviola, M. V. Duya, D. S. Balete, and L. R. Heaney. 2007. Report on a survey of mammals of the Sierra Madre Range, Luzon Island, Philippines. *Banwa* 4:41–68.

Duya, M. R. M., M. V. Duya, P. A. Alviola, D. S. Balete, and L. R. Heaney. 2011. Diversity of small mammals in montane and mossy forests on Mount Cetaceo, Cagayan Province, Luzon. *Fieldiana: Life and Earth Sciences* 2:88–95.

Eduardo, S. L., C. Y. J. Domingo, and B. P. Divina. 2008. Zoonotic parasites of rats in the Philippines. Pp. 157–194 in *Philippine Rats: Ecology and Management*, ed. G. R. Singleton, R. C. Joshi, and L. S. Sebastian. Muñoz City: Philippine Rice Research Institute.

Encarnación, J. 2004. Multiple ophiolite generation preserved in the northern Philippines and the growth of an island arc complex. *Tectonophysics* 392:103–130.

Environmental Science for Social Change. 1999. *Mining Revisited*. Quezon City: Environmental Science for Social Change, Ateneo de Manila University.

Environmental Science for Social Change. 2000. *Decline of the Philippine Forest*. Makati City, Philippines: Bookmark.

Esselstyn, J. A., A. S. Achmadi, and K. C. Rowe. 2012a. Evolutionary novelty in a rat with no molars. *Biology Letters* 8:990–993.

Esselstyn, J. A., and R. M. Brown. 2009. The role of repeated sea-level fluctuations in the generation of shrew (Soricidae: *Crocidura*) diversity in the Philippine Archipelago. *Molecular Phylogenetics and Evolution* 53:171–181.

Esselstyn, J. A., H. J. D. Garcia, M. G. Saulog, and L. R. Heaney. 2008. A new species of *Desmalopex* (Pteropodidae) from the Philippines, with a phylogenetic analysis of the Pteropodini. *Journal of Mammalogy* 89:815–825.

Esselstyn, J. A., and S. M. Goodman. 2010. New species of shrew (Soricidae: *Crocidura*) from Sibuyan Island, Philippines. *Journal of Mammalogy* 91:1467–1472.

Esselstyn, J. A., S. P. Maher, and R. M. Brown. 2011. Species interactions during diversification and community assembly in an island radiation of shrews. *PLoS ONE* 6:e21885.

Esselstyn, J. A., C. H. Oliveros, R. G. Moyle, A. T. Peterson, J. A. McGuire, and R. M. Brown. 2010. Integrating phylogenetic and taxonomic evidence illuminates complex biogeographic patterns along Huxley's modification of Wallace's Line. *Journal of Biogeography* 37:2054–2066.

Esselstyn, J. A., J. L. Sedlock, F. A. Anwarali Khan, B. J. Evans, and L. R. Heaney. 2012b. Single-locus species delimitation: A test of the mixed Yule-coalescent model, with an empirical application to Philippine round-eared bats. *Proceedings of the Royal Society B: Biological Sciences* 279:3678–3686.

Esselstyn, J. A., R. M. Timm, and R. M. Brown. 2009. Do geological or climatic processes drive speciation in dynamic archipelagos? The tempo and mode of diversification in Southeast Asian shrews. *Evolution* 63:2595–2610.

Esselstyn, J. A., P. Widmann, and L. R. Heaney. 2004. The mammals of Palawan Island, Philippines. *Proceedings of the Biological Society of Washington* 117:271–302.

Fabre, P. H., M. Pages, G. G. Musser, Y. S. Firiana, J. Fjeldsa, A. Jennings, K. A. Jonsson, J. Kennedy, J. Michaux, G. Semiadi, N. Supriatna, and K. M. Helgen. 2013. A new genus of rodent from Wallacea (Rodentia: Muridae: Murinae: Rattini), and its implications for biogeography and Indo-Pacific Rattini systematics. *Zoological Journal of the Linnean Society* 169:408–447.

FAO. 2008. *Forest Faces: Hopes and Regrets in Philippine Forestry*. Bangkok: Food and Agriculture Organization of the United Nations.

Feiler, A. 1999. Ausgestorbene Saugetiere, Typusexemplare und bemerkenswerte Lokalserien von Saugetieren aus der Sammlung des Staatlichen Museums für Tierkunde Dresden (Mammalia). *Zoologische Abhandlungen Staatliches Museum für Tierkunde Dresden* 50:401–414.

Fernando, E. S., M. H. Suh, J. Lee, and D. K. Lee. 2008. *Forest Formations of the Philippines*. Seoul: GeoBook.

Fooden, J. 2006. Comparative review of *fascicularis*-group species of macaques (Primates: *Macaca*). *Fieldiana: Zoology*, n.s., 107:1–43.

Francis, C. M. 2008. *A Guide to the Mammals of Southeast Asia*. Princeton, NJ: Princeton University Press.

Francis, C. M., and J. E. Hill. 1986. A review of the Bornean *Pipistrellus* (Mammalia: Chiroptera). *Mammalia* 50: 43–55.

Garcia, M. G., and A. H. Deocampo. 1995. Wild pig meat trading in Iligan City, Mindanao. *Sylvatrop* 5:29–34.

Gaubert, P., and P. Cordeiro-Estrela. 2006. Phylogenetic systematics and tempo of evolution of the Viverrinae (Mammalia, Carnivora, Viverridae) within feliformians: Implications for faunal exchanges between Asia and Africa. *Molecular Phylogenetics and Evolution* 41:266–278.

Geronimo-Catane, S. 1994. Mode of emplacement of two debris-avalanche deposits at Banahao Volcano, Southern Luzon, Philippines. *Bulletin of the Volcanological Society of Japan* 39:113–127.

Giannini, N. P., F. C. Almeida, N. B. Simmons, and R. Desalle. 2006. Phylogenetic relationships of the enigmatic harpy fruit bat, *Harpyionycteris* (Mammalia: Chiroptera: Pteropodidae). *American Museum Novitates* 3533:1–12.

Giannini, N. P., F. C. Almeida, N. B. Simmons, and K. M. Helgen. 2008. The systematic position of *Pteropus leucopterus* and its bearing on the monophyly and relationships of *Pteropus* (Chiroptera: Pteropodidae). *Acta Chiropterologica* 10:11–20.

Gomez, R. K. S. C., J. C. Ibañez, and S. T. Bastian. 2005. Diversity and community similarity of pteropodids and notes on insectivorous bats in the Arakan Valley Conservation Area, Mindanao. *Sylvatrop* 15:87–102.

Graham, C. H., A. C. Carnaval, C. D. Cadena, K. R. Zamudio, T. E. Roberts, J. L. Parra, C. M. McCain, R. C. K. Bowie, C. Moritz, S. B. Baines, C. J. Schneider, J. VanDerWal, C. Rahbeck, K. H. Kozak, and N. J. Sanders. 2014. The origin and maintenance of montane diversity: Integrating evolutionary and ecological processes. *Ecography* 37:711–719.

Griffin, P. B., and M. B. Griffin. 2000. Agta hunting and sustainability of resource use in northeastern Luzon, Philippines. Pp. 325–335 in *Hunting for Sustainability in Tropical Forests*, ed. J. G. Robinson and E. L. Bennett. New York: Columbia University Press.

Grubb, P., and C. P. Groves. 1983. Notes on the taxonomy of the deer (Mammalia, Cervidae) of the Philippines. *Zoologischer Anzeiger, Leipzig* 210:119–144.

Guillén-Servent, A., C. M. Francis, and R. E. Ricklefs. 2003. Phylogeny and biogeography of the horseshoe bats. Pp. xii–xxiv in *Horseshoe Bats of the World (Chiroptera: Rhinolophidae)*, ed. G. Csorba, P. Ujhelyi, and N. Thomas. Shropshire, UK: Alana Books.

Hall, R. 1998. The plate tectonics of Cenozoic SE Asia and the distribution of land and sea. Pp. 99–131 in *Biogeography and Geological Evolution of SE Asia*. ed. R. Hall and J. D. Holloway. Leiden: Backhuys.

Hall, R. 2012. Late Jurassic-Cenozoic reconstructions of the Indonesian region and the Indian Ocean. *Tectonophysics* 570–571:1–41.

Hall, R. 2013. The palaeogeography of Sundaland and Wallacea since the Late Jurassic. *Journal of Limnology* 72:1–17.

Hamilton, W. 1979. *Tectonics of the Indonesian Region*. Geological Survey Professional Paper 1078. Washington, DC: US Government Printing Office.

Harris, D. B. 2009. Review of negative effects of introduced rodents on small mammals on islands. *Biological Invasion* 11:1611–1630.

Heaney, L. R. 1984. Mammalian species richness on islands on the Sunda Shelf, Southeast Asia. *Oecologia* 61:11–17.

Heaney, L. R. 1985. Zoogeographic evidence for Middle and Late Pleistocene land bridges to the Philippine Islands. *Modern Quaternary Research in Southeast Asia* 9:127–144.

Heaney, L. R. 1986. Biogeography of the mammals of Southeast Asia: Estimates of colonization, extinction, and speciation. *Biological Journal of the Linnean Society* 28:127–165.

Heaney, L. R. 1991. An analysis of patterns of distribution and species richness among Philippine fruit bats (Pteropodidae). *Bulletin of the American Museum of Natural History* 206:145–167.

Heaney, L. R. 1993. Biodiversity patterns and the conservation of mammals in the Philippines. *Asia Life Sciences* 2:261–274.

Heaney, L. R. 2000. Dynamic disequilibrium: A long-term, large-scale perspective on the equilibrium model of island biogeography. *Global Ecology and Biogeography* 9:59–74.

Heaney, L. R. 2001. Small mammal diversity along elevational gradients in the Philippines: An assessment of patterns and hypotheses. *Global Ecology and Biogeography* 10:15–39.

Heaney, L. R. 2004a. Remarkable rats and the origins of island biological diversity. *In the Field* 75:16–18.

Heaney, L. R. 2004b. Conservation biology in oceanic archipelagos. Pp. 345–360 in *Frontiers in Biogeography: New Directions in the Geography of Nature*. ed. M. V. Lomolino and L. R. Heaney. Sunderland, MA: Sinauer Associates.

Heaney, L. R. 2007. Is a new paradigm emerging for oceanic island biogeography? *Journal of Biogeography* 34:753–757.

Heaney, L. R. (ed.). 2011. Discovering diversity: Studies of the mammals of Luzon Island, Philippines. *Fieldiana: Life and Earth Sciences* 2:1–102.

Heaney, L. R. 2013a. Introduction and commentary. Pp. xi–lxxi in A. R. Wallace, *Island Life*. Chicago: University of Chicago Press [reprint of 1880 first edition].

Heaney, L. R. 2013b. Preface to studies of mammalian biodiversity on Luzon Island, Philippines. *National Museum of the Philippines: Journal of Natural History* 1:iii–v.

Heaney, L. R., and A. C. Alcala. 1986. Flat-headed bats (Mammalia, *Tylonycteris*) from the Philippine Islands. *Silliman Journal* 33:117–123.

Heaney, L. R., and D. S. Balete. 2012. Discovering diversity: Newly discovered mammals highlight areas of unique biodiversity in Luzon. *Wildernews* 1:4–9.

Heaney, L. R., D. S. Balete, P. A. Alviola, M. R. M. Duya, and E. A. Rickart. 2013a. The mammals of Mt. Anacuao, NE Luzon Island, Philippines: A test of predictions of Luzon small mammal diversity patterns. *National Museum of the Philippines: Journal of Natural History* 1:1–13.

Heaney, L. R., D. S. Balete, P. A. Alviola, E. A. Rickart, and M. Ruedi. 2012. *Nyctalus plancyi* and *Falsistrellus petersi* (Chiroptera: Vespertilionidae) from northern Luzon, Philippines: Ecology, phylogeny, and biogeographic implications. *Acta Chiropterologica* 14:265–278.

Heaney, L. R., D. S. Balete, and A. T. L. Dans. 1997. Terrestrial mammals. Pp. 139–168 in *Philippine Red Data Book*, ed. Wildlife Conservation Society of the Philippines. Manila: Bookmark.

Heaney, L. R., D. S. Balete, M. L. Dolar, A. C. Alcala, A. T. L. Dans, P. C. Gonzales, N. R. Ingle, M. V. Lepiten, W. L. R. Oliver, P. S. Ong, E. A. Rickart, B. R. Tabaranza Jr., and R. C. B. Utzurrum. 1998. A synopsis of the mammalian fauna of the Philippine Islands. *Fieldiana: Zoology*, n.s., 88:1–61.

Heaney, L. R., D. S. Balete, M. R. M. Duya, M. V. Duya, S. A. Jansa, S. J. Steppan, and E. A. Rickart. Submitted. Doubling diversity: A cautionary tale of previously unsuspected mammalian diversity on a tropical oceanic island. *Frontiers of Biogeography*.

Heaney, L. R., D. S. Balete, G. V. Gee, M. V. Lepiten-Tabao, E. A. Rickart, and B. R. Tabaranza. 2003. Preliminary report on the mammals of Balbalasang, Kalinga Province, Luzon. *Sylvatrop* 13:51–62.

Heaney, L. R., D. S. Balete, and E. A. Rickart. 2013c. Models of oceanic island biogeography: Changing perspectives on biodiversity dynamics in archipelagoes. *Frontiers of Biogeography* 5:249–257.

Heaney, L. R., D. S. Balete, E. A. Rickart, P. A. Alviola, M. R. M. Duya, M. V. Duya, M. J. Veluz, L. VandeVrede, and S. J. Steppan. 2011a. Seven new species and a new subgenus of forest mice (Rodentia: Muridae: *Apomys*) from Luzon Island. *Fieldiana: Life and Earth Sciences* 2:1–60.

Heaney, L. R., D. S. Balete, E. A. Rickart, and S. A. Jansa. 2014b. Three new species of *Musseromys* (Muridae, Rodentia), the endemic Philippine tree mouse from Luzon Island. *American Museum Novitates*, 3802:1–28.

Heaney, L. R., D. S. Balete, E. A. Rickart, R. C. B. Utzurrum, and P. C. Gonzales. 1999. Mammalian diversity on Mount Isarog, a threatened center of endemism on southern Luzon Island, Philippines. *Fieldiana: Zoology*, n.s., 95:1–62.

Heaney, L. R., D. S. Balete, E. A. Rickart, M. J. Veluz, and S. Jansa. 2009. A new genus and species of small "tree mouse" (Rodentia, Muridae) related to the Philippine giant cloud-rats. Pp. 205–229 in *Systematic Mammalogy: Contributions in Honor of Guy G. Musser*, ed. R. S. Voss and M. D. Carleton. *Bulletin of the American Museum of Natural History* 331. New York: American Museum of Natural History.

Heaney, L. R., D. S. Balete, E. A. Rickart, M. J. Veluz, and J. Sarmiento. 2004. Welcome surprises from Mt. Banahaw. *Haring Ibon* 16:10–16.

Heaney, L. R., D. S. Balete, R. G. B. Rosell-Ambal, M. J. Veluz, and E. A. Rickart. 2013b. The non-volant small mammals of Mt. Banahaw–San Cristobal National Park, Luzon, Philippines: Distribution and ecology of a highly endemic fauna. *National Museum of the Philippines: Journal of Natural History* 1:45–60.

Heaney, L. R., D. S. Balete, J. Sarmiento, and P. A. Alviola. 2006c. Losing diversity and courting disaster: The mammals of Mt. Data National Park. *Haring Ibon* 25:15–23.

Heaney, L. R., D. S. Balete, M. J. Veluz, S. J. Steppan, J. A. Esselstyn, A. W. Pfeiffer, and E. A. Rickart. 2014a. Two new species of Philippine forest mice (*Apomys*, Muridae, Rodentia) from Lubang and Luzon Islands, with a redescription of *Apomys sacobianus* Johnson, 1962. *Proceedings of the Biological Society of Washington* 126:395–413.

Heaney, L. R., A. Diesmos, B. R. Tabaranza Jr., N. A. Mallari, and R. Brown. 2000. Beacon of hope: A first report from Kalinga Province, in the northern Central Cordillera. *Haring Ibon* 2:14–18.

Heaney, L. R., M. L. Dolar, D. S. Balete, J. A. Esselstyn, E. A. Rickart, and J. L. Sedlock. 2010. *Synopsis of Philippine Mammals*. Field Museum of Natural History, www.fieldmuseum.org/philippine_mammals/.

Heaney, L. R., P. C. Gonzales, R. C. B. Utzurrum, and E. A. Rickart. 1991. The mammals of Catanduanes Island: Implications for the biogeography of small land-bridge islands in the Philippines. *Proceedings of the Biological Society of Washington* 104:399–415.

Heaney, L. R., and P. D. Heideman. 1987. Philippine fruit bats, endangered and extinct. *Bats* 5:3–5.

Heaney, L. R., P. D. Heideman, E. A. Rickart, R. C. B. Utzurrum, and J. S. H. Klompen. 1989. Elevational zonation of mammals in the central Philippines. *Journal of Tropical Ecology* 5:259–280.

Heaney, L. R., and N. A. D. Mallari. 2000. A preliminary analysis of current gaps in the protection of threatened Philippine terrestrial mammals. *Sylvatrop* 10:28–39.

Heaney, L. R., P. J. Piper, and A. S. Mijares. 2011b. The first fossil record of endemic murid rodents from the Philippines: A Late Pleistocene cave fauna from northern Luzon. *Proceedings of the Biological Society of Washington* 124:234–247.

Heaney, L. R., and D. S. Rabor. 1982. Mammals of Dinagat and Siargao Island, Philippines. *Occasional Papers of the Museum of Zoology, University of Michigan* 699:1–30.

Heaney, L. R., and J. C. Regalado Jr. 1998. *Vanishing Treasures of the Philippine Rain Forest*. Chicago: Field Museum.

Heaney, L. R., and E. A. Rickart. 1990. Correlations of clades and clines: Geographic, elevational, and phylogenetic distribution patterns among Philippine mammals. Pp. 321–332 in *Vertebrates in the Tropics*, ed. G. Peters and R. Hutterer. Bonn: Alexander Koenig Zoological Research Institute and Zoological Museum.

Heaney, L. R., and T. E. Roberts. 2009. New perspectives on the long-term biogeographic dynamics and conservation of Philippine fruit bats. Pp. 17–58 in *Ecology, Evolution, and Conservation of Island Bats*. ed. T. H. Fleming and P. Racey. Chicago: University of Chicago Press.

Heaney, L. R., and M. Ruedi. 1994. A preliminary analysis of biogeography and phylogeny of *Crocidura* from the Philippines. Pp. 357–377 in *Advances in the Biology of Shrews*, ed. J. E. Meritt, G. L. Kirkland, and R. K. Rose. Special Publication 18. Pittsburgh: Carnegie Museum of Natural History.

Heaney, L. R., B. R. Tabaranza Jr., D. S. Balete, E. A. Rickart, and N. R. Ingle. 2006b. The mammals of Mt. Kitanglad Nature Park, Mindanao, Philippines. *Fieldiana: Zoology*, n.s., 112:1–63.

Heaney, L. R., B. R. Tabaranza Jr., D. S. Balete, and N. Rigertas. 2006a. Synopsis and biogeography of the mammals of Camiguin Island, Philippines. *Fieldiana: Zoology*, n.s., 106:28–48.

Heaney, L. R., and R. C. B. Utzurrum. 1991. A review of the conservation status of Philippine land mammals. *ASPB Communication* 3:1–13.

Heaney, L. R., E. K. Walker, B. R. Tabaranza Jr., and N. R. Ingle. 2000. Mammalian diversity in the Philippines: An assessment of the adequacy of current data. *Sylvatrop* 10:6–27.

Heaney, L. R., J. S. Walsh, and A. T. Peterson. 2005. The roles of geological history and colonization abilities in genetic differentiation between mammalian populations in the Philippine Archipelago. *Journal of Biogeography* 32:229–247.

Heideman, P. D. 1989. Delayed development in Fischer's pygmy fruit bat, *Haplonycteris fischeri*, in the Philippines. *Journal of Reproduction and Fertility* 85:363–382.

Heideman, P. D. 1995. Synchrony and seasonality of reproduction in tropical bats. *Symposia of the Zoological Society of London* 67:141–165.

Heideman, P. D., J. A. Cummings, and L. R. Heaney. 1993. Observations on reproductive timing and early embryonic development in an Old World fruit bat, *Otopteropus cartilagonodus* (Megachiroptera). *Journal of Mammalogy* 74:621–630.

Heideman, P. D., and L. R. Heaney. 1989. Population biology and estimates of abundance of fruit bats (Pteropodidae) in Philippine submontane rainforest. *Journal of Zoology* (London) 218:565–586.

Heideman, P. D., L. R. Heaney, R. L. Thomas, and K. R. Erickson. 1987. Patterns of faunal diversity and species abundance of non-volant small mammals on Negros Island, Philippines. *Journal of Mammalogy* 68:884–888.

Heideman, P. D., and K. S. Powell. 1998. Age-specific reproductive strategies and delayed embryonic development in an Old World fruit bat, *Ptenochirus jagori*. *Journal of Mammalogy* 79:295–311.

Heideman, P. D., and R. C. B. Utzurrum. 2003. Seasonality and synchrony of reproduction in three species of nectarivorous Philippine bats. *BMC Ecology* 3, www.biomedcentral.com/1472-6785/3/11/.

Helgen, K. M., D. Kock, R. K. S. C. Gomez, N. R. Ingle, and M. H. Sinaga. 2007. Taxonomy and natural history of the Southeast Asian fruit-bat genus *Dyacopterus*. *Journal of Mammalogy* 88:302–318.

Hollister, N. 1913. A review of the Philippine land mammals in the United States National Museum. *Proceedings of the United States National Museum* 46:299–341.

Hoogstraal, H. 1951. Philippine Zoological Expedition, 1946–1947: Narrative and itinerary. *Fieldiana: Zoology* 33:1–86.

Huke, R. E. 1963. *Shadows on the Land: An Economic Geography of the Philippines*. Manila: Bookmark.

Hutterer, R. 2005. Records of shrews from Panay and Palawan, Philippines, with the description of two new species of *Crocidura* (Mammalia: Soricidae). *Lynx* 38:5–20.

Inger, R. F. 1954. Philippine Zoological Expedition 1946–1947: Systematics and zoogeography of the Philippine Amphibia. *Fieldiana: Zoology* 33:181–531.

Ingle, N. R. 1992. The natural history of bats on Mt. Makiling, Luzon Island, Philippines. *Silliman Journal* 36:1–26.

Ingle, N. R. 1993. Vertical stratification of bats in a Philippine rainforest. *Asia Life Sciences* 2:215–222.

Ingle, N. R. 2003. Seed dispersal by wind, birds, and bats between Philippine montane rainforest and successional vegetation. *Oecologia* 134:251–261.

Ingle, N. R., and L. R. Heaney. 1992. A key to the bats of the Philippine Islands. *Fieldiana: Zoology*, n.s., 69:1–44.

Jansa, S. A., F. K. Barker, and L. R. Heaney. 2006. The pattern and timing of diversification of Philippine endemic rodents: Evidence from mitochondrial and nuclear gene sequences. *Systematic Biology* 55:73–88.

JAXA/EORC. 2014. *Tropical Cyclone Database*, http://sharaku.eorc.jaxa.jp/TYP_DB/index_e.shtml.

Johnson, D. H. 1962. Two new murine rodents. *Proceedings of the Biological Society of Washington* 75:317–319.

Jones, D. P., and T. H. Kunz. 2000a. *Pteropus hypomelanus*. *Mammalian Species* 639:1–6.

Jones, D. P., and T. H. Kunz. 2000b. *Pteropus vampyrus*. *Mammalian Species* 642:1–6.

Jotterand-Bellomo, M., and P. Schauenberg. 1988. Les chromosomes du rat de Cuming, *Phloeomys cumingi* Waterhouse, 1839 (Mammalia: Rodentia). *Genetica* 76:181–190.

Justiniano, R., J. J. Schenk, D. S. Balete, E. A. Rickart, J. A. Esselstyn, L. R. Heaney, and S. J. Steppan. 2015. Testing diversification models of endemic Philippine forest mice (*Apomys*) with nuclear phylogenies across elevational gradients reveal repeated colonization of isolated mountain ranges. *Journal of Biogeography* 42:51–64.

Kellogg, R. 1945. Two new Philippine rodents. *Proceedings of the Biological Society of Washington* 58:121–124.

Kingston, T., G. Jones, Z. Akbar, and T. H. Kunz. 1999. Echolocation signal design in Kerivoulinae and Murininae (Chiroptera: Vespertilionidae) from Malaysia. *Journal of Zoology* (London) 249:359–374.

Kingston, T., G. Jones, A. Zubaid, and T. H. Kunz. 2000. Resource partitioning in rhinolophoid bats revisited. *Oecologia* 124:332–342.

Kingston, T., B. L. Lim, and Z. Akbar. 2006. *Bats of Krau Wildlife Reserve*. Bangi: Penerbit Universiti Kebangsaan Malaysia.

Kitchener, D. J., and Maharadatunkamsi. 1991. Description of a new species of *Cynopterus* (Chiroptera: Pteropodidae) from Nusa Tenggara, Indonesia. *Records of the Western Australian Museum* 15:307–363.

Klingener, D., and G. K. Creighton. 1984. On small bats of the genus *Pteropus* from the Philippines. *Proceedings of the Biological Society of Washington* 97:395–403.

Knittel-Weber, C., and U. Knittel. 1990. Petrology and genesis of the volcanic rocks on the eastern flank of Mount Malinao, Bicol Arc (southern Luzon, Philippines). *Journal of Southeast Asian Earth Sciences* 4:267–280.

Kock, D. 1969a. *Dyacopterus spadiceus* (Thomas 1890) auf den Philippinen (Mammalia, Chiroptera). *Senckenbergiana Biologica* 50:1–7.

Kock, D. 1969b. Eine Bemerkenswerte neue Gattung und Art Flughunde von Luzon, Philippinen (Mammalia, Chiroptera). *Senckenbergiana Biologica* 50:329–338.

Kock, D. 1981. *Philetor brachypterus* auf New-Britannien und den Philippinen (Mammalia: Chiroptera: Vespertilionidae). *Senckenbergiana Biologica* 61:313–319.

Kowal, N. E. 1966. Shifting cultivation, fire, and pine forest in the Cordillera Central, Luzon, Philippines. *Ecological Monographs* 36:389–419.

Ku, Y., C. Chen, S. Song, Y. Iizuka, and J. J. Shen. 2009. A 2 Ma record of explosive volcanism in southwestern Luzon: Implications for the timing of subducted slab steepening. *Geochemistry, Geophysics, Geosystems* 10:Q06017, doi:10.1029/2009GC002486.

Kummer, D. M. 1991. *Deforestation in the Postwar Philippines*. Geography Research Paper No. 234. Chicago: University of Chicago Press.

Kuramoto, T., H. Torii, H. Ikeda, H. Endo, W. Rerkamnuaychoke, and T. Oshida. 2012. Mitochondria DNA sequences of Finlayson's squirrel found in Hamamatsu, Shizuoka Prefecture, Japan. *Mammal Study* 37:63–67.

Lapitan, P. G., E. S. Fernando, M. H. Suh, R. U. Fuentes, Y. K. Shin, N. M. Pampolina, M. L. Castillo, R. P. Cereno, J. H. Lee, S. Han, T. B. Choi, and D. K. Lee. 2010. *Biodiversity and Natural Resources Conservation in Protected Areas of Korea and the Philippines*. Seoul: ASEAN-Korea Environmental Cooperation Unit.

Largen, M. J. 1985. Taxonomically and historically significant specimens of mammals in the Merseyside County Museums, Liverpool. *Journal of Mammalogy* 66:412–418.

Laurence, W. F., G. R. Clements, S. Sloan, C. S. O'Connell, N. D. Mueller, M. Goosem, O. Venter, D. P. Edwards, B. Phalan, A. Balmford, R. Van Der Ree, and I. B. Arrea. 2014. A global strategy for road building. *Nature* 513:229–232.

Lawrence, B. L. 1939. Collections from the Philippine Islands: Mammals. *Bulletin of the Museum of Comparative Zoology* 86:28–73.

Lekagul, B., and J. McNeely. 1977. *Mammals of Thailand*. Bangkok: Kurusapha Ladprao Press.

Leones, M. A. B., and A. Navarro. 2012. Conservation areas, a treasure trove of biodiversity. *Wildernews* 1:10–13.

Lepiten, M. V. 1995. The mammals of Siquijor Island, central Philippines. *Sylvatrop* 5:1–17.

Lomolino, M. V., B. R. Riddle, R. J. Whittaker, and J. H. Brown. 2010. *Biogeography*. Sunderland, MA: Sinauer Associates.

Lucchini, V., E. Meijaard, C. H. Diong, C. P. Groves, and E. Randi. 2005. New phylogenetic perspectives among species of south-east Asian wild pig (*Sus* sp.) based on mtDNA sequences and morphometric data. *Journal of Zoology* (London) 266:25–35.

MacArthur, R. H., and E. O. Wilson. 1963. An equilibrium theory of insular zoogeography. *Evolution* 17:373–387.

MacArthur, R. H., and E. O. Wilson. 1967. *The Theory of Island Biogeography*. Monographs in Population Biology 1. Princeton, NJ: Princeton University Press.

Mallari, N. A. D., and A. Jensen. 1993. Biological diversity in Northern Sierra Madre, Philippines: Its implication for conservation and management. *Asia Life Sciences* 2:101–112.

Mallari, N. A. D., B. R. Tabaranza Jr., and M. J. Crosby. 2001. *Key Conservation Sites in the Philippines*. Makati City, Philippines: Bookmark.

Manalo, E. B. 1956. The distribution of rainfall in the Philippines. *Philippine Geographical Journal* 4:104–147.

Marshall, J. T. 1977. Rats and mice, Family Muridae. Pp. 397–487 in *Mammals of Thailand*, ed. B. Lekagul and J. A. McNeely. Bangkok: Kurusapha Ladprao Press.

Marshall, J. T. 1986. Systematics of the genus *Mus*. *Current Topics in Microbiology and Immunology* 127:12–18.

McCain, C. M. 2005. Elevational gradients in diversity of small mammals. *Ecology* 86:366–372.

Md. Nor, S. 1996. The mammalian fauna on the islands at the northern tip of Sabah, Borneo. *Fieldiana: Zoology*, n.s., 83:1–51.

Medway, L. 1969. *The Wild Mammals of Malaya*. London: Oxford University Press.

Meldahl, K. H. 2011. *Rough-Hewn Land: A Geologic Journey from California to the Rocky Mountains*. Berkeley: University of California Press.

Mickleburgh, S. P., P. A. Racey, and A. M. Hutson, eds. 1992. *Old World Fruit Bats: An Action Plan for the Family Pteropodidae*. Gland, Switzerland: International Union for the Conservation of Nature.

Mijares, A. S. B. 2008. The Late Pleistocene to Early Holocene foragers of northern Luzon. *Ippa Bulletin* 28:99–107.

Mijares, A. S. [B.], F. Détroit, P. Piper, R. Grun, P. Bellwood, M. Aubert, G. Champion, N. Cuevas, A. De Leon, and E. Dizon. 2010. New evidence for a 67,000-year-old human presence at Callao Cave, Luzon, Philippines. *Journal of Human Evolution* 59:123–132.

Mildenstein, T. L., S. C. Stier, C. E. Nuevo-Diego, and L. S. Mills. 2005. Habitat selection of endangered and endemic large flying-foxes in Subic Bay, Philippines. *Biological Conservation* 126:93–102.

Miller, G. S., Jr. 1907. The families and genera of bats. *Bulletin of the United States National Museum* 57:1–282.

Miller, G. S., Jr. 1910. Descriptions of two new genera and sixteen new species of mammals from the Philippine Islands. *Proceedings of the United States National Museum* 38:3391–3404.

Miller, R. W., A. M. Stuart, R. C. Joshi, P. B. Banks, and G. R. Singleton. 2008. Biology and management of rodent communities in complex agroecosystems—rice terraces. Pp. 25–36 in *Philippine Rats: Ecology and Management*, ed. G. R. Singleton, R. C. Joshi, and L. S. Sebastian. Muñoz City: Philippine Rice Research Institute.

Mona, S., E. Randi, and M. Tommaseo-Ponzetta. 2007. Evolutionary history of the genus *Sus* inferred from cytochrome b sequences. *Molecular Phylogenetics and Evolution* 45:757–762.

Morley, R. J., and J. R. Flenley. 1987. Late Cainozoic vegetational and environmental changes in the Malay Archipelago. Pp. 50–59 in *Biogeographical Evolution of the Malay Archipelago*, ed. T. C. Whitmore. Oxford: Clarendon Press.

Mould, A. 2012. Cave bats of the central west coast and southern section of the northwest Panay Peninsula, Panay Island, the Philippines. *Journal of Threatened Taxa* 4:2993–3028.

Mudar, K. M., and M. S. Allen. 1986. A list of bats from northeastern Luzon, Philippines. *Mammalia* 50:219–225.

Musser, G. G. 1973. Zoogeographical significance of the ricefield rat, *Rattus argentiventer*, on Celebes and New Guinea and the identity of *Rattus pesticulus*. *American Museum Novitates* 2511:1–30.

Musser, G. G. 1977. *Epimys benguetensis*, a composite, and one zoogeographic view of rat and mouse faunas in the Philippines and Celebes. *American Museum Novitates* 2624:1–15.

Musser, G. G. 1981. Results of the Archbold Expeditions, no. 105: Notes on systematics of Indo-Malayan murid rodents, and descriptions of new genera and species from Ceylon, Sulawesi, and the Philippines. *Bulletin of the American Museum of Natural History* 168:225–234.

Musser, G. G. 1982a. Results of the Archbold Expeditions, no. 107: A new genus of arboreal rat from Luzon Island in the Philippines. *American Museum Novitates* 2730:1–23.

Musser, G. G. 1982b. Results of the Archbold Expeditions, no. 108: The definition of *Apomys*, a native rat of the Philippine Islands. *American Museum Novitates* 2746:1–43.

Musser, G. G. 1982c. Results of the Archbold Expeditions, no. 110: *Crunomys* and the small-bodied shrew-rats native to the Philippine Islands and Sulawesi (Celebes). *Bulletin of the American Museum of Natural History* 174:1–95.

Musser, G. G. 1987. The mammals of Sulawesi. Pp. 73–93 in *Biogeographical Evolution of the Malay Archipelago*, ed. T. C. Whitmore. Oxford: Clarendon Press.

Musser, G. G., and M. D. Carleton. 2005. Superfamily Muroidea. Pp. 894–1531 in *Mammal Species of the World: A Taxo-*

nomic and Geographic Reference, ed. D. Wilson and D. M. Reeder. Baltimore: Johns Hopkins University Press.

Musser, G. G., and P. W. Freeman. 1981. A new species of *Rhynchomys* (Muridae) from the Philippines. *Journal of Mammalogy* 62:154–159.

Musser, G. G., L. K. Gordon, and H. Sommer. 1981. Species limits in the Philippine murid, *Chrotomys*. *Journal of Mammalogy* 63:514–521.

Musser, G. G., and L. R. Heaney. 1992. Philippine rodents: Definitions of *Tarsomys* and *Limnomys* plus a preliminary assessment of phylogenetic patterns among native Philippine murines (Murinae, Muridae). *Bulletin of the American Museum of Natural History* 211:1–138.

Musser, G. G., L. R. Heaney, and B. R. Tabaranza Jr. 1998. Philippine rodents: Redefinition of known species of *Batomys* (Muridae, Murinae) and description of a new species from Dinagat Island. *American Museum Novitates* 3237:1–51.

Musser, G. G., and C. Newcomb. 1983. Malaysian murids and the giant rat of Sumatra. *Bulletin of the American Museum of Natural History* 174:327–598.

Newhall, C. G., and R. S. Punongbayan (eds.). 1996. *Fire and Mud: Eruptions and Lahars of Mount Pinatubo, Philippines*. Quezon City: Philippine Institute of Volcanology and Seismology.

Nowak, R. M. 1999. *Walker's Mammals of the World*. Baltimore: Johns Hopkins University Press.

Nunn, P. D. 2009. Oceanic islands. Pp. 689–696 in *Encyclopedia of Islands*, ed. R. G. Gillespie and D. A. Clague. Berkeley: University of California Press.

Oliver, W. L. R. 1992. The taxonomy, distribution, and status of Philippine wild pigs. *Silliman Journal* 36:55–64.

Oliver, W. L. R., C. R. Cox, P. C. Gonzales, and L. R. Heaney. 1993a. Cloud rats in the Philippines—preliminary report on distribution and status. *Oryx* 27:41–48.

Oliver, W. L. R., C. R. Cox, and C. P. Groves. 1993b. The Philippine warty pigs (*Sus philippensis* and *Sus cebifrons*). Pp. 145–155 in *Pigs, Peccaries, and Hippos: Status Survey and Conservation Action Plan*, ed. W. L. R. Oliver. Gland, Switzerland: International Union for the Conservation of Nature.

Oliver, W. [L. R.], and L. R. Heaney. 2008. *Sus philippensis*. In *IUCN 2013: IUCN Red List of Threatened Species*, version 2013.2, www.iucnredlist.org.

Oliver, W. [L. R.], J. MacKinnon, P. Ong, and J. C. Gonzales. 2008. *Rusa marianna*. In *IUCN 2013: IUCN Red List of Threatened Species*, version 2013.2, www.iucnredlist.org.

Ong, P., L. E. Afuang, and R. G. Rosell-Ambal (eds.). 2002. *Philippine Biodiversity Conservation Priorities: A Second Iteration of the National Biodiversity Strategy and Action Plan*. Quezon City: Department of the Environment and Natural Resources.

Ong, P. S., M. Pedregosa, and M. D. de Guia. 1999. Wildlife inventory of the University of the Philippines (UP) and the Atenao de Manila University Campus, Diliman, Quezon City, Luzon, Philippines. *Science Diliman* 11:6–20.

Ong, P. S., and E. A. Rickart. 2008. Ecology of native and pest rodents in the Philippines. Pp. 101–115 in *Philippine Rats: Ecology and Management*, ed. G. R. Singleton, R. C. Joshi, and L. S. Sebastian. Muñoz City: Philippine Rice Research Institute.

Ozawa, A., T. Tagami, E. L. Listanco, C. B. Arpa, and M. Sudo. 2004. Initiation and propagation of subduction along the Philippine Trench: Evidence for the temporal and spatial distribution of volcanoes. *Journal of Asian Earth Sciences* 23:105–111.

Paguntalan, L. M. J., and P. G. C. Jakosalem. 2008. Conserving threatened and endemic fruit bats in isolated forest patches in Cebu (with notes on new records and rediscoveries). *Silliman Journal* 48:81–94.

Pasicolan, S. A. 1993. Biology of the northern Luzon slender-tailed cloud rat (*Phloeomys pallidus*) in captivity. *Asia Life Sciences* 2:223–226.

Patou, M., R. Debruyne, A. P. Jennings, A. Zubaid, J. J. Rovie-Ryan, and G. Veron. 2008. Phylogenetic relationships of the Asian palm civets (Hemigalinae & Paradoxurinae, Viverridae, Carnivora). *Molecular Phylogenetics and Evolution* 47:883–892.

Patou, M., A. Wilting, P. Gaubert, J. A. Esselstyn, C. Cruaud, A. P. Jennings, J. Fickel, and G. Veron. 2010. Evolutionary history of the *Paradoxurus* palm civets—a new model for Asian biogeography. *Journal of Biogeography* 37:2077–2097.

Patrick, L. E., E. S. McCulloch, and L. A. Ruedas. 2013. Systematics and biogeography of the arcuate horseshoe bat species complex (Chiroptera, Rhinolophidae). *Zoologica Scripta* 42:1–38.

Pavey, C. R., and J. C. Burwell. 1977. The diet of the diadem leaf-nosed bat *Hipposideros diadema*: Confirmation of a morphologically based prediction of carnivory. *Journal of Zoology* 243:295–303.

Payne, J., C. M. Francis, and K. Phillipps. 1985. *A Field Guide to the Mammals of Borneo*. Kota Kinabalu, Sabah, Malaysia: Sabah Society.

Peterson, A. T., and L. R. Heaney. 1993. Genetic differentiation in Philippine bats of the genera *Cynopterus* and *Hap-*

lonycteris. *Biological Journal of the Linnean Society* 49:203–218.

Philippine Bureau of Mines. 1963. *Geological Map of the Philippines* [9 sheets]. Manila: Department of Agriculture and Natural Resources, Bureau of Mines.

Piper, P. J., H. Hung, F. Z. Campos, P. Bellwood, and R. Santiago. 2009. A 4000-year-old introduction of domestic pigs into the Philippine Archipelago: Implications for understanding routes of human migration through Island Southeast Asia and Wallacea. *Antiquity* 83:687–695.

Piper, P. J., J. Ochoa, E. Robles, H. Lewis, and V. Paz. 2011. The palaeozoology of Palawan Island, Philippines. *Quaternary International* 233:142–158.

Pitra, C., J. Fickel, E. Meijaard, and C. P. Groves. 2004. Evolution and phylogeny of Old World deer. *Molecular Phylogenetics and Evolution* 33:880–895.

Posa, M. R. C., A. C. Diesmos, N. S. Sodhi, and T. M. Brooks. 2008. Hope for threatened tropical biodiversity: Lessons from the Philippines. *BioScience* 58:231–40.

Rabor, D. S. 1955. Notes on mammals and birds of the central northern Luzon highlands, Philippines, part 1: Notes on mammals. *Silliman Journal* 2:193–218.

Rabor, D. S. 1977. *Philippine Birds and Mammals*. Quezon City: University of Philippines Press.

Rabor, D. S. 1986. *Guide to Philippine Flora and Fauna*. Vol. 11, *Birds and Mammals*. Quezon City: Ministry of Natural Resources and University of the Philippines.

Rahbeck, C. 1995. The elevational gradient of species richness: A uniform pattern? *Ecography* 18:200–205.

Reginaldo, A. A., V. F. Ballesteros, Ma. Princess A. V. Gonzales, and C. M. Austria. 2013. Small non-volant mammals in forest patches of Baguio City, Luzon, Philippines. *Asia Life Sciences* 22:131–139.

Reginaldo, A. A., and A. P. de Guia. 2014. Species richness and patterns of occurrence of small non-flying mammals of Mt. Sto Tomas, Luzon Island, Philippines. *Philippine Science Letters* 7:37–44.

Reiter, J., and E. Curio. 2001. Home range, roost switching, and foraging area in a Philippine fruit bat (*Ptenochirus jagori*). *Ecotropica* 7:109–113.

Rickart, E. A. 1993. Diversity patterns of mammals along elevational and disturbance gradients in the Philippines: Implications for conservation. *Asia Life Sciences* 2:251–260.

Rickart, E. A. 2003. Chromosomes of Philippine mammals (Insectivora, Dermoptera, Primates, Rodentia, Carnivora). *Proceedings of the Biological Society of Washington* 116:699–709.

Rickart, E. A., D. S. Balete, P. A. Alviola, M. J. Veluz, and L. R. Heaney. Submitted. The mammals of Mt. Amuyao: An assessment of the highly endemic fauna of the Central Cordillera, Luzon Island, Philippines. *Mammalia*.

Rickart, E. A., D. S. Balete, and L. R. Heaney. 2007. Habitat disturbance and the ecology of small mammals in the Philippines. *Journal of Environmental Science and Management* 10:34–41.

Rickart, E. A., D. S. Balete, R. J. Rowe, and L. R. Heaney. 2011b. Mammals of the northern Philippines: Tolerance for habitat disturbance and resistance to invasive species in an endemic fauna. *Diversity and Distributions* 17:530–541.

Rickart, E. A., and L. R. Heaney. 1991. A new species of *Chrotomys* (Muridae) from Luzon Island, Philippines. *Proceedings of the Biological Society of Washington* 104:387–398.

Rickart, E. A., and L. R. Heaney. 2002. Further studies on the chromosomes of Philippine rodents (Muridae: Murinae). *Proceedings of the Biological Society of Washington* 115:473–487.

Rickart, E. A., L. R. Heaney, D. S. Balete, P. A. Alviola, M. R. M. Duya, M. V. Duya, R. G. Rosell-Ambal, and J. L. Sedlock. 2013. The mammals of Mt. Natib, Bataan Province, Philippines. *National Museum of the Philippines: Journal of Natural History* 1:31–44.

Rickart, E. A., L. R. Heaney, D. S. Balete, and B. R. Tabaranza Jr. 1998. A review of the genera *Crunomys* and *Archboldomys* (Rodentia, Muridae, Murinae) with descriptions of two new species for the Philippines. *Fieldiana: Zoology*, n.s., 89:1–24.

Rickart, E. A., L. R. Heaney, D. S. Balete, and B. R. Tabaranza Jr. 2011a. Small mammal diversity along an elevational gradient in northern Luzon, Philippines. *Mammalian Biology* 76:12–21.

Rickart, E. A., L. R. Heaney, S. M. Goodman, and S. Jansa. 2005. Review of the Philippine genera *Chrotomys* and *Celaenomys* (Murinae) and description of a new species. *Journal of Mammalogy* 86:415–428.

Rickart, E. A., L. R. Heaney, P. D. Heideman, and R. C. B. Utzurrum. 1993. The distribution and ecology of mammals on Leyte, Biliran, and Maripipi Islands, Philippines. *Fieldiana: Zoology*, n.s., 72:1–62.

Rickart, E. A., L. R. Heaney, and M. J. Rosenfeld. 1989a. Chromosomes of ten species of Philippine fruit bats (Chiroptera: Pteropodidae). *Proceedings of the Biological Society of Washington* 102:520–531.

Rickart, E. A., L. R. Heaney, and B. R. Tabaranza Jr. 2002. Review of *Bullimus* (Muridae: Murinae) and description of a new species from Camiguin Island, Philippines. *Journal of Mammalogy* 83:421–428.

Rickart, E. A., L. R. Heaney, and R. C. B. Utzurrum. 1991. Distribution and ecology of small mammals along an elevational transect in southeastern Luzon, Philippines. *Journal of Mammalogy* 72:458–469.

Rickart, E. A., P. D. Heideman, and R. C. B. Utzurrum. 1989b. Tent-roosting by *Scotophilus kuhlii* (Chiroptera: Vespertilionidae) in the Philippines. *Journal of Tropical Ecology* 5:433–436.

Rickart, E. A., J. A. Mercier, and L. R. Heaney. 1999. Cytogeography of Philippine bats (Mammalia: Chiroptera). *Proceedings of the Biological Society of Washington* 112:453–469.

Rickart, E. A., and G. G. Musser. 1993. Philippine rodents: Chromosomal characteristics and their significance for phylogenetic inference among 13 species (Rodentia: Muridae: Murinae). *American Museum Novitates* 3064:1–34.

Ricklefs, R. E., and G. C. Cox. 1972. Taxon cycles in the West Indian avifauna. *American Naturalist* 106:195–219.

Ringenbach, J. C., N. Pinet, J. F. Stéphan, and J. Delteil. 1993. Structural variety and tectonic evolution of strike-slip basins related to the Philippine fault system, northern Luzon, Philippines. *Tectonics* 12:187–203.

Roberts, T. E. 2006a. History, ocean channels, and distance determine phylogeographic patterns in three widespread Philippine fruit bats (Pteropodidae). *Molecular Ecology* 15:2183–2199.

Roberts, T. E. 2006b. Multiple levels of allopatric divergence in the endemic Philippine fruit bat *Haplonycteris fischeri* (Pteropodidae). *Biological Journal of the Linnean Society* 88:329–349.

Robins, J. H., P. A. McLenachan, M. J. Phillips, L. Craig, H. A. Ross, and E. Matisoo-Smith. 2008. Dating of divergences within the *Rattus* genus phylogeny using whole mitochondrial genomes. *Molecular Phylogenetics and Evolution* 49:460–466.

Rowe, K. C., A. S. Achmadi, and J. A. Esselstyn. 2014. Convergent evolution of aquatic foraging in a new genus and species (Rodentia: Muridae) from Sulawesi Island, Indonesia. *Zootaxa* 3815:541–564.

Rowe, K. C., M. L. Reno, D. M. Richmond, R. M. Adkins, and S. J. Steppan. 2008. Pliocene colonization and adaptive radiations in Australia and New Guinea (Sahul): Multilocus systematics of the Old Endemic rodents (Muroidea: Murinae). *Molecular Phylogenetics and Evolution* 47:84–101.

Ruedas, L. A., J. R. Demboski, and R. V. Sison. 1994. Morphological and ecological variation in *Otopteropus cartilagonodus* Kock 1969 (Mammalia: Chiroptera: Pteropodidae) from Luzon, Philippines. *Proceedings of the Biological Society of Washington* 107:1–16.

Ruedi, M., N. Friedli-Weyeneth, E. C. Teeling, S. J. Puechmaille, and S. M. Goodman. 2012. Biogeography of Old World emballonurine bats (Chiroptera: Emballonuridae) inferred with mitochondrial and nuclear DNA. *Molecular Phylogenetics and Evolution* 64:204–211.

Ruedi, M., B. Stadlemann, Y. Gager, E. J. P. Douzery, C. M. Francis, L. K. Lin, A. Guillén-Servent, and A. Cibois. 2013. Molecular phylogenetic reconstructions identify East Asia as the cradle for the evolution of the cosmopolitan genus *Myotis* (Mammalia, Chiroptera). *Molecular Phylogenetics and Evolution* 69:437–449.

Sanborn, C. C. 1952. Philippine Zoological Expedition 1946–1947: Mammals. *Fieldiana: Zoology* 33:89–158.

Sax, D. F., J. J. Stachowicz, and S. D. Gaines. 2005. *Species Invasions: Insights into Ecology, Evolution, and Biogeography*. Sunderland, MA: Sinauer Associates.

Scheffers, B. R., R. T. Corlett, A. Diesmos, and W. F. Laurance. 2012. Local demand drives a bushmeat industry in a Philippine forest preserve. *Tropical Conservation Science* 5:121–129.

Schenk, J. J., K. C. Rowe, and S. J. Steppan. 2013. Ecological opportunity and incumbency in the diversification of repeated continental colonizations by muroid rodents. *Systematic Biology* 62:837–864.

Schnitzler, H.-U., and E. K. V. Kalko. 2001. Echolocation by insect-eating bats. *Bioscience* 51:557–569.

Schnitzler, H.-U., and J. Ostwald. 1983. Adaptations for the detection of fluttering insects by echolocation in horseshoe bats. Pp. 801–827 in *Advances in Vertebrate Neuroethology*, ed. J. P. Ewert. New York: Plenum Press.

Sedlock, J. L. 2001. Inventory of insectivorous bats on Mount Makiling, Philippines, using echolocation call signatures and a new tunnel trap. *Acta Chiropterologica* 3:163–178.

Sedlock, J. L., N. R. Ingle, and D. S. Balete. 2011. Enhanced sampling of bat assemblages: A field test on Mount Banahaw, Luzon. *Fieldiana: Life and Earth Sciences* 2:96–102.

Sedlock, J. L., R. P. Jose, J. M. Vogt, L. M. J. Paguntalan, and A. B. Carino. 2014b. A survey of bats in a karst landscape in the central Philippines. *Acta Chiropterologica* 16:197–211.

Sedlock, J. L., F. Kruger, and E. L. Clare. 2014a. Island bat diets: Does it matter more who you are or where you live? *Molecular Ecology* 23:3684–3694.

Sedlock, J. L., and S. E. Weyandt. 2009. Genetic divergence between morphologically and acoustically cryptic bats:

Novel niche partitioning or recent contact? *Journal of Zoology* 279:388–395.

Sedlock, J. L., S. E. Weyandt, L. Cororan, M. Damerow, S. Hwa, and B. Pauli. 2008. Bat diversity in tropical forest and agro-pastoral habitats within a protected area in the Philippines. *Acta Chiropterologica* 10:349–358.

Simmons, N. 2005. Order Chiroptera. Pp. 312–525 in *Mammal Species of the World: A Taxonomic and Geographic Reference*, ed. D. Wilson and D. M. Reeder. Baltimore: Johns Hopkins University Press.

Singleton, G. R., R. C. Joshi, and L. S. Sebastian (eds.). 2008. *Philippine Rats: Ecology and Management*. Muñoz City: Philippine Rice Research Institute.

Smith, A. T., and Y. Xie. 2008. *A Guide to the Mammals of China*. Princeton, NJ: Princeton University Press.

Stanley, W. T. (ed.). 2011. Studies of montane vertebrates of Tanzania. *Fieldiana: Life and Earth Sciences* 4:1–116.

Start, A., and A. G. Marshall. 1975. Nectarivorous bats as pollinators of trees in West Malaysia. Pp. 141–150 in *Tropical Trees: Variation, Breeding, and Conservation*, ed. J. Burley and B. T. Styles. London: Academic Press.

Steppan, S. J., C. Zawadski, and L. R. Heaney. 2003. Molecular phylogeny of the endemic Philippine rodent *Apomys* (Muridae) and the dynamics of diversification in an oceanic archipelago. *Biological Journal of the Linnean Society* 80:699–715.

Stevenson, J., F. Siringan, J. Finn, D. Madulid, and H. Heijnis. 2010. Paoay Lake, northern Luzon, the Philippines: A record of Holocene environmental change. *Global Change Biology* 16:1672–1688.

Stibig, H.-J., and J.-P. Malingreau. 2003. Forest cover of insular Southeast Asia mapped from recent satellite images of coarse spatial resolution. *Ambio* 32:469–475.

Stier, S. C., and T. L. Mildenstein. 2005. Dietary habits of the world's largest bats: The Philippine flying foxes, *Acerodon jubatus* and *Pteropus vampyrus lanensis*. *Journal of Mammalogy* 86:719–728.

Stinus-Remonde, M., and C. Vertucci (eds.). 1999. *Minding Mining: Lessons from the Philippines*. Quezon City, Philippine International Forum.

Struebig, M. J., T. Kingston, A. Zubaid, S. C. Le Comber, A. Mohd-Adnan, A. Turner, J. Kelly, M. Bożek, and S. J. Rossiter. 2009. Conservation importance of limestone karst outcrops for Palaeotropical bats in a fragmented landscape. *Biological Conservation* 142:2089–2096.

Stuart, A. M., C. V. Prescott, and G. R. Singleton. 2008. Biology and management of rodent communities in complex agroecosystems-lowlands. Pp. 37–55 in *Philippine Rats: Ecology and Management*, ed. G. R. Singleton, R. C. Joshi, and L. S. Sebastian. Muñoz City: Philippine Rice Research Institute.

Stuart, A. M., C. V. Prescott, G. R. Singleton, R. C. Joshi, and L. S. Sebastian. 2007. The rodent species of the Ifugao rice terraces, Philippines—target or non-target species for management? *International Journal of Pest Management* 53:139–146.

Suarez, R. K., and P. E. Sajise. 2010. Deforestation, swidden agriculture, and Philippine biodiversity. *Philippine Science Letters* 3:91–99.

Tan, J. M. L. 1995. *A Field Guide to Whales and Dolphins in the Philippines*. Manila: Bookmark.

Taylor, E. H. 1934. *Philippine Land Mammals*. Monograph, Bureau of Science 30. Manila: Bureau of Printing.

Thomas, O. 1895. Preliminary diagnoses of new mammals from northern Luzon, collected by Mr. John Whitehead. *Annals and Magazine of Natural History, London*, ser. 6, 15:160–164.

Thomas, O. 1898. On the mammals collected by Mr. John Whitehead during his recent expedition to the Philippines, with field notes by the collector. *Transactions of the Zoological Society of London* 14:377–412.

Thorington, R. W., Jr., J. L. Koprowski, M. A. Steele, and J. F. Whatton. 2012. *Squirrels of the World*. Baltimore: Johns Hopkins University Press.

Timm, R. M., and E. C. Birney. 1980. Mammals collected by the Menage Scientific Expedition to the Philippine Islands and Borneo, 1890–1893. *Journal of Mammalogy* 61:566–571.

US National Oceanographic and Atmospheric Administration (NOAA). 2014. *Global Mapping of Cloudiness Using MODIS Imagery*. Cloud Climatology Project, www.nssl.NOAA.gov/projects/pacs/web/MODIS.

Utzurrum, R. C. B. 1992. Conservation status of Philippine fruit bats (Pteropodidae). *Silliman Journal* 36:27–45.

Utzurrum, R. C. B. 1995. Feeding ecology of Philippine fruit bats: Patterns of resource use and seed dispersal. *Symposia of the Zoological Society of London* 67:63–77.

Utzurrum, R. C. B. 1998. Geographic patterns, ecological gradients, and the maintenance of tropical fruit bat diversity: The Philippine model. Pp. 342–353 in *Bat Biology and Conservation*, ed. T. H. Kunz and P. A. Racey. Washington, DC: Smithsonian Institution Press.

Van der Geer, A., G. Lyras, J. de Vos, and M. Dermitzakis. 2010. *Evolution of Island Mammals: Adaptation and Extinction of Placental Mammals on Islands*. Oxford: Wiley-Blackwell.

van der Ploeg, J., A. B. Masipiqueña, and E. C. Bernardo (eds.). 2003. *The Sierra Madre Mountain Range: Global Relevance, Local Realities; Papers Presented at the 4th Regional Conference on Environment and Development*. Tuguegarao City, Philippines: Cagayan Valley Program on Environment and Development.

van der Ploeg, J., M. van Weerd, A. B. Masipiqueña, and G. A. Persoon. 2011. Illegal logging in the Northern Sierra Madre Natural Park, the Philippines. *Conservation and Society* 9:202–215.

van Weerd, M., J. P. Guerrero, B. A. Tarun, and D. C. Rodriguez. 2003. Flying foxes of the Northern Sierra Madre Natural Park. In *The Sierra Madre Mountain Range: Papers Presented at the 4th Regional Conference on Environment and Development*, ed. J. van der Ploeg, A. B. Masipiqueña, and E. C. Bernardo. Tuguegarao City, Philippines: Cagayan Valley Program on Environment and Development.

van Weerd, M., J. P. Guerrero, B. A. Tarun, and D. C. Rodriguez. 2009. Flying foxes of the Northern Sierra Madre Natural Park. In *The Sierra Madre Mountain Range: Papers Presented at the 4th Regional Conference on Environment and Development*, ed. J. Van der Ploeg, A. B. Masipiquena, and E. C. Bernardo. Tuguegarao City, Philippines: Cagayan Valley Program on Environment and Development.

van Whye, J., and K. Rookmaaker. 2013. *Alfred Russel Wallace, Letters from the Malay Archipelago*. Oxford: Oxford University Press.

Veron, G., M. Willsch, V. Dacosta, M.-L. Patou, A. Seymour, C. Bonillo, A. Couloux, S. T. Wong, A. P. Jennings, J. Fickel, and A. Wilting. 2014. The distribution of the Malay civet *Viverra tangalunga* (Carnivora: Viverridae) across Southeast Asia: Natural or human-mediated dispersal? *Zoological Journal of the Linnean Society* 170:917–932.

Vitug, M. D. 1993. *The Politics of Logging: Power from the Forest*. Manila: Philippine Center for Investigative Journalism.

Vogel, T. A., T. P. Flood, L. C. Patino, M. S. Wilmot, R. P. R. Maximo, C. B. Arpa, C. A. Arcilla, and J. A. Stimac. 2006. Geochemistry of silicic magmas in the Macolod Corridor, SW Luzon, Philippines: Evidence of distinct, mantle-derived, crustal sources for silicic magmas. *Contributions to Mineral Petrology* 151:267–281.

Vondra, C. F., M. E. Mathisen, D. R. Burggraf Jr., and E. P. Kvale. 1981. Plio-Pleistocene geology of northern Luzon, Philippines. In *Hominid Sites: Their Geological Settings*, ed. G. Rapp Jr. and C. F. Vondra. *AAAS Selected Symposiums* 63:255–310.

Wallace, A. R. 1859 [1860]. On the zoological geography of the Malay Archipelago. *Journal of the Proceedings of the Linnean Society: Zoology* 4:172–184 [communicated by Charles Darwin and read on 3 November 1859; published 1860].

Wallace, A. R. 1880. *Island Life*. London: Macmillan [Chicago: University of Chicago Press, 2013, reprint of 1880 edition].

Walpole, P. 2010. Figuring the forest figures: Understanding forest cover data in the Philippines and where we might be proceeding. *Environmental Science for Social Change*, http://download.essc.org.ph/forest/ESSC-PWalpole _Figuring forest figures_reduced_.pdf.

Wei, L., N. Han, L. Zhang, K. M. Helgen, S. Parsons, S. Zhou, and S. Zhang. 2008. Wing morphology, echolocation calls, diet, and emergence time of black-bearded tomb bats (*Taphozous melanopogon*, Emballonuridae) from southwest China. *Acta Chiropterologica* 10:51–59.

Wells, N. A. 2003. Some hypotheses on the Mesozoic and Cenozoic paleoenvironmental history of Madagascar. Pp. 16–34 in *The Natural History of Madagascar*, ed. S. M. Goodman and J. P. Benstead. Chicago: University of Chicago Press.

Whitehead, J. 1899. Field-notes on birds collected in the Philippine Islands in 1893–1896 [4 parts]. *Ibis* 5:81–111, 210–246, 381–399, 485–501.

Whittaker, R. J., and J. M. Fernández-Palacios. 2007. *Island Biogeography: Ecology, Evolution, and Conservation*. Oxford: Oxford University Press.

Whittaker, R. J., K. A. Triantis, and R. J. Ladle. 2008. A general dynamic theory of oceanic island biogeography. *Journal of Biogeography* 35:977–994.

Whittaker, R. J., K. A. Triantis, and R. J. Ladle. 2010. A general dynamic theory of oceanic island biogeography: Extending the MacArthur-Wilson theory to accommodate the rise and fall of volcanic islands. Pp. 88–115 in *The Theory of Island Biogeography Revisited*, ed. J. B. Losos and R. E. Ricklefs. Princeton, NJ: Princeton University Press.

Wildlife Conservation Society of the Philippines. 1997. *Philippine Red Data Book*. Manila: Bookmark.

Wiles, G. J., D. W. Buden, and D. J. Worthington. 1999. History of introduction, population status, and management of Philippine deer (*Cervus mariannus*) on Micronesian islands. *Mammalia* 63:193–215.

Wilson, E. O. 1961. The nature of the taxon cycle in the Melanesian ant fauna. *American Naturalist* 95:169–193.

Wolfe, J. A. 1988. Arc magmatism and mineralization in North Luzon and its relationship to subduction at the East Luzon and North Manila Trenches. *Journal of Southeast Asian Earth Sciences* 2:79–93.

Wolfe, J. A., and S. Self. 1983. Structural lineaments and Neogene volcanism in southwestern Luzon. Pp. 157–172 in *The Tectonic and Geological Evolution of Southeast Asian Seas and Islands: Part 2*, ed. D. E. Hayes. Geophysical Monograph 27. Washington, DC: American Geophysical Union.

World Bank Group. 2014. *Climate Change Knowledge Portal*, http://sdwebx.worldbank.org/climateportal/index.cfm?page=country_historical_climate&ThisRegion=Asia&ThisCCode=PHL/ [accessed 6 June 2014].

Wyatt, K. B., P. F. Campos, M. T. P. Gilbert, S.-O. Kolokotronis, W. H. Hynes, R. DeSalle, P. Daszak, R. D. E. MacPhee, and A. D. Greenwood. 2008. Historical mammal extinctions on Christmas Island (Indian Ocean) correlates with introduced infectious disease. *PLoS ONE* 3:e3602.

Yang, T. F., J. Tien, C. Chen, T. Lee, and R. S. Punongbayan. 1995. Fission-track dating of volcanics in the northern part of the Taiwan-Luzon Arc: Eruption ages and evidence for crustal contamination. *Journal of Southeast Asian Earth Sciences* 11:81–93.

Yumul, G. P., Jr., C. B. Dimalanta, V. B. Maglambayan, and E. J. Marquez. 2008. Tectonic setting of a composite terrane: A review of the Philippine island arc system. *Geosciences Journal* 1:7–17.

Yumul, G. [P.], Jr., C. Dimalanta, K. Queano, and E. Marquez. 2009. Philippines, geology. Pp. 732–738 in *Encyclopedia of Islands*, ed. R. G. Gillespie and D. A. Clague. Berkeley: University of California Press.

Zachos, J., M. Pagani, L. Sloan, E. Thomas, and K. Billups. 2001. Trends, rhythms, and aberrations in the global climate 65 Ma to Present. *Science* 292:686–693.

Zubaid, A. 1990. Food and roosting habits of the black-bearded tomb bat, *Taphozous melanopogon* (Chiroptera, Emballonuridae) from peninsular Malaysia. *Mammalia* 54:159–162.

Index

Primary species accounts are shown in boldface.

LAWRENCE R. HEANEY is Negaunee Curator of Mammals at the Field Museum of Natural History. He received his MA and PhD from the University of Kansas and has conducted research on Philippine mammals since 1981. DANILO S. BALETE is a research associate at the Field Museum. He grew up in the Bicol region of Luzon, joined the Field Museum's Philippine Mammal Project in 1989, and received his MS degree from the University of Illinois–Chicago. ERIC A. RICKART is curator of vertebrates at the Natural History Museum of Utah. He received his MA degree from the University of Kansas and PhD from the University of Utah, and he began studies of Luzon mammals in 1987. VELIZAR SIMEONOVSKI is a research associate and resident artist at the Field Museum; he began making drawings for this volume in 2003. He received his MS degree in Zoology of Vertebrates from the Sofia University "St. Clement Ohridski" in Sofia, Bulgaria. ANDRIA NIEDZIELSKI is a research assistant at the Field Museum. She received her Bachelor of Fine Arts degree from the School of the Art Institute of Chicago and has worked on the Mammals of Luzon Project since 2010.